THE FRONTIERS COLLECTION

THE FRONTIERS COLLECTION

Series Editors:

A.C. Elitzur L. Mersini-Houghton M.A. Schlosshauer M.P. Silverman R. Vaas H.D. Zeh

The books in this collection are devoted to challenging and open problems at the forefront of modern science, including related philosophical debates. In contrast to typical research monographs, however, they strive to present their topics in a manner accessible also to scientifically literate non-specialists wishing to gain insight into the deeper implications and fascinating questions involved. Taken as a whole, the series reflects the need for a fundamental and interdisciplinary approach to modern science. Furthermore, it is intended to encourage active scientists in all areas to ponder over important and perhaps controversial issues beyond their own speciality. Extending from quantum physics and relativity to entropy, consciousness and complex systems – the Frontiers Collection will inspire readers to push back the frontiers of their own knowledge.

Other Recent Titles

Weak Links
Stabilizers of Complex Systems from Proteins to Social Networks
By P. Csermely

The Biological Evolution of Religious Mind and Behaviour
Edited by E. Voland and W. Schiefenhövel
Particle Metaphysics

A Critical Account of Subatomic Reality
By B. Falkenburg

The Physical Basis of the Direction of Time
By H.D. Zeh

Mindful Universe
Quantum Mechanics and the Participating Observer
By H. Stapp

Decoherence and the Quantum-To-Classical Transition
By M. Schlosshauer

The Nonlinear Universe
Chaos, Emergence, Life
By A. Scott

Symmetry Rules
How Science and Nature are Founded on Symmetry
By J. Rosen

Quantum Superposition
Counterintuitive Consequences of Coherence, Entanglement, and Interference
By M.P. Silverman

Series home page – springer.com

Dean Cvetkovic • Irena Cosic
Editors

States of Consciousness

Experimental Insights into Meditation, Waking, Sleep and Dreams

Editors
Dean Cvetkovic
RMIT University
School of Electrical
and Computer Engineering
PO Box 2476V
3001 Melbourne Victoria
Australia
dean.cvetkovic@rmit.edu.au

Irena Cosic
RMIT University
College of Science, Engineering
and Health
PO Box 2476V
3001 Melbourne Victoria
Australia
irena.cosic@rmit.edu.au

ISSN 1612-3018
ISBN 978-3-642-18046-0 e-ISBN 978-3-642-18047-7
DOI 10.1007/978-3-642-18047-7
Springer Heidelberg Dordrecht London New York

Library of Congress Control Number: 2011931470

Cover design: eStudio Calamar S.L.

Printed on acid-free paper

Springer is part of Springer Science+Business Media (www.springer.com)

Dedicated to my partner and family for endless love, inspiration and support.
Dedicated to everyone who is searching to explain that elusive conscious experience that has once left a memorable mark in our lives, changed us and continues to do so...

Dean Cvetkovic

Preface

A search for a deeper understanding and exploration of consciousness has long been and still is widely studied by scientists and philosophers alike. There are countless examples of individuals making significant changes to their lives in following a brief encounter with state of consciousness. Naturally, a philosopher would want to define this state of consciousness and a scientist would want to measure it. Both philosophers and scientists have the common goal of wanting to explain consciousness. Whether philosophical and scientific approaches will succeed in capturing this elusive "thing" that we so often refer to as consciousness remains to be seen.

The first few chapters present introductory theory and insights into various states and disorders of consciousness. In the opening chapter, I introduce consciousness, its history, various philosophical and scientific theories, contemporary technological advances, natural medical phenomena, solid experimental findings that reveal correlations between physiological processes and consciousness, and finally an overview of the following chapters. Other chapters present various psychophysiological, neurocognitive, neuroscientific and neurobiological theories and models of waking, sleeping, dreaming and meditation, tested with advanced neuroimaging and engineering biomarkers. These chapters highlight the need to utilise the knowledge of consciousness in order to develop corrective treatments for certain disorders and pathological problems of consciousness. The remaining chapters describe the more specific altered states of consciousness, such as hypnagogic phenomena and transcendental meditation, which are generated both internally, within the human body, and externally, as environmental stimuli.

While the main states of consciousness are considered to be wakefulness, sleep and dreaming, there are multiple specific states that originate across these three states and along its borders. Altered states of consciousness are dynamic transitional processes, where a subject continually enters a new state and leaves the old state. One may become aware of these altered states of consciousness where one's own conscious experiences and perception are characterised by electrophysiological, cognitive and behavioural changes.

The book is not aiming to define or, identify all possible states of consciousness, nor is it claiming to present exact ways to measure the level of these states of consciousness. On the contrary, the idea of writing and compiling this book emerged from my long fascination with certain intricacies of the phenomenon of consciousness. Whether these intricacies can be identified as states, contents, properties, levels or processes of consciousness is debatable. But what remains important is that they exist and are classified as either normal or disordered consciousness. The invited chapters are meant to serve both as a basic introduction and an in-depth research on some well-known and rare states of consciousness.

We hope that this book will provide inspiration to people from all walks of life, from professionals to students at all levels of education. We also hope that the book will encourage readers to explore certain aspects of consciousness using scientific approaches in order to gain insight in own conscious experiences.

Melbourne, Australia
June 2011

Dean Cvetkovic

Contents

Chapter 1
Introduction to States of Consciousness

Dean Cvetkovic

Abstract The problem of consciousness is mostly regarded as identical to the mind-body problem. According to Chalmers' philosophical arguments, the hard problem of consciousness lies in establishing and explaining the link between physical processes and conscious experiences, via psychological processes. A brief history of various theories of consciousness is given and a selection of theories are tested against Zeman's three fundamental intuitions and Chalmers' controversial zombie argument. The hard problem of consciousness is further described using Levine's notion of an explanatory gap between physical matter and conscious experience, through the first and third persons. Various states, contents, levels and processes of consciousness are summarised, including Damasio and Meyer's dual perspective for defining consciousness. Tart's three definitions do not entirely describe altered states of consciousness. While the challenge of finding the core function of human and animal sleep remains unknown when tested under the null hypothesis, studies on the neural correlates of consciousness during meditation have revealed neuroplasticity effects. The synchrony of gamma brain oscillations reflecting various styles of meditation or attention, also known as the binding problem, may be related to conscious experiences. This binding problem with gamma brain oscillatory synchronization also arises in relation to sensory awareness or perception, affecting the perception of time and hallucinatory experiences in various disorders of consciousness such as severe schizophrenic and déjà vu (in healthy or epileptic) patients. In conjunction with medication treatments, music therapy is often useful in accelerating the healing process in most such disorders of consciousness. It is still unknown how this sensory awareness to music is perceived in medicated patients suffering from disorders of consciousness. More clinically elusive are near death experiences, in which consciousness persists independently of brain function, where there is no scientific basis for such consciousness to exist and no physiological or psychological model that can explain it. Near death

D. Cvetkovic (✉)
School of Electrical and Computer Engineering, RMIT University, Melbourne, VIC, Australia
e-mail: dean.cvetkovic@rmit.edu.au

D. Cvetkovic and I. Cosic (eds.), *States of Consciousness*, The Frontiers Collection,
DOI 10.1007/978-3-642-18047-7_1, © Springer-Verlag Berlin Heidelberg 2011

experiences can be regarded as a special state of consciousness, which provides further evidence that the consciousness problem may be very close to the mind-body problem that originates in Descartes' classic theory of dualism and is transformed into Chalmers' contemporary theory of natural dualism. The final section of this chapter offers an overview of all invited chapters.

> I know my processes are just electronic circuits, but how does this explain my experience of thought and perception? (Hofstadter 1985, p. 186)

1.1 Psychological and Phenomenal Consciousness

The best way we can get closely acquainted with consciousness is to experience it using all our senses, thoughts, feelings, emotions and perceptions. The subjective experience is so strong in shaping a human character that we refer to it as human psychology. When we think of psychology we often ignore physiology, and this fact reminds us that consciousness may relate to the mind-body problem. Memory is very powerful, yet it is different for everyone. Memories can retain information from all the senses that we know to exist. Our senses of smell, taste, touch, hearing and vision can all translate our environmental experiences into information that is stored in our memory. These experiences may be visual, auditory, tactile or olfactory, or relate to taste, temperature, pain, other body sensations (orgasms, itches, etc.), mental imagery, conscious thoughts, emotions or the sense of self ("I"). How does the brain maintain a sense of self? While conscious experiences generate new neural pathways, by constant wiring and rewiring, the self-image or understanding of "I" remains elusive. However, we do understand that the "I" can act as a reference point for ordering our thoughts, emotions and experiences, and it can create boundaries between internal and external events to form experiences. If these experiences were pleasant when they were first sensed (e.g. hearing a popular song while walking along the beach in one's early childhood) and later the same sensations occur (e.g. years later the same song is heard), the sense is triggered, information is extracted from memory and the experiences are relived, so that mental states or emotions are induced and physical actions may be realised (e.g. by dancing). In extreme cases, these experiences may be peak experiences, or "sudden ecstatic moments of great happiness, awe, and of a feeling of unity that gives way to serenity and contemplation" (Maslow 1971, p. 235). Even the wistful memories of the smell of mother's cooked pancakes can induce a yearning for the past, resulting in experiences of nostalgia. It was only recently revealed that such nostalgic experiences can elevate moods, increase self-esteem and improve our overall well-being. Indeed, "like armour shielding the mind from dark thoughts, nostalgia protects against psychological onslaughts in the future" (Gebauer and Sedikides 2010, p. 35).

While the mind can easily differentiate the information coming from the various sensory modalities (vision, hearing, taste etc.), the mind finds it challenging to distinguish emotions, memories, feelings and thoughts from each other (Peat 1987). If a person feels sad or happy, he/she may not be able to find the main cause of that reaction. The main cause may be not the external sensory information that triggered the state, but an internal feeling initiated by a thought or by fragmentary memories leading to the creation of a series of inner thoughts. Such external and internal mental processes can often result in an inability from a first-person or a third-person perspective to recognise or distinguish the causes. Why am I feeling this way all of a sudden? The phenomenal state of mind is the conscious experience. The psychological state of mind is the explanatory basis of behaviour, and is studied in cognitive science. The phenomenal state is characterised by the way it feels (or how it feels) and the psychological state is characterised by what it does. Chalmers suggested that for every phenomenal state there may also be a psychological state and that one cannot be differentiated from the other. Both states are part of the mental concept, which may be described as having a double life.

Scientific methods have difficulty in observing the phenomenon of consciousness. David Chalmers suggests that the problem of consciousness lies on the border between philosophy and science and that if one is to study this problem scientifically, one must also understand the philosophy and vice versa. Chalmers distinguishes the "easy" and "hard" problems of consciousness (Chalmers 1996). While most easy problems still continue to challenge us scientifically, the philosophical and scientific aspects of the hard problem remain untouched. The easy problems mainly deal with the neural correlates of consciousness and the *how* question (i.e. the physiological processes in the brain and how psychological processes respond under the influence of internal or external sensory stimuli). By contrast, the hard problem is concerned with the question of why these physiological and psychological processes live a double life as experiences.

Chalmers also addressed the mind-body problem when he distinguished easy and hard problems. The hardest problem is this: "how could a physical system give rise to conscious experience?" (Chalmers 1996, p. 25). Chalmers argues that the relation between the physical process and the conscious (phenomenal) experience depends on the dual link between the physical and psychological, and psychological and phenomenal. It is well understood that there is a link at the level of how physical processes can have psychological properties. However, it remains unknown "why and how these psychological properties are accompanied by phenomenal properties" and "understanding the link between psychological and the phenomenal is crucial to understand conscious experience" (Chalmers 1996, p. 25). Pain is a good example to illustrate how we distinguish phenomenal from psychological mental concepts. Pain experience cannot be measured objectively, but subjectively it can be rated (e.g. by the level of unpleasant phenomenal quality). Pain also causes a psychological effect when the person suffering it assumes that this unpleasantness is generated by an injury or damage to the organism, which then leads to other reactions.

An example of a dual mental concept is that of perception, which can be considered to be psychological or phenomenal, or to combine both components. The psychological component of perception concept is processed in the cognitive system, influenced by environmental stimulation. By contrast, the phenomenal perception consists of the perceived conscious experience. The sensation can be regarded as the phenomenal component and the perception as the psychological component. But the concepts of perception and sensation can lean towards each other or blur together. Sensation is regarded a state responsible for feeling. This makes it impossible to objectively or even subjectively measure the experience of colour sensation. It is believed from both scientific and personal exploration (via physics, biology, neuroscience and social and cultural studies) that we can consciously experience colour. The conscious experience of any colour needs to be understood first in the very same social and cultural environment that we live in, through perceptual learning. Karl H. Pribram, psychologist and psychiatrist, presented an example of how some cultures perceive and experience some colours differently to other cultures (Pribram 2004). For example, people in the northern national state XYZ (where "XYZ" is used to preserve anonymity) were unable to distinguish red from green, or red from yellow or black, mainly because the colour red was rarely experienced in their environment. But the people of XYZ were able to distinguish many shades of green. People of other cultures working in XYZ at the time were unable to distinguish these shades of green.

Pribram states that conscious experiences are "initially emergent from brain processes produced by input generated by the brain's control over its physiological, physical, chemical, and socio-cultural environment. When changes occur in that environment, changes are produced in the brain processes. Only when these peripheral changes become implemented in the brain's memory do the resulting experiences become accessible to further processing" (Pribram 2004, p. 24).

1.2 History, Philosophy and Theories of Consciousness

There are many philosophical theories of consciousness, such as: dualism, behaviourism, idealism, functionalism, identity (personal), phenomenalism, phenomenology, emergentism, mysticism, externalism, physicalism and others. Only a few of these widely known theories are described in this chapter.

Mind-body dualism was formally introduced in modern philosophy by René Descartes (1596–1650), through the statement *Cogito ergo sum* ("I think therefore I am"). Descartes did not distinguish consciousness from the mind-body nexus. He defined body as extended physical material (space-filling) and mind as unextended (did not take up space) non-physical *Res cogitans* ("thinking thing"). Dualism dwells in both physical and non-physical worlds. Cartesian dualism sought to explain how an independent mental substance can influence physical processes, which gave meaning to the familiar idea of mind over matter. A contemporary philosophical perspective by John Searle is that dualism is wrong and that only changes in the brain state can create conscious states (Searle 1992). Bishop George Berkeley (1685–1753) was another philosopher of consciousness, an *idealist* who attempted to show that we only experience percepts, thoughts and feelings and that an external world (explained by physics) is imagination and illusion. This philosophical theory was never finally discredited. William James (1842–1910) defined his own philosophical theory of mind, known as *functionalism*. Functionalist theory explains that a mental state does not depend on any physical internal properties but on the way it functions or the role it plays in the system (mind). In 1890, from that theory of mind, James introduced the stream of consciousness as a process, stream or a flow of thoughts of which one is aware (James 1890). He described four major characteristics of this stream of consciousness: every thought belongs to some personal consciousness; consciousness is in the constant change; personal consciousness is continuous and like a stream, flowing from one place to another; and the stream flows towards one particular place that brings stability to the constant change of thought and feeling (Kokoszka 2007).

Pioneering psychologists Sigmund Freud (1856–1939) and Carl Gustav Jung (1875–1961) theorised that consciousness has three levels or categories, respectively. Sigmund Freud contributed immensely to the study of consciousness and altered states of consciousness (ASC) by identifying three levels of consciousness: the conscious level (awareness of oneself and one's environment); the subconscious level (where information remains hidden from consciousness until it presents itself over time or with therapy); and the unconscious level (where information is blocked from consciousness and is extremely difficult to access, requiring years of therapy, but may be accessed through dreams and dream interpretation). To access the subconscious and unconscious levels, Freud proposed psychoanalysis and dream content analysis (where dream symbols represent conscious experiences), respectively. Conscious experiences can be observed and communicated with others, but the unconscious processes are difficult to access. Sigmund Freud devoted his life to developing a technique that was able to bring these unconscious processes up to the level of consciousness, where they could be shared and treated in therapy. Freud believed that there is a subconscious mind and there are unconscious beliefs and desires that can trigger certain related behaviours. Modern hypnosis or self-hypnosis therapies are used to "correct" such unconscious desires, through voluntary relaxation and auto-suggestive techniques.

Jung developed his own theory of mind and of natural and altered states of consciousness, extending on Freud's theory of unconsciousness. Jung identified three categories of consciousness: personal consciousness (conscious awareness at one point of time); the personal unconscious (memories at the edge of conscious recall, similar to Freudian repressed memories); and the collective unconscious (universal human reactions based on pre-existing forms known as archetypes that influence individual experience). Jung believed that there is a strong directional link from unconscious to consciousness. The unconscious acts as a memory organizer and feeds the relevant information to consciousness whenever is needed.

These advances in psychology (rather than in philosophy) set the psychological and phenomenal concepts of mind apart. Behaviourist psychologists ignored the phenomenal concept and highlighted the psychological concept. Some behaviourists denied that consciousness had anything to do with psychology and even denied that consciousness existed. From the 1940s, there was a shift from the phenomenal to the psychological concept of mind, mainly in the work of Gilbert Ryle (1900–1976), who believed that mental states can be analysed in terms of certain behaviour. In the 1960s, mental concepts began to be analysed functionally, by investigating the interaction of mental states and their causes and effects.

For Adam Zeman, any theory of consciousness must obey three fundamental "intuitions" (as he calls them) (1) consciousness is a robust phenomenon which deserves to be explained rather than being explained away; (2) consciousness is bound up with our physical being; (3) consciousness makes a difference (Zeman 2001, p. 1282). Each of the three theories of consciousness – dualism, identity and functionalism – clashes with at least one of these intuitions.

Chalmers has proposed naturalistic property dualism. This theory has some but not all aspects of dualist theory. Chalmers does not assume that there is a mental substance or "thinking thing" but builds on fundamental laws aligned with modern scientific results. He accepts that these fundamental laws are unknown and may be incomprehensible to humans, but says they are still naturalistic. In modern science, where physical theory states that "for every physical event, there is a physical sufficient cause" (Chalmers 1996, p. 125), the elements of religion, spiritualism or supernaturalism are considered nonscientific and those elements should not be associated with dualism.

Since experiences or other elements may appear as phenomenal properties of consciousness, they may be correlated with physical properties to establish dualism. The fundamental laws of consciousness may be psychophysiological laws that specify the dependency of phenomenal properties on physical properties. Chalmers expressed this by saying that while physical theory "gives a theory of physical processes", psychophysiological theory "tells us how those processes give rise to experiences" (Chalmers 1996, p. 128). Chalmers concluded that "once we have a fundamental theory of consciousness to accompany a fundamental theory in physics, we may truly have a theory of everything" (Chalmers 1996, p. 127). Responding to Chalmers' explanation of his natural dualism, Zeman concluded that he tends "to be driven to the conclusion that conscious experience is a beautiful but functionally irrelevant embellishment of physical processes" (Zeman 2001, p. 1284). For Zeman, dualism fails his third intuition that consciousness makes a difference.

An identity theory was introduced by experimental psychologist Edwin G. Boring (1886–1968) in 1933. The corresponding philosophical theory of mind was formally established in the 1950s by Ullin T. Place and Herbert Feigl. Place and Feigl claimed that mental states and processes are the same as states and processes of the brain. For example, the identity theory treats the experience of hearing a melody or seeing a baby blue colour as identical with brain processes and not as mere correlates of brain processes. Behaviouristic, materialistic, physicalistic and reductionist approaches have been applied to explain the identity theory of mind. Testing the identity theory against his intuitions, Zeman infers that its main implication is that conscious events originate in the brain and have a functional role for behaviour, which fails to satisfy his first intuition that "the properties of experience are robust phenomena in need of explanation" (Zeman 2001, p. 1283).

He concludes that conscious experiences cannot be reduced to the neural structures and processes on which they depend.

Functionalism is another philosophy of mind which explains that a mental state does not depend on the physical internal properties but on the way it functions or the role it plays in the system (mind). Based on this functionalist theory, Daniel Dennett proposed two cognitive models of consciousness. He called the first one a box-and-lines model (Dennett 1978) by analogy with an engineering functional block diagram that can be used to explain the functionality of a system. In Dennett's case, the system is the mind, which consists of a flow of information (lines) between modules (boxes). The modules are a perceptual unit, a short-term memory store, control systems and a "public relations" module. This model may explain our ability to report the contents of our internal states and to control our behaviour using our perceptual information. Dennett called the second model a "pandemonium" model, since it consists of multiple cognitive agents (demons) that compete to control mental processing (Dennett 1991). Instead of the main control module, this model includes multiple channels forming a highly parallel information system. This model has been adapted for applications in neuroscience, artificial intelligence, etc. Dennett's model may be applied in artificial intelligence to model how visual experiences arise from countless acts of discrimination and classification. The advantages of these cognitive models are that they focus on attention, explain reportability and show the influence of information on control of behaviour. However, most of these cognitive models, as Chalmers argues, basically ignore the hard problem of consciousness, of "why should there be conscious experience in the vicinity of these capacities" (Chalmers 1996, p. 114). In other words, they fail to explain the role of phenomenal consciousness in describing how our minds generate the conscious experience that we enjoy. As Zeman says, "Why should consciousness be like this?" (Zeman 2001, p. 1283). Dennett's models explain our human ability to report verbally on our mental states while ignoring their physical neural properties. The cognitive models can be applied to describe psychological consciousness, but the functionalist theory of mind fails to explain phenomenal consciousness. They fail to meet Zeman's intuition that consciousness deserves not to be explained away.

Considering that science and philosophy strive to explain everything in physical terms, there is a tendency to explain consciousness in a similar way. A thought experiment called the zombie argument shows where physicalists and dualists have different views. Chalmers, a natural dualist, proposed the concepts of a zombie and a zombie world, where a zombie is defined to be someone or something that is physically identical to a conscious being, like "me", but has no conscious experience. The zombie world is a world where people are all zombies but is otherwise identical to ours. A zombie lacks only conscious experience. Chalmers called an unconscious clone a zombie twin and explained that our zombie twin is psychologically and physically identical to us. This zombie can have enough function to perceive any external information, have psychological senses, be awake, report the contents of internal states, and maintain focused attention towards an external environment. As Chalmers says, "none of this functioning will be accompanied by any real conscious experience. There will be no phenomenal feel. There is nothing it is like to be a zombie." (Chalmers 1996, p. 95).

Physicalists often criticise this Chalmers zombie argument, claiming that it is not possible and believe that zombies are organisms that may have consciousness but not in the reality. Dennett, disagrees with Chalmers on this zombie argument, stating that pain, for example, cannot be replicated from a human mental state, without considering its behavioural or physiological differences. Chalmers on the other hand, believes that even though zombies are probably not naturally existent with the current laws of nature, zombie scenario is logically possible. Chalmers also claims that the brain physiology despite its complexity, cannot "conceptually entail consciousness" (Chalmers 1996, p. 98). Despite its controversy and different views in the philosophical community about the zombie argument, conscious experiences may still be challenging to explain.

1.3 Problems of Consciousness

In defining consciousness, the neuroscientists Antonio Damasio and Kasper Meyer (2009) identified a dual perspective (which is not a dualist theory). One perspective is internal and first-person (subjective, cognitive and mental) and the other perspective is external and third-person (behavioural, that of an objective observer). Lutz and Thompson contrasted these two perspectives by noting that first-person subjective reports are problematic and often biased and inaccurate (Lutz and Thompson 2003). In the early 1980s, Levine (1983) introduced the notion of an explanatory gap between conscious experience and its physical substrate, between the respective first-person (subjective experience) and third-person (body and behaviour) reports. Bridging this explanatory gap could help to solve the problem of consciousness. Varela (1996, 1999) proposed to bridge the gap through research in neurophenomenology that would develop towards the study of consciousness. The main aim of neurophenomenology is to generate and refine first-person data by means of a phenomenological exploration experience in order to interpret and quantify the physiological and behavioural processes that are relevant to consciousness. This explanatory gap is not supposed to be a different problem to Chalmers' (1996, p. 25) hard problem of consciousness, that of how a physical system could give rise to conscious experience, but a "methodological substitute" for the hard problem. The methodological problem is "how to relate first-person phenomenological accounts of experience to third-person cognitive-neuroscientific accounts" (Lutz and Thompson 2003, p. 47). The notion of an explanatory gap is not intended to address the hard problem, nor is anyone claiming to have bridged the gap. It provides only a pathway for scientific research. The reason for proposing this pathway is to bridge the related gap between the areas of neuroscience (third person) and psychology and philosophy (first person). Lutz has suggested that this neurophenomenological approach may complement current behavioural and cognitive research in areas such as: brain plasticity of human experience; time consciousness (Varela 1999); lucid dreaming; etc. Lutz also distinguished a number of kinds of consciousness (e.g. creature consciousness, background/state consciousness, transitive/intransitive consciousness, access consciousness, phenomenal consciousness, introspective consciousness, and pre-reflective self-consciousness), some of which are debatable and controversial (Lutz and Thompson 2003, p. 34–35).

A world renowned neuroscientist, E. Roy John (1924–2009) summarised the problem of consciousness thus: "perceptual awareness involves the integration of distributed synchronous activity representing fragments of sensation into unified global perception. How this statistical information is transformed into a personal subjective experience is the problem of consciousness." (John 2002, p. 3).

1.4 States, Contents, Levels and Processes of Consciousness

Karl H. Pribram, psychologist and psychiatrist, identified three modes of conscious experience: states, contents and processes. Conscious states are influenced by the biochemical and biophysical substrates of wakefulness, sleep and dreaming. Conscious contents are better known under perception. Conscious processing binds state with perceived content and content with state.

Steven Laureys, a neurologist, has defined consciousness clinically as having two main components: *awareness* of self and environment (content of consciousness) and *arousal* (alertness or vigilance) or wakefulness (level of consciousness) (Laureys and Tononi 2009). Awareness refers to conscious perception and consists of cognition, intentions, experiences stored in memory (from the past) and the present. Awareness of self is a mental process entangled in the socio-cultural environment but not necessarily dependent on or influenced by external stimuli. Awareness of the environment, on the other hand, is conscious perception of the environment through the sensory modalities. The functionality of consciousness is unknown and it is also unknown if it exists in the first place. Modern cognitive science has so far addressed sensation, perception, attention, and emotion. But psychiatrist and dream researcher John A. Hobson says it does not address the details in "characterizing each aspect of mentation (i.e. the process or result of mental activity)" (Hobson 2009, p. ix).

Chalmers identified psychological consciousness as consisting of: *awareness* (our ability to process information about the world and deal with it in rational fashion), *introspection* (the process by which we can become aware of the contents of our internal states), *reportability* (our ability to report the contents of our mental states), *self-consciousness* (our ability to think about ourselves or our awareness of our existence as individuals and of our distinctness from others, which is limited to humans and few animal species), *attention*, *voluntary control* and *knowledge* (Chalmers 1996, p. 26). Our awareness is accompanied by phenomenal consciousness, which is often unintentionally ignored because psychological consciousness gets all the attention.

Damasio and Meyer, with their dual perspective, identified the external perspective as the signs of human consciousness in wakefulness, background emotion, attention and purposeful behaviour. The background emotion, which differs from normal or primal emotion (e.g. anger, fear, sadness, joy and happiness) and social emotions (our "de facto moral compass" (Goleman 2007, p. 131), including embarrassment, compassion and guilt), can be expressed and observed (by an external observer) from body language (body posture, movement etc.). Interestingly, it was recently discovered that guilt dwells in the front and back cingulum of the cerebral cortex. Attention directed towards a mental object and events is evidence of the presence of consciousness. These mental objects and events can be external (e.g. phone ringing and your name being called) or internal (i.e. thoughts). In cases of schizophrenia, these perceived internal objects and events are distorted by hallucinatory internal voices, which is why consciousness experience in schizophrenic people is not considered as normal but might be seen as partly reflective behaviour. In all, Damasio summarised the external perspective by linking thoughts, emotion and behaviour: "conscious human behaviour exhibits a continuity of emotions induced by a continuity of thoughts" (Damasio and Meyer 2009, p. 5). Damasio's internal perspective can be identified as the human conscious mental state when it represents objects and events in relation to itself. In that case, the person is a "perceiving agent" (Damasio and Meyer 2009, p. 6) who is "generating the appearance of an owner and observer of the mind, within that very same mind" (Damasio and Meyer 2009, p. 5; Damasio 1999/2000). The mental events are created from a series or scenario of mental images which are generated by neural activity and integrated via sensory modalities in space and time. However, there are unknowns in how humans or animals perceive time linked to these internal conscious events.

Is it possible for anyone to consciously experience both the external and internal (dual) perspectives within themselves? Or are we able only to experience one of the perspectives at a time? Damasio defined consciousness as a "momentary creation of neural patterns which describe a relation between the organism, on the one hand, and an object or event, on the other" and extended the definition by adding that "the creation of self neural patterns is accompanied by characteristic observable behaviours" (Damasio and Meyer 2009, p. 6). The central problem in the study of consciousness is the self state: the sense of self ("I"), subjectivity and its subjective process.

According to Damasio, consciousness can be divided into two kinds, *core* consciousness and *extended* consciousness. Core consciousness is a simpler kind of consciousness, which "establishes the relationship between an object and the organism" (Damasio and Meyer 2009, p. 12). Extended consciousness is a more complex kind of consciousness, which "enriches the relationship by creating additional links between the object and the organism, not just with respect to the presence of the latter in the here and now, but also to its past and anticipated future" (Damasio and Meyer 2009, p. 12). Extended consciousness depends on core consciousness. It consists of multi-level mental information processing, depends on memory and is enhanced by language, which represent the contents or perception.

By contrast, core consciousness is independent of memory or language. Core consciousness depends on the neural activity generated in the brain mainly during wakefulness. The neurobiology of core consciousness "requires the discovery of a composite neural map which brings together in time the pattern for the object, the pattern for the organism, and establishes the relationship between the two" (Damasio and Meyer 2009, p. 8; Damasio 1999/2000). Damasio hypothesises that both extended and core consciousness dwell in the posteromedial cortex and in thalamocortical interaction (Llinas and Ribary 1993).

1.5 Altered State of Consciousness

The waking state of consciousness is typically experienced by our central nervous system responses to external stimuli, which are sensed by our sensory modalities (i.e. smell, taste, touch, hearing, vision). This waking state of consciousness relies on the way these stimuli interact with our mind and body. However, if the level of stimulus is increased or decreased so as to affect the internal responses and cause us to feel or sense the change, we know we are not in a "normal" waking state of consciousness, but in an altered state of consciousness (ASC), which may be defined as "a qualitative alteration in the overall pattern of mental functioning such that the experiencer feels his [or her] consciousness is radically different from the 'normal' way it functions" (Tart 1972, p. 1203).

Over the years, the ASC has been defined in a number of ways: as a "changed pattern of subjective experiences" (Kihlstrom 1997); or a "reflective awareness of changed pattern of subjective experiences" (Tart 1972); or a "changed pattern of subjective experiences and physiological responses" (Shapiro 1977). Tart pointed out that the three definitions do not entirely describe altered states. The first definition relies on subjective experiences and is insufficient. The second definition requires one to be aware of what is normal or altered in subjective experience, and most of the time this distinction between the normal or altered may not be clear. What is a normal experience? The third definition may give rise to problems in defining the exact neural correlates and mechanisms that may influence (or be influenced by) the subjective experiences. But it can at least assist in measuring what is normal and what is altered experience. This is where contemporary science is continuing to make progress in consciousness studies.

Tart attempted to identify mind through awareness and consciousness components (Tart 1975). Consciousness encompasses awareness in the complexity of our mind. Assigning a precise logical definition to all three keywords (consciousness, awareness and mind) may never be possible, because, as Tart explained, logic is only one component or product of the functioning of the whole mind, and this one part may not define the whole mind. For example, to describe awareness and consciousness, Tart considers the perception of sound to describe awareness and complexity to recognise that sound and identify the instrument or object that generated it.

The level of stimulus can influence the ASC, by the reduction of sensory input (meditation or isolation) or increase of sensory input through repetitive stimulation (chanting, drumming, dancing, etc.), sensory overload (e.g. dancing, music, lights, crowd or a combination of all of these), or various mental and emotional states. (Ludwig 1969). There are other stimulus factors which may also influence the ASC, but they do not depend on the sensory systems. These other stimulus factors contribute to alteration of body biochemistry or neurophysiology through sleep deprivation, stress, mental burn-out, fasting, physical and mental exercise, substance intake (e.g. drug abuse) and others. Trying to re-create an ASC whenever we wish may require years to master (e.g. via meditation and yoga) and it can lead to healthy or unhealthy lifestyles. The only time we may experience an ASC is by noticing the difference while experiencing a waking state of consciousness. The key to noticing these differences between waking and altered states of consciousness is through self-awareness. Even this self-awareness can be influenced by perception (so that it is not what it seems) and distorted by many psychophysiological and psychotic disorders (e.g. schizophrenia, a distortion of reality characterised by hearing voices and seeing unreal images).

For Dietrich, "any modification of information processing, from the sensory level to the prefrontal cortex, alters the content of consciousness" (Dietrich 2003, p. 237). Dietrich proposed that to induce each altered state of consciousness, a particular behavioural technique would need to be utilized. Inducing an ASC would generally decrease the activity in the prefrontal cortex or cause prefrontal cortex deregulation or transient hypofrontality, and in turn produce minor changes in the contents of consciousness. Such ASCs as dreaming, daydreaming, meditation, hypnosis, "runner's high" (endurance or marathon running experience) and drug-induced states, all have a cognitive aspect that is controlled by deregulation of the prefrontal cortex. It is not just the prefrontal cortex or neural structure that is necessary for consciousness. There are functional neuroanatomical layers that are responsible for an ASC. If one neural layer (according to Dietrich's model) were damaged, then it would alter a particular state of consciousness.

Recent research indicates that romantic rejection and similar psychological pain can alter the heart rhythm. This romantic rejection can be traumatic: the feeling of being rejected can slow the heart rate and even stop the heart momentarily. It has been suggested that some brain regions are origins for processing the physical and emotional pain and that the pain affects the autonomic nervous system. What is more important is that this incident can trigger depression in some predisposed people. Others may have post-experience of an ASC by having precognitive dreams, strong intuitive feelings, etc. However, it not known why such ASCs occur during these post-traumatic periods. Conceivably, they "heal" the pain of rejection by providing "hope" to a person that he or she will be fine and perhaps reconnect with their loved one or find another romance. Perhaps it is simply a distorted physiological effect that triggers this ASC. If a human body is capable of self-sufficient healing at the molecular level (as living cells continually die and new cells replace them), is there a psychological and phenomenological process or mechanism to heal a broken heart from the trauma of romantic rejection?

Drowsiness is a transition between the waking state and the sleep state. It may not necessarily involve turning one state off or on to make room for the other state. The waking state may drift in and out of an early sleep stage a couple of times, creating the state or process of drowsiness. But how does that process translate to how one would feel during that transition? For example, an experienced professional truck driver can fall asleep momentarily while driving and let go of the steering wheel, causing the truck to steer off the road and crash. It is understood that when a driver falls asleep, the muscle tone decreases, the eyelids close, the jaw and neck muscles relax, and the hands loosen their grip on the steering wheel. However, what is difficult to comprehend is that the same driver would be aware of falling asleep and loosening his or her grip on the steering wheel and yet continue to drift further into the state of drowsiness. In that state of deep drowsiness, one would have an momentary feeling that nothing is important and let go of the conscious control that is present only during wakefulness. In reality, the truck driver would normally be trying to make a freight delivery on time and that background state of stress would be sufficient to ensure alertness and keep the driver awake while driving. But the pleasant sensation accompanying the attitude that nothing matters while drifting from wakefulness to sleep (a transition from one state of consciousness to another) becomes the primary working mode instead of responsible driving. Often, a sleep debt or tiredness may help to make a driver fall asleep, but in other circumstances, it may only be boredom in traffic that stops the brain from staying alert. The details are irrelevant here. What is important is the power of transitional consciousness and state switching to control our lives and change our behaviour.

While sleep debt, tiredness and traffic boredom can cause a driver to slip in and out of and the waking-sleeping transition, a daydreaming ASC might be working in the background mode and the same driver may not be aware of it while driving. That driver would eventually arrive at the destination after few hours, navigating through traffic, and would not be able to recall how he or she consciously drove all that time. This experience of what may be an ASC proves that the awareness and processing of external sensory information does not decrease but is on the contrary increased when the driver's mind wanders elsewhere. It is not yet known whether these simultaneous state-switching operations (i.e. conscious driving while navigating through traffic, slow drifting to sleep and daydreaming) are generated by the same or different cortex layers or how these layers interact with each other. We do not know which cognitive function and neural mechanism decides when and whether an ASC works in the primary working mode or background mode.

1.6 Human and Animal Sleep

Wakefulness and sleep are equally fascinating states of consciousness, especially the unique ASC, known as sleep onset, the hypnagogic state or the "wake to sleep" transition (the author describes this scientific research in more detail in Chap. 7). Sleep research pioneer Nathaniel Kleitman (1895–1999) claimed that conscious experiences are different while awake and asleep. In fact, one can be conscious while asleep and unconscious while awake. This observation may contradict the general understanding that one is not conscious while asleep, because the dream experiences can tap into the unconscious states. Sleepwalking is a sleep disorder, which occurs during deep stages of sleep (low consciousness) when one is consciously active (walking, talking, etc.) much like when awake. René Descartes was fascinated with philosophical ideas that resulted from images that appeared to him during the hypnagogic state. The mental activity at sleep onset can lead to hypnagogic hallucinations, resulting in either terrifying or creative experiences. It has been suggested that creativity and insight in sleep (compared to wakefulness) generally involve seeing information in new combinations and sequences, presented to consciousness (Broughton 1982). There have been cases where sleep (either nocturnal slumber or diurnal napping) has lead to the creation of musical or poetic compositions. Dreaming in sleep has solved problems, as in the case of at least two Nobel prizes. One was for discovering the elusive structure of the benzene molecule, which was in the form of a ring, in fact a hexagon, which August Kekulé visualized in his dream as a coiled snake. Another Nobel prize resulting from a dream was for Otto Loewi's experimental demonstration of neurochemical transmission via acetylcholine and involving epinephrine (adrenaline). If John Allan Hobson explains dreaming by microscopic disorientation and defines it as delirium, how do solutions to problems and insights appear to a dreamer?

Sleep "grogginess" or sleep inertia is the abnormal excessive confusion that can occur during the transition from sleep to wakefulness, a transition known as the sleep offset process (as opposed to sleep onset, i.e. the wakefulness to sleep transition). This sleep inertia is a natural altered state of consciousness, which requires further study. It is reported that sleep offset affects sleep inertia, which in turn affects mood behaviour. Moods can be positive or negative, depending on the amount of sleep and the moment of sleep offset (i.e. whether the subject is awakened from light, deep or REM sleep). Generally, if a subject is awakened from deep sleep, the sleep inertia is longer and greater. However, it is still unknown how mood is affected as a result of this prolonged sleep inertia.

The core function of sleep is unknown when tested under the null hypothesis. If the null hypothesis were correct, studies would be able to find that: there are animals that do not sleep at all; there are animals that do not need recovery sleep after a prolonged awake period; and the effects of sleep deprivation do not result in death. The nature of sleep is not universal across all known living species, but "sleep is present and strictly regulated in all animal species that have been carefully studies so far" (Cirelli and Tononi 2008, p. 1607). For some reptiles, amphibians and fish, scientific investigations suggest that they might simply rest instead of sleeping the way mammals and birds do. Flies' cerebral electrical activity is similar when awake and asleep to mammals. Bullfrogs' respiratory responses to stimuli suggest that they do not sleep, but more studies are needed to confirm this. Some coral reef fish sleep while they swim at night. However, there is no evidence to suggest that any species do not sleep at all. It is known that all terrestrial (excluding marine) mammals normally exhibit simultaneous electroencephalographic (EEG) activity in both sides of the brain (left and right hemispheres). However, there are exceptions to this simultaneous EEG activity in both sides of the brain for humans, whales and dolphins. Bottlenose dolphins sleep by alternately turning off the left and right hemispheres of their brains. If the left side of their brain is asleep, the right is awake, and vice versa. This behaviour is known as unihemispheric sleep. The main reason for this hemispheric switching to regulate wake-sleep cycles in dolphins is that dolphins are constantly on the move in order to survive and protect their offspring from predators, such as sharks. This movement is unceasing from a dolphin's birth to its death. Humans are never consistently mobile and active (sleepwalking occurs in shorter episodes but is never continuous). More importantly, humans are never able to switch their brain sides in order to alternate their wake-sleep cycles. This may be one of the reasons for regarding dolphins as the most intelligent marine species. It remains unknown whether dolphins ever experience drowsiness.

Sleep has qualitative (intensity or depth) and quantitative (duration) measures. Humans and rodents experience the deepest sleep just after sleep onset. Humans experience sleep debt by staying awake for prolonged periods of time and then need a restorative sleep. This sleep debt reflects on the qualitative and quantitative measures of their sleep. While some people regularly need 8–10 h doses of sleep, others need fewer hours. The amount of sleep necessary for a human to function properly varies from one person to another. Studies have suggested that reducing regular sleep by a few hours can lead to significant risk of overall mortality. Studies on sleep-deprived rats have resulted in their death after 2–4 weeks. Death due to sleep deprivation was also evident in flies, cockroaches and humans. Humans can develop an extremely rare degenerative brain disease, called fatal familial insomnia. This disease, which prevents humans from falling asleep and maintaining a normal sleep, can last for several months and eventually leads to death. However, it is still not clear whether death is caused directly by sleep deprivation or indirectly by forced arousal, stress (associated with sleep deprivation), heart disease, stroke, diabetes, cancer, etc. One experiment revealed that a human can remain awake for 11 days. Unverified reports have claimed that some humans have remained awake for weeks and still functioned properly.

1.7 Meditation and Neuroplasticity

More research is required to characterise the nature of the differences among types of meditation. The phenomenological differences suggest that these various meditative states may be associated with different cerebral electrical activity or EEG oscillatory signatures. Much remains unknown about the meditation or mental training process and its impact on the brain. Some studies have recently linked meditation with plasticity of the brain or neuroplasticity. The term neuroplasticity is used in the neuroscience community to describe the ability of the neural network to change by growing new neurons or dendrites and synapses. Some fascinating studies have suggested that neuroplasticity can be enhanced under the mental training of meditation (Berger et al. 2007; Poldrack 2002). Some of the studies report a progressive increase in theta and alpha EEG bands, where the transition from a neutral state to a meditative state requires 5–15 s (depending on the meditation subject) and is characterised by gamma EEG band synchronisation. Functional magnetic resonance imaging (fMRI) experiments were conducted on Buddhist monks in clinical settings and the results reveal that changes in the cortical evoked response to visual stimuli and amplitude and synchrony of gamma EEG band oscillations reflect the influence of various styles of meditation on attention. There are further suggestions that these changes in neural connectivity are triggered by the gamma EEG band synchronisation (Lutz et al. 2004).

1.8 Neural Correlates of Consciousness

Chalmers asked how one performs experiments to detect a correlation between some neural process and consciousness (Chalmers 1996, p. 115). Often, the main parameters of consciousness would consist of: focus of attention, subjective description of particular internal state, control of behaviour and even emotion. These parameters would then be correlated with any neural processes upon its activation or occurrence. The importance of studying the correlation between neural processes and parameters of consciousness is undisputed but the link between the psychology and conscious experience remains unknown.

Since the early 1990s, neuroscientists have been investigating synchronised gamma EEG activity (35–45 Hz) in relation to the binding problem. The activity may be related to conscious experience and seems able to bind certain functional information by synchronising the frequency and phase of the gamma waves carrying it. The binding problem is that of how different sensory input, such as colour, shape and location can be bound and experienced together as a single event. Functional information may include the matching of memory contents and perceptual contents, grouping of letters to form a word or words to form a sentence, learning, etc. The hypothesis is that information is bound and stored in working memory and finally integrated into the contents of consciousness. Two renowned neuroscientists, Francis Crick and Christof Koch, were the first to demonstrate the link between temporal binding and consciousness by bridging binding and sensory awareness (Crick and Koch 1990). Experiments show that gamma EEG activity is related to visual sensory binding. Evidently there is a mechanism that processes the functional information along a common synchronised pathway to generate the large functional states that underlie cognition. In other words, gamma EEG thalamocortical resonances are correlates of cognition. Similar research has revealed that the same 40 Hz waves of electromagnetic activity occur during wakefulness and dreaming (REM activity), despite the evident differences between these two states. The dreaming state may be described as "hyperattentiveness in which sensory input cannot address the machinery that generates conscious experience" (Llinas and Ribary 1993, p. 2081). This is one of a number of theories where awareness has physiology and anatomy. Another study suggests that gamma EEG band synchrony may play a major role in the processes of visual consciousness (arousal, selective attention, working memory, etc.) and higher order consciousness (motivation, symbolic processing, action planning, etc.) (Engel and Singer 2001). The latter study revealed a single object response of gamma oscillatory synchronisation within and between visual cortical regions.

It is believed that subjective time is experienced as continuous. However, the neurophysiological processes that characterise consciousness are discontinuous in time (around the 80 ms epoch) and define a "travelling moment of perception" (Allport 1968). The significance of perception and what is actually perceived within our central nervous system can be explained by means of the illusions or tricks that out mind plays on us through our own sensory systems. For example, we perceive the flickering of lights at frequencies above 50–60 Hz as a steady light, and do so every day with our house lighting. As soon as the flicker frequency decreases, we begin to distinguish dark from light phases. This subjective visual sensation occurs somewhere below 50 Hz at a frequency called the critical fusion frequency. Similarly, our auditory sensation alters from a tone to an intermittent sound at 30–35 Hz (Wever and Lawrence 1954).

Tononi identified two properties of conscious experience: integration, or the inability to subdivide consciousness; and differentiation, the ability to select out of a wide range of different conscious states within a short time (Tononi 1998). In terms of integration of conscious experience, it is false to assume that certain local neurons or regions are more responsible for the occurrence of consciousness than other neurons. The general property of conscious experience is that the selection among integrated states must occur within few hundred ms in order to create a contemporaneous conscious experience. A group of neurons can influence conscious experience only if it can perform a certain functionality or task in less than a second. These groups of neurons may interact between posterior thalamocortical (associated with perceptual characterisation), anterior (associated with memory) and other regions. Tononi's claim contradicts the 1950s hypothesis by Wilder Penfield that stimulated brain structures outside or within the thalamocortical system, have no direct influence on conscious experience (Penfield 1958). Experimental findings indicate that in order to generate a conscious sensory experience, a high-frequency somatosensory (sensory modality for touch, temperature and pain) stimulus needs to be delivered to the thalamus for some 500 ms, and for at least 150 ms to trigger sensory detection without conscious awareness (Libet 1993). These times in ms allow a sufficient number of oscillatory cycles for synchronisation of the EEG activity. During the low-amplitude high-frequency EEG activity of the awake state and of REM sleep, the firing of neurons is not globally synchronous, which is consistent with reports of vivid dreams when awake. By contrast, people do not recall any vivid dreams or conscious experiences during deep sleep, characterised by high-amplitude low-frequency EEG activity, where neurons are synchronous and interactive in the thalamocortical system. Similar states of unconsciousness are generated during seizures, where neurons are highly interactive and synchronous across the whole brain and the subject becomes unconscious. Such experimental evidence confirms that certain conscious experiences are correlated with alterations in neural activity that are driven by external or internal (images, dreams or memories) stimuli.

Our conscious experience of time is not generated by biophysical and biochemical neural processes alone. It is unknown how the human or animal perception of time is linked to experience and how that perception can be altered. The scientific focus has been to ask whether perception is continuous or segmented into episodes, as on an old film strip. Many visual illusions are part of our everyday experience. One such illusion is the "wagon wheel" effect where if a wheel is observed rotating at a certain speed, its spokes may be visually perceived as not moving or moving backwards. When this illusion was tested, it was found that when all subjects reported that the wheel was moving in the opposite direction, the EEG activity in their right inferior parietal lobe (associated with perception of visual location) was oscillating at 13 Hz. The results suggest that our visual information is processed in discrete frames and certainly not continuously. This frame-based visual perception may have to do with the brain processing different objects independently. Similar studies of visual perception showed that photic threshold stimuli are detected or perceived as theta and low alpha EEG phase responses (Busch et al. 2009). For the neural system to process sensory information in the auditory, visual or tactile modes, the information needs to be segmented into frames of 30–50 ms. Sensory information spread over longer than about 50 ms is not detected within one frame. Ernst Poppel described these frames as building blocks of consciousness. Other studies have revealed that if subjects are stimulated by a burst of auditory and visual tones and flashes, this may be perceived as a longer or shorter intermittent tone or flash. Reports also indicate that such auditory and visual stimuli may entrain gamma and other EEG activity, and that the faster the oscillations (in the gamma band), the more building blocks of consciousness or sensory perceptions are processed in our neural system.

1.9 Schizophrenia and Its Neural Correlates

Schizophrenic patients typically perceive time differently, suggesting that their internal clock might be altered. But what alters that internal clock in the first place? The term "schizophrenia" comes from the Greek language and means split mind. It was first recognised a century ago by a psychiatrist Eugene Bleuler, who described this mind splitting as a symptom of "erronous, irrational thinking and the inability to experience normal emotions" (Kidman 2007, p. 1). A split personality, where patients can turn into dangerous and violent people, is not a symptom of schizophrenia. Unfortunately, this is one of the social stigmas that has been associated with this illness for a long time. The primary symptoms (positive and negative) of schizophrenia include: hallucinations, hearing voices (one or more), seeing visions, having imaginary conversations and thoughts which command them to do things, hearing sounds and even music, having delusions (imagining that people can read their thoughts or that they are religious figures), and thought disorder and sleep abnormalities. The secondary symptoms are chronic are often follow from the primary symptoms. They include emotional blunting (inability to express emotion by normal means), loss of motivation and interest in work and social life, social isolation and withdrawal. All this explains why schizophrenia is considered a debilitating psychotic disorder. How are these voices, music (which may not be recalled from memory), commanding thoughts and so on generated within one's mind?

Hallucinations are conceptualized as perceptual experiences related to attentional factors, which predominantly occur during the waking state of consciousness. There have been suggestions that schizophrenic hallucinations are caused by peripheral sensory impairment and even elevated random neural activity. But if they are caused by random neural activation, how can this explain the structured commanding voices and thoughts (which tell the patient what to do) that are experienced as verbal hallucinations? Hallucinations are generated in the thalamocortical part of the brain during the state of arousal and attention. The sensory input information gets distorted on its way to (or within) the thalamocortical region. How are these hallucinations developed as part of a mental illness in the first place? One suggestion is that "biological vulnerability and psychological influences" (Behrendt 2003, p. 433) are psychological factors that explain the verbal hallucinations in mental illness patients. Verbal hallucinations may (unconsciously, as Freud suggested) reflect interpersonal ability, social phobias and problems, anxiety and a desire to have social experiences. Some researchers maintain that a person who is socially isolated and has a "biological vulnerability" or predisposition would be more inclined to experience verbal hallucinations and develop psychotic illness. The patients generate verbal hallucinations to compensate for deficits in social interactions and experiences. An alternative suggestion is that hallucinations are caused by a poor integration of sensory input with memory.

Studies have revealed that hallucinatory experiences in schizophrenic patients are correlated with coherent EEG gamma-band oscillations. During hallucinatory episodes, activity in the thalamocortical region is synchronized without any sensory input to drive that synchronisation (Behrendt 2003). Other experiments on schizophrenic patients have shown that stimulation with sensory input (an auditory 40 Hz click) can substantially delay the synchronization and desynchronisation of the EEG in the gamma band (40 Hz). However, this delay might not be evident in healthy patients. Other studies have verified that rhythmic sensory stimulation responses differ between healthy and schizophrenic patients. The schizophrenic patients exhibited reduced responses, highlighting that there might be sensory impairment or degenerative alteration in relation to thalamocortical activity (Herrmann and Demiralp 2005). Most studies on schizophrenic patients, including binding problem studies, reveal a deficiency of gamma EEG oscillations linked to the memory and attention deficits present in schizophrenic patients. However, there are also a few studies that reveal an increase in gamma EEG oscillations under the induction of somatosensory hallucinations. This reduction and increase in gamma EEG oscillations depends on whether the schizophrenic patients have positive or negative symptoms. A reduction of gamma activity appears with negative symptoms and an increase in gamma activity with positive symptoms.

1.10 Déjà Vu

The déjà vu experience is believed by some people to be a part of dreaming or supernatural premonition. Because of its relation to memory, déjà vu encompasses ingredients of dreams, recalling scenes from dreaming or waking states and real or imaginary scenes. Unfortunately, déjà vu is commonly observed in epileptic and schizophrenic patients. Déjà vu occurs when a momentarily experienced scene seems familiar, as if the patient has seen it before or as if history is repeating itself, which can be disturbing if it occurs frequently (especially in epileptic patients). In any case, a person having a déjà vu experience cannot recall that familiar scene because it might not have been experienced in the first place and it might not be stored in memory. Gamma EEG oscillations were increased in both healthy and schizophrenic patients while they hallucinated known sounds. Perhaps the reason for an increase in EEG gamma oscillations during any perception is to be found in the mechanisms linking sensory input and brain processes (i.e. memory and attention). For schizophrenic hallucinations, the sensory input is absent and the perception is generated within the brain process on its own. These mechanisms are triggered to distinguish the perceived information (identifying objects, sounds, scenes, etc.) from background noise or unrelated information. A model has been proposed for déjà vu and hallucinations in normal, epileptic and schizophrenic conditions based on gamma EEG oscillations (Herrmann and Demiralp 2005). There have been many hypotheses for déjà vu, such as distortion in time perception; dissociation between familiarity and recall where a sense of familiarity is not linked to experience in a real history; and a strong link between recall, familiarity and emotion. Emotional arousal can amplify the sensation of déjà vu and the sense of reliving a history that in fact never happened in the first place. Results have revealed that epilepsy affects the amygdala brain region, which is responsible for emotion, as well as the perirhinal cortex, responsible for feelings of familiarity. In consequence, some observers report an influence of binding emotional stimuli and emotional perception on the decrease or increase of gamma EEG oscillations. Déjà vu is a cognitive process operating in a background mode (like daydreaming) but it does not include the sensations of recollection, daydreaming or awake reality. It is surely one of the most intriguing conscious experiences ever experienced by a healthy person.

1.11 Music Therapy

Music therapy is a psychosocial therapy that can complement psychopharmalogical medication by treating symptoms that medications cannot reach. Music therapy can engage the schizophrenic patient in music experiences that provide a supportive atmosphere and stimulate behavioural changes, perceptual-motor behaviour, social interaction and interpersonal skills (Unkefer and Thaut 2005).

Why does music have this effect? Music and song live in the core of human lives. Music is a powerful medium that can alter mood and behaviours. Music can induce changes in human emotions. For that reason, it is used in clinics as a therapeutical method to treat many psychological, mental and physiological disorders. Neuroscience has always attempted to reveal what is going inside our brain when we listen to music. Music connects us with others and ourselves by creating a socio-cultural cohesion that is larger than us. Music is often stored in our memory and linked to our history of awareness of self and the environment, which is also linked to conscious perception. Music can affect our whole mind and body via the experience of euphoria. It is still unknown how musical features trigger this special human feeling and why we respond to certain music genres or songs more than others. Music can "reach the 'inner self' and function as a source of emotional support, as the patient deals with the trauma of the disease" (Purdie 1997, p. 212). The ongoing research since the 1990s into music therapy has been aimed at discovering how the neural activity of rhythmic information processing interacts with the rhythms and sounds in music. Music is complex in time, unpredictable and non-sequential, and it arouses attention through anticipation of the next musical tone and note. Music comprises pitch, timbre, vibrato, intensity, beat tracking, tempo or velocity, contour or intonation, rhythm variation or timing, and melody and harmonic features, all of which can drive and entrain neural, respiratory and cardiac activity. Studies suggest that music can arouse and excite the spinal motor neurons via auditory neurons at the brainstem and spinal chord level. It is the brainstem that connects and synchronises the neural activity with respiratory and cardiac activity. Music provides a complex and strong temporal stimulus that can induce emotions. At times, musical rhythms can also provide "temporal structures through metrical organization, predictability and patterning" and can regulate "physiological and behavioural functions via entrainment mechanisms" (Thaut 2008, p. 83). Music is referred to as the language of the emotions, and music sounds the way moods feel. "Music is structured in terms of tension and release, motion and rest, fulfilment and change" (Winner 1982, p. 211). These emotions can be described as "an inferred complex sequence of reactions to a stimulus and includes cognitive evaluations, subjective changes, autonomic and neural arousal, impulses to action, and behaviour designed to have an effect on the stimulus that initiated the complex sequence" (Plutchik 1984, p. 217). Music can actively engage attention and memory systems, stimulate attention on multiple levels and in many ways facilitate cognitive and behavioural changes. Music has been utilised in auditory attention, perception and memory training. It is well known that positive mood states enhance memory while listening to music and enhanced memory recall. Music can form the neurobiological basis of perception and learning. Studies have reported that brain regions (Broca's and Wernicke's areas) used to process music are also involved in processing language. Children who are training to play music undergo changes in the structure of their brains, as shown by developed auditory, motor and visuo-spatial areas, and by increased volume of the corpus callosum (which provides the information path between the left and right hemispheres).

A study was conducted to see whether listening to music was able to synchronise the EEG activity of musicians and non-musicians (Bhattachrya and Petsche 2001). This EEG synchronisation while listening to music was assumed to be related to long-term memory representation of music and its access to that memory. It was found that only the musicians exhibited a significant increase in gamma EEG band synchronisation while listening to music, which explains how their ability to learn and play music is related to their memory. We can perhaps relate this finding about EEG synchronisation to our understanding of the popular idea that someone (a musician or non-musician) is in tune with the music or has an ear for music. A new group of subjects who are not musicians but have those abilities should perhaps be investigated in the future.

1.12 Near Death Experiences and Consciousness

Scientific studies of near death experience (NDE) continue to perplex the medical and neurophysiological communities in relation to human consciousness and the mind-body problem (Van Lommel et al. 2001). It is hard to see how someone could have a conscious experience while their body is clinically dead (Sabom 1998). It has been suggested that NDE is a special state of consciousness that includes out-of-body experiences during subjective reports of various life-threatening circumstances such as cardiac arrest, coma following brain injury, etc. There have been studies investigating the relationship between physiological, psychological, pharmacological (and other) aspects of these conscious experiences of clinical death (Van Lommel et al. 2001). The studies revealed that brain activity in brain-dead patients could not generate these conscious experiences and that there is no scientific evidence for these relationships to exist. Furthermore, they suggested that NDE would be a *changing* state of consciousness in which memories, identity, cognition and emotion functioned independently of the unconscious body.

The interconnections and transitions between the waking and other special states of consciousness, such as NDE, sleep, dream, daydreaming, mind-wandering, trance, dissociation, meditation and hypnosis, offer an explanation for various extraordinary elements that humans report, such as glimpses of creative insights, hypnagogic hallucinations, intuition, visions and prognostic dreams. It seems possible that the combined interconnections and approximate transitions (through fusion) of waking and these special states of consciousness often enables humans (perhaps even animals) to consciously experience these unexplained events, despite no clear scientific evidence on the physiological or psychological basis. Indeed, it seems possible that it is only when these transitions between states of consciousness occur in synchrony with other unknown factors that we as humans first become aware of these extraordinary powers. Whereas sleep paralysis results in hallucinatory experiences in which one is convinced that paranormal or supernatural experiences are real, the NDE and other experiences remain far from being defined as just another hallucination. If there are other factors that work in synchrony with this transition between states of consciousness, what are they and can they be controlled?

Considering that many specific states of consciousness have the potential to be categorised as altered states, it seems that the only time humans are aware of a change in their consciousness is through their own conscious experience along the pathway of a transitional state of consciousness. It also seems that anyone who remains in one state of consciousness may be unable to reach that experience of change. Indeed, these transitional states may only amplify an existing conscious experience so that we actually get to experience it, perhaps as something that cannot be understood or explained. However, the most powerful transitional state of consciousness that one can experience is the NDE, because this transition is from one's life to death. But is death a state of consciousness? The author, at least, has been fascinated with the transition from waking to sleeping, which generally coincides with the transition from day to night.

1.13 Overview of Contributions

This book encompasses many aspects of consciousness studies. Despite the fact that there is now a huge literature on consciousness, the editors have decided to focus on four main areas: introduction to states of consciousness (Chaps. 1–3); sleep and dreaming consciousness (Chaps. 4–7); biomarkers in monitoring altered states of consciousness (Chaps. 7–9); meditation consciousness and sound trance and yogic slow breathing induction (Chaps. 9–11); and misconceptions about what are widely called states of consciousness (Chap. 12).

Olivia Gosseries, Steven Laureys and their colleagues survey and address various disorders of consciousness, characterized by a disrupted relationship between arousal and awareness components. Conscious states such as coma and vegetative disorders clearly exhibit distinctive arousal and awareness characteristics and provide insights to the neural correlates of consciousness. This clinical research group survey contemporary functional neuro-imaging and electrophysiological monitoring techniques, which still limit the quality of clinical diagnosis, prognosis and treatment of many disorders of consciousness.

Gerard A. Kennedy provides an introductory account of states of consciousness and surveys many modes of conscious experiences, as defined by Pribram and Laureys. Furthermore, Kennedy proposes that it may be possible to identify neural and other physiological correlates that underlie consciousness. From a psychological point of view, Kennedy argues that intimate knowledge of the sequencing of "units" of consciousness may allow abnormal patterns of behaviour to be identified from as early as childhood in order to develop proper corrective treatments for many pathological problems and disorders.

In Chap. 4, Michael Schredl and Daniel Erlacher address the question of how dream consciousness is related to the physiology of sleeping. They describe how dream and lucid dream research studies on REM sleep have so far shown strong correlations between dreamed and actual actions regarding central nervous activity, autonomic responses and time aspects. These authors firmly believe that for the phenomenon of dream consciousness and consciousness in general to be further explored and understood, continued research is needed into the psychophysiology of sleep and dream content through a more methodological approach and advanced neuro-imaging technologies.

Agostinho C. da Rosa and João P.M. Rodrigues believe that sleep and dreaming are inseparable phenomena and from this perspective present old and new behavioural and physical interventions on dreaming during the REM-NREM sleep process. The authors describe the relationship of REM-NREM sleep and dreaming neurophysiology with the autonomous nervous system (ANS). They also review dream therapy as part of the analysis of objective dream content and neuro-cognitive theory.

Over the years, investigators have attempted to develop theories of states of consciousness in order to define it scientifically. Some of these theories have been mentioned briefly in this chapter and are widely accepted, but there are also other theories that continue to be debated in many scientific communities. One theory that is widely accepted is the neurobiological theory of dreaming, developed by J.A. Hobson. In Chap. 6, Janette L. Dawson and Russell Conduit present some of the basic principles of theory construction that can provide a framework for evaluating Hobson's neurobiological theory of dreaming. Dawson and Conduit have attempted to highlight the nature of the assumptions that underlie dream theories, the logic of argument, and the validity of methodologies used in collecting the empirical evidence, according to principles of theory construction and validity.

In Chap. 7, Dean Cvetkovic and Irena Cosic (the volume editors) explore and evaluate biomedical engineering biomarker techniques in monitoring the process of falling asleep (sleep onset). Throughout the complex and unpredictable sleep onset process, various electrophysiological, cognitive and behavioural alterations occur, all linked to this transitional state of consciousness. The authors believe that this unique ASC, known as the sleep onset process, can be voluntarily altered and induced using a biofeedback technology. The phase synchrony during the sleep onset process within and between neural, cardio and respiratory activities and the environment provides a "signature of subjective experiences" that is linked to this transitional state of consciousness. While there is gradual progress through technological advances in various human studies, one would hope that by questioning why and how we consciously experience hypnagogic phenomena through a dual perspective (the first-person view and the observer's third-person view), we might also be able to shed a light that may lead us towards a better understanding of consciousness and ourselves.

Complementing the research presented in Chap. 7 on new biomedical engineering biomarker techniques in monitoring the sleep onset process and its transitional state of consciousness, Nada Pop-Jordanova demonstrates in Chap. 8 that an EEG parameter known as "brain rate" can objectively measure a mental arousal. This mental arousal maybe able to characterise the level of consciousness, irrespective of its content, and be utilised as a diagnostic indicator of general mental activation in addition to heart rate, blood pressure or temperature as standard indicators. Referring to her individually adapted biofeedback studies, Pop-Jordanova shows that brain rate can be used to discriminate between various arousal, sleep and attention deficit hyperactivity disorders.

Emil Jovanov argues in Chap. 9 that in order to understand and achieve physiological correlates with ASC or expanded states of consciousness, one can adopt biofeedback tool with slow yogic breathing and chanting techniques. In Chap. 7, it was explained how biofeedback can be achieved by attempting to steer one's physiological rhythms towards the desired state of consciousness of hypnagogia using artificial driven environmental stimuli, such as light and sound. Jovanov explains how a similar biofeedback technique can be adopted to create stimuli through one's own breathing and chanting in order to induce states of expanded consciousness. Jovanov believes that this approach may facilitate improvement of efficiency, creativity and spiritual growth.

While Jovanov argues that slow yogic breathing can induce relevant altered state of consciousness, Frederick Travis asserts in Chap. 10 that unique phenomenological and physiological correlates can be achieved during the practice of transcendental meditation (TM). Travis proposes three categories of meditation practices, namely focused attention, open monitoring and automatic self-transcending meditations. Travis discusses a particular model that integrates meditation experiences with the three background states of consciousness (i.e. waking, sleeping and dreaming) in order to investigate the full spectrum of other states of consciousness. The intriguing aspect of Travis' research through his modelling is that neural correlates suggest that there are certain conscious experiences associated with different combination of transitions between waking, sleeping and dreaming and various EEG transients. Most of all, each of these conscious experiences is also combined with pure consciousness, which Travis says is generated from the practice of automatic self-transcending meditation. Travis' approach of associating certain conscious experiences with the transition between waking and sleeping coincides well with the suggestion made in Chap. 7 that conscious experience exists along the pathway of the transitional wake and sleep states of consciousness or within the hypnagogic phenomenon.

As described in this first chapter, music has therapeutic properties, which can assist throughout a medicated treatment. Whereas Jovanov discussed how chanting can induce certain state of consciousness, in Chap. 11 Jörg Fachner and Sabine Rittner discuss their investigation on how music can induce ASC through a trance mechanism and whether the setting and rituals connected to music are responsible for the induction of ASC. Like Pribram, they argue that conscious experiences can be perceived through the socio-cultural environment, recalling Laureys' claim that the awareness of self is a mental process, entangled in the socio-cultural environment. Fachner and Rittner also explored sound trance induction through neural correlates and hypnotisability and found certain trance reactions in line with EEG activity changes.

Finally, Adam Rock and Stanley Krippner argue that while states of consciousness have been the subjects of extensive talks and publications, the concept is neither defined nor sufficiently understood. Rock and Krippner claim that definitions of states of consciousness typically confuse consciousness and its contents by explicitly stating that a state of consciousness is the content available to conscious awareness. In other words, the terms "states of consciousness" and "altered states of consciousness" rest on a conflation of consciousness and its contents. Rock and Krippner believe that to correct this misconception or fallacy, instead of "altered states of consciousness" one should say "altered pattern of phenomenal properties".

References

Allport DA (1968) Phenomenal simultaneity and perceptual moment hypothesis. Br J Psychol 59:395–406

Behrendt RP (2003) Hallucinations: synchronisation of thalamocortical gamma oscillations underconstrained by sensory input. Conscious Cogn 12:413–451

Berger A, Kofman O, Livneh U, Henik A (2007) Multidisciplinary perspectives on attention and the development of self-regulation. Prog Neurobiol 82(5):256–286

Bhattachrya J, Petsche H (2001) Enhanced phase synchrony in the electroencephalograph gamma band for musicians while listening to music. Phys Rev E 64(012902):1–4

Broughton R (1982) Human consciousness and sleep/waking rhythms: a review and some neuropsychological considerations. J Clin Neuropsychol 4(3):193–218

Busch NA, Dubois J, Van Rullen R (2009) The phase of ongoing EEG oscillations predicts visual perception. J Neurosci 29(24):7869–7876

Chalmers DJ (1996) The conscious mind, in search of a fundamental theory. Oxford University Press, Oxford

Cirelli C, Tononi G (2008) Is sleep essential? PLoS Biol 6(8):1606–1611, e216m

Crick F, Koch C (1990) Towards a neurobiological theory of consciousness. Semin Neurosci 2:263–275

Damasio A (1999/2000) The feeling of what happens: body and emotion in the making of consciousness. Harcourt Brace, New York

Damasio A, Meyer K (2009) Consciousness: an overview of the phenomenon and of its possible neural basis. In: Laureys S, Tononi G (eds) The neurology of consciousness. Academic, Elsevier, pp 3–14

Dennett DC (1978) Toward a cognitive theory of consciousness. In: Dennett DC (ed) Brainstorms. MIT, Cambridge, MA

Dennett DC (1991) Consciousness explained. Little Brown, Boston

Dietrich A (2003) Functional neuroanatomy of altered states of consciousness: the transient hypofrontal hypothesis. Conscious Cogn 12:231–256, 237

Engel AK, Singer W (2001) Temporal binding and neural correlates of sensory awareness. Trends Cogn Sci 5(1):16–25

Gebauer J, Sedikides C (2010) Yearning for yesterday. Scientific American Mind, pp 30–35

Goleman D (2007) Social intelligence. Arrow Books, London

Herrmann CS, Demiralp T (2005) Human EEG gamma oscillations in neuropsychiatric disorders. Clin Neurophysiol 116:2719–2733

Hobson JA (2009) Prologue. In: Laureys S, Tononi G (eds) The neurology of consciousness, cognitive neuroscience and neuropathology. Academic, Elsevier

Hofstadter DR (1985) Who shoves whom around inside the careenium? In metamagical themas. Basic Books, New York

James W (1890) In: Miller GA (ed) The principles of psychology. Harvard University Press, Cambridge, 1983
John ER (2002) The neurophysics of consciousness. Brain Res Rev 39:1–28
Kidman A (2007) Schizophrenia: a guide for families. Biochemical & General Services, Australia, pp 1–49
Kihlstrom JF (1997) Convergence in understanding hypnosis? Perhaps, but perhaps not quite so fast. Int J Clin Exp Hypn 45:324–332
Kokoszka A (2007) States of consciousness: models for psychology and psychotherapy. Springer Science+Business Media LLC, Philadelphia
Laureys S, Tononi G (2009) The neurology of consciousness, cognitive neuroscience and neuropathology. Academic, Elsevier
Levine J (1983) Materialism and qualia: the explanatory gap. Pac Philos Q 64:354–361
Libet B (1993) The neural time factor in consciousness and unconscious events. Ciba Found Symp 174:123–137
Llinas R, Ribary U (1993) Coherent 40-Hz oscillation characterizes dream state in humans. Proc Natl Acad Sci USA 90:2078–2081
Ludwig AM (1969) Altered states of consciousness. In: Tart C (ed) Altered states of consciousness. Wiley, New York, NY, pp 9–22
Lutz A, Thompson E (2003) Neurophenomenology, integrating subjective experience and brain dynamic in the neuroscience of consciousness. J Conscious Stud 10(9–10):31–52
Lutz A, Greischar L, Rawlings NB, Ricard M, Davidson RJ (2004) Long-term meditators self-induce high-amplitude synchrony during mental practice. Proc Nat Acad Sci USA 101(46): 16369–16373
Maslow A (1971) The father reaches of human nature. Harmondsworth, Eng., Penguin
Peat FD (1987) Synchronicity: the bridge between matter and mind. Bantam Books, New York
Penfield W (1958) The excitable cortex in conscious man. Thomas, Springfield, IL
Plutchik R (1984) A general psychoevolutionary theory. In: Scherer K, Ekman P (eds) Approaches to emotion. Erlbaum, Hillsdale, NJ, pp 197–219
Poldrack RA (2002) Neural systems for perceptual skill learning. Behav Cogn Neurosci Rev 1(1):76–83
Pribram KH (2004) Consciousness reassessed. Mind Matter 2(1):7–35
Purdie H (1997) Music therapy in neurorehabilitation: recent developments and new challenges. Crit Rev Phys Rehabil Med 9:205–217
Sabom MB (1998) Light and death: one doctor's fascinating account of near-death experiences: the case of Pam reynolds. In: Sabom MB (ed) Death: the final frontier. Michigan, Zondervan, pp 37–52
Searle JR (1992) The rediscovery of the mind. MIT, Cambridge, MA
Shapiro D (1977) A biofeedback strategy in the study of consciousness. In: Zinberg NE (ed) Alternate States of Consciousness. The Free Press, New York, NY
Tart CT (1972) States of consciousness and state-specific sciences. Science 176:1203–1210
Tart CT (1975) States of consciousness. E.P. Dutton & Co., New York, Introduction to the Web Edition, (1997)
Thaut MH (2008) Rhythm, music, and the brain. Routledge, New York, p 83
Tononi G (1998) Consciousness and complexity. Science 282:1846–1851
Unkefer RF, Thaut MH (eds) (2005) Music therapy in the treatment of adults with mental disorders: theoretical bases and clinical interventions. Barcelona, Gilsum, NH
Van Lommel P, Van Wees R, Meyers V, Elfferich I (2001) Near-death experience in survivors of cardiac arrest: a prospective study in the Netherlands. Lancet 358:2039–2045
Varela FJ (1996) Neurophenomenology: a methodological remedy to the hard problem. J Conscious Stud 3(4):330–350
Varela FJ (1999) The specious present: a neurophenomenology of time consciousness. In: Petitot J et al (eds) Naturalizing phenomenology. Stanford University Press, Stanford, CA
Wever EG, Lawrence M (1954) Psychological acoustics. Princeton University Press, Princeton
Winner W (1982) Invented worlds. Harvard University Press, Cambridge, MA
Zeman A (2001) Consciousness. Brain 124:1263–1289

Chapter 2
Disorders of Consciousness: Coma, Vegetative and Minimally Conscious States

Olivia Gosseries, Audrey Vanhaudenhuyse, Marie-Aurélie Bruno, Athena Demertzi, Caroline Schnakers, Mélanie M. Boly, Audrey Maudoux, Gustave Moonen, and Steven Laureys

Abstract Consciousness can be defined by two components: arousal and awareness. Disorders of consciousness (DOC) are characterized by a disrupted relationship between these two components. Coma is described by the absence of arousal and, hence, of awareness whereas the vegetative state is defined by recovery of arousal in the absence of any sign of awareness. In the minimally conscious state, patients show preserved arousal level and exhibit discernible but fluctuating signs of awareness. The study of DOC offers unique insights to the neural correlates of consciousness. We here review the challenges posed by the clinical examination of DOC patients and discuss the contribution of functional neuroimaging and electrophysiological techniques to the bedside assessment of consciousness. These studies raise important issues not only from a clinical and ethical perspective (i.e. diagnosis, prognosis and management of DOC patients) but also from a neuroscientific standpoint, as they enrich our current understanding of the emergence and function of the conscious mind.

Keywords Brain injury • Coma • Consciousness • Electrophysiology • Functional neuroimaging • Locked-in syndrome • Minimally conscious state • Prognosis • Treatment • Vegetative state

---

O. Gosseries • A. Vanhaudenhuyse • M.-A. Bruno •
A. Demertzi • C. Schnakers • M.M. Boly • A. Maudoux
Coma Science Group, Neurology Department and Cyclotron Research Centre, University of Liège, Sart Tilman-B30, 4000 Liège, Belgium

G. Moonen
Department of Neurology, University of Liège, Liège, Belgium

S. Laureys (✉)
Coma Science Group, Neurology Department and Cyclotron Research Centre, University of Liège, Sart Tilman-B30, 4000 Liège, Belgium
and
Department of Neurology, University of Liège, Liège, Belgium
e-mail: steven.laureys@ulg.ac.be

D. Cvetkovic and I. Cosic (eds.), *States of Consciousness*, The Frontiers Collection,
DOI 10.1007/978-3-642-18047-7_2,

2.1 Introduction

In recent years, resuscitation techniques have led to a considerable increase in the number of patients who survive severe brain injuries. Some patients recover in the first days after the accident while others die quickly. Others, however, recover more slowly through different stages before fully or partially recovering consciousness. Patients in altered states of consciousness present major challenges concerning the diagnosis, prognosis and daily care. Indeed, detecting signs of consciousness is not always easy and, in some cases, they may remain unnoticed. Some of these signs are of prognostic importance. Prognosis, in turn, may influence therapeutic choices. The interdependency of these factors cannot be ignored: the announcement of a favorable outcome may imply the commitment of the medical team, while a bad prognosis may jeopardize the patient's potential recovery. End-of-life decisions, aggressive therapy with analgesic drugs and discussion about options of euthanasia often arise and, in many cases, lead to difficult debates among the medical staff, and sometimes more widely through the media and society.

2.1.1 Two Components of Consciousness

Clinically defined, consciousness encompasses two main components: arousal and awareness (Zeman 2001). At the bedside, arousal (also called vigilance or alertness) is observed by looking at the presence of eye opening. At a neuroanatomical level, the level of arousal (and in particular of sleep-wake cycles) is mainly supported by the brainstem (which is the region between the brain and the spinal cord), and the thalami (which are the nuclei in the center of the brain) (Schiff 2008; Lin 2000).

Awareness, the second component of consciousness, refers to conscious perception which includes cognition, experiences from the past and the present, and intentions. At a clinical level, awareness is mostly inferred by command following (e.g. "squeeze my hand", "close your eyes"). At a neuroanatomical level, awareness is underpinned by the cerebral cortex, which is a thin mantle of gray matter covering the surface of each cerebral hemisphere, and mainly through a wide frontoparietal network (see Sect. 3.2.1). Awareness can be further divided into awareness of the environment and awareness of self. Awareness of the environment can be defined as the conscious perception of one's environment through the sensory modalities (e.g. visual, auditory, somesthetic or olfactory perception) whereas awareness of self is a mental process that does not require the mediation of the senses and is not related to external stimuli for its presence (as shown by mind wandering, daydreaming, inner speech, mental imagery, etc.). Awareness of self also refers to the knowledge of our own social and cultural history as well as our family membership.

To be aware, we need to be awake but when awake, we are not necessarily aware. Consciousness depends on the interaction between the activity of the cerebral cortex, the brainstem and thalamus. When one of these systems is disrupted, consciousness gets impaired. Thus, consciousness is not an all-or-nothing phenomenon but lies on a continuum of states (Wade 1996). The various states of consciousness include wakefulness, deep sleep and paradoxical sleep (dreaming sleep, i.e. rapid eye movement sleep, REM sleep), anesthesia, coma, vegetative state and the minimally conscious state (Fig. 2.1). The boundaries between these different states are not always sharp but often are progressive transitions.

But how do these disorders of consciousness (DOC) occur? As illustrated in Fig. 2.2, after an acute brain injury that could be of traumatic (i.e. motor vehicle accident, falling, etc.) or non-traumatic (i.e. stroke, anoxia, etc.) etiology, patients may lose consciousness and fall into a coma. In most cases where the damage is severe, patients die within a few days. From the moment the patients open their eyes, they move out of a coma and, if still unresponsive, evolve into a vegetative state. Typically, the vegetative patient (VS) gradually recovers awareness and enters a minimally conscious state (MCS). This is often followed by a period of transient post-traumatic amnesia where the patient remains confused and amnesic. In most cases, the patient recovers within a few weeks, but in some cases, they may remain in a state of no awareness or minimal consciousness for several months or even years or decades. Another exceptional condition the locked-in syndrome (LIS), where the patient awake from the coma fully conscious but is unable to move or communicate, except by eye movements.

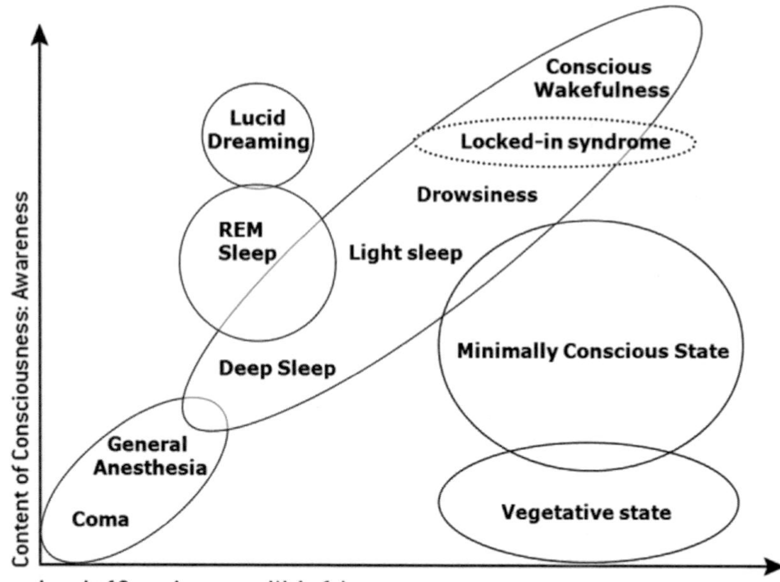

Fig. 2.1 Illustration of the two major components of consciousness: the level of consciousness (arousal or wakefulness) and the content of consciousness (awareness) in normal physiological states, where the level and the content of consciousness are generally positively correlated, and in pathological states or pharmacological coma (adapted from Laureys 2005)

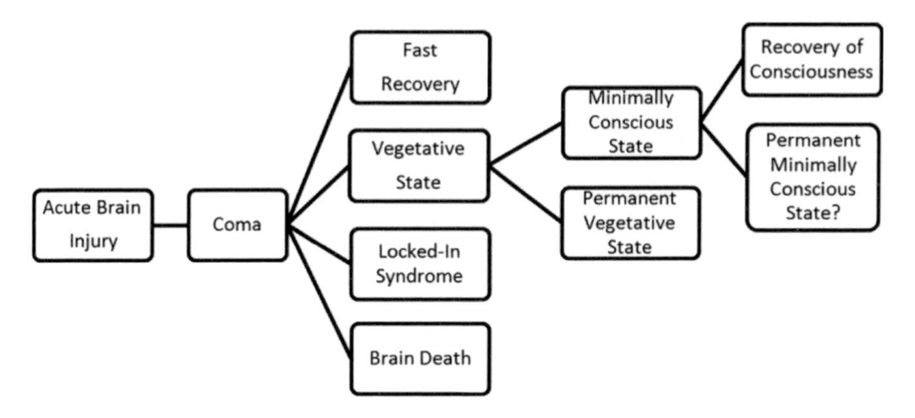

Fig. 2.2 Different conditions may follow acute brain injury. Classically, coma lasts for a couple of days, and once the patients open their eyes they evolve into a vegetative state. Then they may enter a minimally conscious state after showing some signs of consciousness, and eventually they recover full consciousness. In rare cases, a person may develop locked-in syndrome, a nearly complete paralysis of the body's voluntary motor responses

2.1.2 *Clinical Entities*

2.1.2.1 Brain Death

Brain death is characterized by the irreversible loss of all reflexes of the brainstem and the demonstration of continuing cessation of brain function and respiration in a persistently comatose patient (Laureys and Fins 2008). There should be an evident cause of coma, and confounding factors, such as hypothermia (low temperature), drugs intoxication, electrolyte, and endocrine disturbances should be excluded (Wijdicks 2001; Laureys 2005). Repeating the evaluation after 6 h is advised, but this time period is considered arbitrary (The Quality Standards Subcommittee of the American Academy of Neurology 1995). The absence of electrical brain activity by electroencephalogram (EEG) or the absence of cerebral blood flow may also serve as conformational tests (Laureys et al. 2004a).

2.1.2.2 Coma

Coma is a state of non-responsiveness in which the patients lie with eyes closed and cannot be awakened even when intensively stimulated (Plum and Posner 1983). Comatose patients are characterized by a lack of sleep–wake cycles (Teasdale and Jennett 1974) and they have neither verbal production nor response to command but can present reflexive responses to painful stimulation. In these patients, there is no awareness of self or of the environment. The autonomous functions such as breathing and thermoregulation are reduced and the patients require respiratory assistance.

Global brain metabolism (i.e. energy use) is also diminished by 50–70% of normal (Laureys 2005). Coma results from a diffuse cortical or white matter damage, or from a brainstem lesion (Vanhaudenhuyse et al. 2009). Coma must last at least 1 h to be distinguished from syncope, concussion, or other states of transient unconsciousness. The prognosis is often made within 3 days; if the etiology is traumatic, half of the patients who have no chance to recover will die during this short period (Schnakers et al. 2004). Prolonged comas are rare but can last 2–5 weeks and then progress to brain death, a vegetative state, or more rarely a locked-in syndrome.

2.1.2.3 Vegetative State

The vegetative state (VS), or newly called unresponsive wakefulness syndrom (Laureys et al, 2010), is defined by eyes opening, either spontaneously or after stimulation. The sleep–wake cycles are characterized by alternating phases of eye opening. The autonomous functions are preserved and breathing occurs usually without assistance. Patients in a vegetative state exhibit no intelligible verbalization, no voluntary response and no signs of awareness of self or the environment (The Multi-Society Task Force on PVS 1994). The vegetative patient is awake but not aware, which shows that both components of consciousness can be completely separated. If patients are still in a vegetative state a month after brain injury, they are said to be in a *persistent vegetative state*. If patients with a non-traumatic etiology remain in this state for more than 3 months, or more than 1 year for patients with a traumatic etiology, they are said to be in a *permanent vegetative state* (American Congress of Rehabilitation Medicine 1995; Jennett and Plum 1972). The term “persistent” refers to a chronic phase and implies an unfavorable prognosis about the possibility of improvement. This terminology confuses the diagnosis and the prognosis, which induces a risk that certain therapies, such as a transfer to a rehabilitation center, are denied to patients diagnosed as in a persistent vegetative state. Similarly, the term *permanent* implies near zero probability of recovery and can therefore give rise to decisions about the cessation of medication and nutrition. It is preferable to avoid using these two terms (often both abbreviated as PVS) and rather mention the duration and cause of the vegetative state.

The brainstem functions of a vegetative patient are preserved, but cortical (including frontal and parietal cortex) and thalamic injuries are present. Brain metabolism is diminished by 40–50% of normal values (Laureys et al. 2000a). The vegetative patient is able to perform a variety of movements, such as grinding teeth, blinking and moving eyes, swallowing, chewing, yawning, crying, smiling, grunting or groaning, but these are always reflexive movements and unrelated to the context. Motor behavior is reduced to a few stereotyped or reflexive movements and is inadequate compared to the intensity of the stimulation. Typical vegetative patients do not track with their eyes a moving object or their image in a mirror.

2.1.2.4 Minimally Conscious State

The minimally conscious state (MCS) (American Congress of Rehabilitation Medicine 1995; Giacino et al. 2002) is a more recently introduced entity and is characterized by primary and inconsistent signs of consciousness of self and the environment. Although patients are unable to communicate functionally, they can sometimes respond adequately to verbal commands and make understandable verbalizations. Emotional behaviors, such as smiles, laughter or tears may be observed. MCS patients may track a moving object, mirror or person (Giacino et al. 2002). Although these responses may be erratic, they must be reproducible in order to conclude that the action is intentional. These voluntary actions are quite distinct from reflexive movements if they are maintained for a sufficient period of time or repeated.

The overall cerebral metabolic activity is reduced by 20–40% (Laureys et al. 2004b). The autonomous functions are preserved and the thalamocortical and corticocortical connections are partly restored (Laureys et al. 2000b). The minimally conscious state may be transitory, chronic or permanent, such as the vegetative state.

2.1.2.5 Emergence of the Minimally Conscious State

Once patients are able to communicate in a functional way, they are said to have emerged from the minimally conscious state. They can therefore use multiple objects in an appropriate manner and their communication systems are adequate and consistent (Giacino et al. 2002). Because this entity is as recent as the minimally conscious state, validation and further research of other diagnostic criteria are still needed.

2.1.2.6 Locked-In Syndrome

Locked-in syndrome (LIS), also known as pseudocoma, is a complete paralysis of the body resulting from a lesion in the brainstem (American Congress of Rehabilitation Medicine 1995). Oral and gestural communications are impossible but patients are often able to blink and move the eyes. Despite the fact that the patients cannot move, their sensations are still intact and they are fully aware of their environment and themselves (Laureys et al. 2005a).

The only way for these patients to communicate with their environment is through eye movement and sometimes later also with the tip of a finger (Bauby 1997). Indeed, they generally recover some control of their fingers, toes or head. The LIS patient is able to answer questions by a simple code such as blinking once for "yes" and twice for "no", or looking up for "yes" and down for "no". Many means of communication have been developed to allow better communication, such as the use of the alphabet based on letter frequency used in English (i.e. E-T-A-O-I-

N-S-R-H-L-D-C-U-M-F-P-G-W-Y-B-V-K-X-J-Q-Z) where the patient blinks when the interlocutor pronounces the desired letter. Here it is necessary to begin over again for each letter in order to form words and sentences. The use of a brain computer interface has also recently become an option and allows a LIS survivor to control his or her environment, use a word processor, operate a telephone or access the Internet and use email (Gosseries et al. 2009).

Nearly 90% of LIS cases are of vascular etiology but they can also be traumatic. Cognitive functions are fully preserved if the lesion is only restricted to the brainstem. If additional cortical lesions are present, the cognitive functions associated with these cortical areas may be affected (Schnakers et al. 2008). Contrary to what we might expect, the quality of life reported by chronic LIS patients is not that much lower than the general population (Bruno 2011) and the demand for euthanasia, albeit existing, is infrequent (Bruno et al. 2008).

2.1.3 Prognosis

The prognosis for survival and recovery from coma, VS, MCS or LIS is still difficult to establish at the individual level. Certain factors, however, increase the chances of recovery. The young age of the patient, a traumatic etiology and the short duration of the state are linked with a better outcome (The Multi-Society Task Force on PVS 1994). Additionally, patients who are in a MCS for 1 month after brain injury have better chances of recovery than patients who are in a VS 1-month post-injury. Life expectancy of most vegetative patients varies between 2 and 5 years, a few patients stay more than 10 years in this state whereas the average survival time for LIS patients is about 6 years. Of notice is the fact that exceptionally some patients can recover even many years after their trauma. Indeed, the American Terry Wallis, who suffered a car accident in 1984, recovered from the minimally conscious state in 2003, 19 years later, he started talking (Wijdicks 2006). Similarly, a Polish man suffered a brain trauma in 1988 and was able to communicate only 19 years later, in 2007.

Clinical and paraclinical assessments can also be used to establish a prognosis, such as the evaluation of the brainstem reflexes, the sensory evoked potential (SEP), the cognitive auditory evoked potential (such as P300 and mismatch negativity, MMN), and the serum marker neuron-specific enolase (NSE). Some of these examinations are described in a later section.

2.2 Clinical Examination

In neurological rehabilitation, the distinction between a vegetative and a minimally conscious state is of great importance, because of the implications in terms of prognosis and treatment decisions, but also at the medico-legal and ethical level.

The main method – known as the gold standard – for detecting signs of consciousness is behavioral observation (Schnakers et al. 2004). The assessment of consciousness in DOC patients is essential from admission throughout hospital discharge in order to obtain information on their cognitive progress and to define appropriate care. Clinical evaluations are therefore used to assess the awareness of self and the environment of the patient. However, no technique is yet available to measure consciousness directly. We can identify its presence but it is much more difficult to prove its absence.

2.2.1 Misdiagnosis in Disorders of Consciousness

Although behavioral assessments are essential in evaluating consciousness, they are sometimes difficult to complete. A patient in a minimally conscious state can be diagnosed as being in a vegetative state, just as a LIS patient can be easily confused with a vegetative state. Indeed, voluntary movements may be wrongly interpreted as reflex movements and motor responses may be very limited due to a paralysis of all limbs (quadriplegia). Motor responses can also be quickly exhaustible and therefore not reproducible (Schnakers et al. 2004). The level of arousal can also fluctuate and patients may become drowsy or even fall asleep while evaluating them. All these boundaries lead to diagnostic errors. Studies have shown that 20–40% of patients diagnosed as vegetative showed signs of consciousness when assessed with sensitive and reliable standardized tools (consciousness scales) (Schnakers et al. 2006, Andrews et al. 1996, Childs and Mercer 1996, Schnakers et al. 2009).

The differential diagnosis requires repeated behavioral assessments by trained medical staff. The risk of misdiagnosis increases if the staff is unfamiliar with the clinical signs of these states. The controversy of some behaviors as reflecting consciousness or not just comes to add further perplexity. For example, blinking to visual threat should not be considered as a sign of consciousness (Vanhaudenhuyse et al. 2008a) whereas visual pursuit clearly should (Giacino et al. 2002). Moreover, the latter can be tested with different tools but it has been revealed that the best means for assessing visual pursuit is the use of a mirror which (by presenting the patient's own face) has the important ability to grab attention (Vanhaudenhuyse et al. 2008b). To avoid events of misdiagnosis, it is necessary that well-experienced personnel use standardized assessments, such as scales and individual testing, in order to objectify the clinical observations.

2.2.2 Consciousness Scales

Many standardized behavioral scales are used in the assessment of consciousness of brain injured patients: the Glasgow Coma Scale (GCS) (Teasdale and Jennett

1974), the Glasgow Liège Scale (Born 1988a), the Coma Recovery Scale-Revised (Giacino et al. 2004), the Full Outline of Unresponsiveness (Wijdicks et al. 2005), the Wessex Head Injury Matrix (Shiel et al. 2000), the Coma-Near Coma scale (Rappaport 2000), the Western Neuro-Sensory Stimulation Profile (Ansell and Keenan 1989), and the Sensory Modality Assessment and Rehabilitation Technique (Gill-Thwaites 1997) are among the most used. Some scales are an aid for diagnoses in the early hours of patients' admission in the intensive care unit (e.g. the GCS and FOUR) while others are rather used throughout the recovery (e.g. CRS-R). Here we review the scales that are used most frequently in clinical practice.

2.2.2.1 Glasgow Coma Scale and Glasgow Liège Scale

The Glasgow Coma Scale (GCS) is the scale of reference used internationally, due to its short and simple administration. It is mainly used in intensive care settings. The GCS measures eye, verbal and motor behaviors. However, the verbal response is impossible to assess in the case of intubation or tracheotomy (patients with artificial respiratory help making speech impossible). Additionally, there may be some concern as to what extent eye opening is sufficient for assessing brainstem function (Laureys et al. 2002a). The total score varies between 3 and 15. In acute stages, brain damage is described as serious if the score is less than or equal to 8 and moderate if the score is between 9 and 12 (Deuschl and Eisen 1999). The Glasgow Liège Scale (GLS) is an extended version of the GCS which includes the standardized evaluation of brainstem reflexes (Born 1988b).

2.2.2.2 Full Outline of Unresponsiveness

The Full Outline of Unresponsiveness (FOUR) is a more recent scale that has been proposed to replace the GCS as it detects more subtle neurological changes (Wijdicks et al. 2005). The scale is named after the number of subscales (eye, motor, brainstem reflexes and respiration) as well as after the maximum score that each subscale can take (four). The assessment takes only a few minutes to administer. It does not include a verbal response, and can therefore be used to assess artificially ventilated or intubated patients. The FOUR is particularly suitable for diagnosing vegetative state, locked-in syndrome and brain death. It also allows to differentiate between VS and MCS patients as it assesses visual pursuit, one of the first signs of recovery of consciousness (Giacino et al. 2002).

2.2.2.3 Coma Recovery Scale-Revised

The Coma Recovery Scale-Revised (CRS-R) also is a recent clinical tool that has been specifically developed to disentangle VS from MCS patients, but also MCS patients from patients who recovered their ability to communicate functionally (Giacino et al. 2004). Indeed, it is the only scale to explicitly incorporate the diagnostic criteria of vegetative and minimally conscious state. It consists of six subscales: auditory, visual, motor and oromotor/verbal functions as well as communication and arousal. The 23 items are ordered according to their degree of complexity; the lowest item on each subscale represents reflexive activity while the highest item represents behaviors that are cognitively mediated. Scoring is based on the presence or absence of operationally defined behavioral responses to specific sensory stimuli (e.g. if the item of visual pursuit is present, the patient's state is diagnosed as minimally conscious). The brainstem reflexes are also measured but not scored. The assessment takes between 10 and 60 min, depending on the patient's responsiveness. In many research centers, the CRS-R is regarded as the gold standard for the behavioral assessment of severely brain injured patients. The scale has been translated and validated in several languages (Schnakers et al 2008) and is freely available (see http://www.comascience.org).

2.2.2.4 Wessex Head Injury Matrix

The objective of this scale is to create a transition between the assessment of coma in acute stages and the realization of neuropsychological tests that are applied much later. The evaluation is based on observations for presence or absence of behaviors. The WHIM has been designed to pick up minute indices demonstrating recovery and it covers a wide range of daily life functions. It assesses motor and cognitive skills, social interactions, the level of wakefulness and the auditivo-verbal, visual-motor and tactile modalities (Majerus and Van der Linden 2000). Compared to the CRS-R, this 62 item matrix assesses the patient without giving any diagnosis, since it does not incorporate the criteria of VS and MCS. It is more useful in assessing MCS patients who show minimal improvement, and in setting goals for rehabilitation from the outset of coma.

2.2.3 Individual Bedside Assessment

Another way to assess severely brain injured patients, which is complementary to the standardized scales, is a quantitative assessment based on the principles of single-subject experimental design. This method identifies whether a specific behavior of interest can be performed in response to command and whether the reliability of this behavior can change over time either spontaneously or in response to treatment (Whyte et al. 1999). The presence of command-following is crucial evidence of consciousness and facilitates differentiation between MCS and VS patients. The ability to follow a command is also important in the rehabilitation process because it means the patient can participate in therapies.

It can also be the starting point of communication. Indeed, specific movements can be used as a means of communication such as "make a thumbs up" for yes and "shake your head" for no.

Quantitative assessment can also lead to conclusions about a patient's visual function that are not readily apparent by clinical observation (Whyte and DiPasquale 1995). Attention deficits, blindness, gaze preference, monocular pathology and impairment in one or both visual fields can be observed objectively through this method by showing the patient a blank card or a photograph, either individually or at the same time, in both visual fields. Systematic visual orienting that involves visual discrimination is evidence of cortical function (i.e. if the patient looks at a picture more often than a blank card). The awareness of these deficits may help to avoid confounding the assessment of a patient's cognitive functions and can also help to adapt the therapy. For example, if a patient presents a left-sided neglect, the therapist should be positioned primarily on the right side to optimize patient's responsiveness.

2.3 Complementary Examination

Brain imaging and electrophysiology techniques are objective ways to investigate residual brain functions in disorders of consciousness. They may show the extent of brain damage for diagnostic, prognostic and therapeutic purposes and can also be used in experimental research. The imaging techniques can therefore lead to a better understanding of the behavioral clinical observations. To simplify, we present two main types of methods which are used for cognitive neuroimaging studies: metabolic or hemodynamic measurements (e.g. positron emission tomography, PET; functional magnetic resonance imaging, fMRI) and electrical measurements (e.g. electroencephalography, EEG; event related potentials, ERP). The PET technique can measure changes in the brain's metabolism using a radioactive tracer (labeled glucose) that is injected into the blood and is accumulated by active areas of the brain (using energy from the glucose). The fMRI indirectly measures regional increases in blood flow by analyzing the magnetic resonance properties of hemoglobin, which varies depending on the blood's oxygenation (energy use). These *functional* neuroimaging techniques should be distinguished from *structural* imaging such as X-ray CT or conventional MRI (offering imaging of the brain without telling anything about their functioning). Electrical measurements collect signals that are related to the intracellular electric current of the brain. PET and fMRI techniques have good spatial resolution but their temporal resolution is poor, whereas EEG methods have excellent temporal resolution but spatial resolution is relatively low and signals from deeper areas are difficult to identify (Laureys and Boly 2008).

2.3.1 Electroencephalography

Electroencephalography (EEG) records continuously and non-invasively the spontaneous electrical brain activity through electrodes placed on the scalp. The EEG well identifies the level of vigilance (Fang et al. 2005) and detects functional cerebral anomalies such as seizures. It can also be used to confirm the clinical diagnosis of brain death (Guérit et al. 2002). The utility of the EEG has been demonstrated to predict poor recovery of patients with brain anoxic and traumatic injury (Zandbergen et al. 1998). However, most of the EEG patterns are not specific (Young 2000) and do not allow reliable differentiation between conscious and unconscious brain processing. The interpretation of raw EEG signals also requires considerable expertise and training. More automated measures deriving from the EEG are therefore welcome, such as the EEG bispectral index measurements.

2.3.1.1 Bispectral Index

The bispectral index (BIS) measures the depth of sedation in anesthesia (Struys et al. 1998) and allows distinction between the different phases of normal sleep (Nieuwenhuijs et al. 2002). BIS values range between 0 and 100: when the subject is awake the values approach 100, whereas when the subject is under general anesthesia, values are around 40–50. BIS values also gradually increase when patients move out from coma to recovery (Schnakers et al. 2008a). However, BIS is a nonspecific measure of consciousness and does not systematically differentiate MCS from VS patients, even if it seems to show prognostic utility (Schnakers et al. 2005b).

2.3.1.2 Event-Related Potentials

The event-related potential (ERP) technique objectively examines sensory and cognitive functions at the patient's bedside by averaging the EEG activity according to the onset of a repeated stimulus (e.g. noise or visual flash). ERPs reveal the time course of information processing from low-level peripheral receptive structures to high-order associative cortices (Vanhaudenhuyse et al. 2008c). Short-latency ERPs, or exogenous ERP components (ranging from 0 to 100 ms), correspond to the passive (automatic) reception of external stimuli whereas cognitive ERPs, or endogenous ERP components (obtained after 100 ms), often reflect cognitive neuronal activity. The ERPs provide neurological markers, where the absence of early ERPs is a good predictor of a bad outcome (i.e. absence of primary cortical responses on somatosensory ERPs) and the presence of cognitive ERPs a good predictors of a favorable outcome (i.e. P300 and Mismatch Negativity responses) (Daltrozzo et al. 2007).

Short-Latency ERPs

Sensory evoked potentials (SEPs) are short-latency ERPs that are routinely used in intensive care. They measure the connection from the body to the brain (called the ascending pathways, which involve the spinal cord, the brainstem, and the primary sensory cortex). Practically, electrical stimulations are elicited from the wrist and the responses are recorded at the level of the nerves, spinal cord, brainstem and cortical levels. Bilateral absence of the cortical response (N20) among patients in coma, especially in anoxic patients who had a lack of oxygen in the brain, after for example cardiac arrest, is strongly associated with poor outcome, but preserved SEPs do not necessarily herald recovery (Cant et al. 1986; Laureys et al. 2005b).

Long-Latency Cognitive ERPs

Mismatch negativity (MMN) is a cognitive ERPs response elicited after approximately 100–200 ms by any change in a sequence of monotonous auditory stimuli in inattentive subjects (Naatanen and Alho 1997). It assesses the residual brain activity and more specifically the integrity of echoic memory, a memory that permits a sound to be remembered in the 2 or 3 s after it is heard. The presence of MMN has prognostic value in predicting recovery after coma (Kane et al. 1996; Fischer et al. 2004; Naccache et al. 2005; Qin et al. 2008).

The auditory evoked potentials P300 response is another ERP wave which is also elicited (around 300 ms after the stimulus) when subjects detect a rare and unpredictable target stimulus in a regular train of standard stimuli (Sutton et al. 1965). It assesses the integrity of acoustic and semantic discrimination. The presence of P300 and MMN is associated with a favorable clinical outcome but their absence does not necessarily imply a poor prognosis as these components can also be absent in some healthy controls, as well as in a significant number of patients who later recover consciousness (van der Stelt and van Boxtel 2008).

The P300 wave can also be observed in response to the patient's own name in VS and MCS patients, when they hear their own name in a sequence of unfamiliar names in a passive condition (Perrin et al. 2006). When asking patients to perform a cognitive task such as counting the number of times they hear their own name, the P300 to the own name stimuli increases (Schnakers et al. 2008b). This permits a demonstration that patients with apparently no behavioral sign of consciousness may be conscious (i.e. show command following) (Schnakers et al. 2009, p. 4588). The P300 ERP technique is also being used to permit EEG-based communication (i.e. in Brain Computer Interface technology) (Sellers et al. 2006), which could allow LIS patients to communicate through their electrical brain activity without moving a single muscle.

2.3.2 *Functional Neuroimaging*

2.3.2.1 Resting State

PET studies have shown that global cerebral metabolism during deep sleep and general anesthesia is diminished by about half of normal values (Maquet et al. 1997; Alkire et al. 1999). The brain metabolism of VS patients is also reduced by 50–60% (Laureys et al. 1999; Levy et al. 1987). However, when the patient has clinically recovered, brain metabolism does not always return back to normal (Fig. 2.3) (Laureys et al. 2000c).

Some brain regions appear to be more important than others for the emergence of consciousness. At rest, patients in a vegetative state show systematic impairment of metabolism in the frontoparietal network that includes polymodal associative cortices (bilateral prefrontal regions, Broca's area, parietotemporal and posterior parietal areas, and precuneus) (Laureys et al. 2006a).

These regions are essential in various functions that are necessary for consciousness, such as attention, memory, and language (Baars et al. 2003). Conscious perception is also linked to the functional connectivity between this frontoparietal network and deeper centers of the brain, such as the thalamus (Fig. 2.4). In vegetative patients, the long-distance connections between different cortical areas as well as between the cortex and the thalamus seem disconnected. The recovery of patients in a vegetative state is linked to the restoration of this frontoparietal network and its connections (Laureys et al. 2000b).

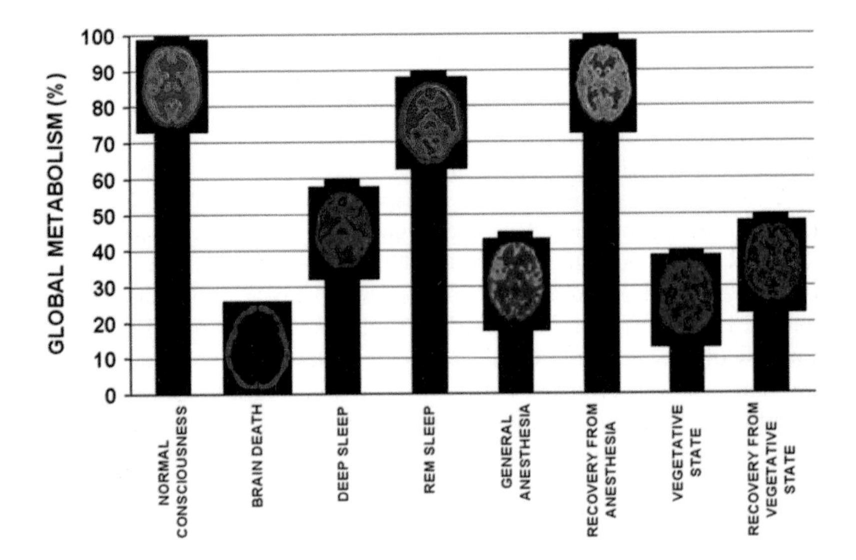

Fig. 2.3 Global cerebral metabolism in various states (adapted from Laureys et al. 2004a)

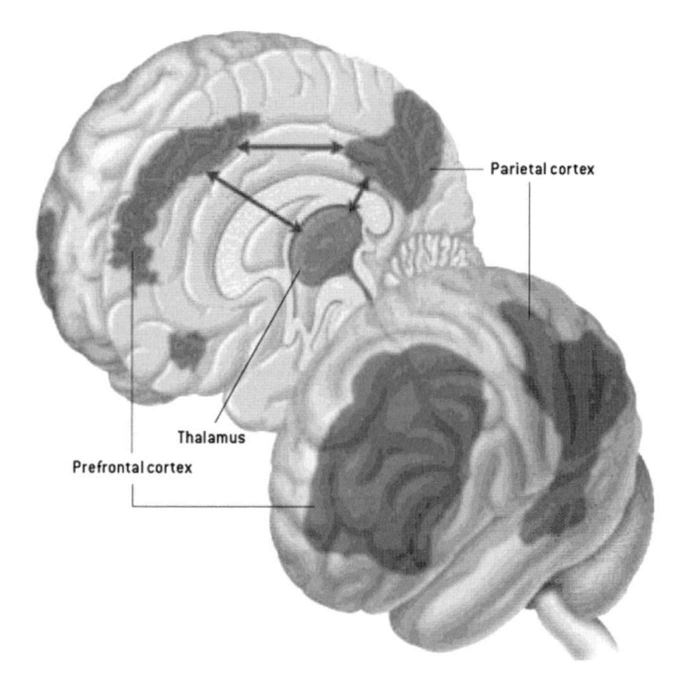

Fig. 2.4 Brain regions encompassing prefrontal and parietal multi-modal associative areas (known as the frontoparietal network) are crucial for consciousness. Conscious perception is also linked to the functional connectivity between this frontoparietal network and deeper centers of the brain such as the thalamus. The vegetative state is characterized by a metabolic dysfunction of this widespread cortical network shown in dark grey (adapted from Laureys 2007)

2.3.2.2 External Stimulation

Despite a massively reduced resting metabolism, primary cortices still seem to be activated during external stimulation in vegetative patients, whereas hierarchically higher-order multimodal association areas are not. When painful stimuli are administered to vegetative patients, only the brainstem, the thalamus and the primary somatosensory cortex are activated, and the latter is isolated and disconnected from the other brain areas, in particular the frontoparietal network (Boly et al. 2005; Laureys et al. 2002b). These findings support the idea that patients in a vegetative state do not consciously perceive pain as do healthy people. In contrast to VS, MCS patients, similar to control subjects, show activation of the complete pain matrix (thalamus, primary and secondary somatosensory, frontoparietal, and anterior cingulate cortices) and show a preserved functional connectivity between these areas (Boly et al. 2008). These results provide evidence for a preserved conscious pain perception capacity in MCS patients, strongly suggesting that these patients should receive pain treatment when needed.

Similarly, in response to auditory stimuli, brain activity in vegetative patients is limited to the primary auditory cortex while polymodal areas of higher order do not become active and remain functionally disconnected (Boly et al. 2004; Laureys et al. 2000d). This primary brain activation does not seem enough to lead to conscious perception and memory formation. In contrast, patients in a minimally conscious state activate higher-order cortical areas. More specifically, a recent study showed a selective impairment in backward connectivity from frontal to temporal cortices in vegetative patients whereas minimally conscious patients present a similar pattern to healthy subjects with preserved feedforward and top-down processes (Boly et al. 2011) Auditory stimuli with emotional content, such as baby cries or the patient's own name, induce even more extensive brain activation than sounds without meaning (Fig. 2.5) (Laureys et al. 2004b; Boly et al. 2005). This implies that content is important when talking to patients in a minimally conscious state.

Similarly, in response to presentation of the patient's own name uttered by a familiar voice, the primary auditory cortices of five VS patients were activated, but none of these patients recovered. In contrast, two other VS patients showed atypical activation of both primary cortex and higher-level associative cortex, and they improved clinically to MCS 3 months after their scan (Di et al. 2007). Another fMRI study showed that MCS patients demonstrated similar responses to healthy volunteers when listening to passive language with personalized narratives. However, when the narratives were presented as a time-reversed signal (without linguistic content) MCS patients demonstrated markedly reduced responses, suggesting again reduced engagement for linguistically meaningless stimuli (Schiff et al. 2005).

Another fMRI study based on mental imagery tasks has been proposed to identify signs of consciousness in non-communicative patients (Boly et al. 2007). Despite the clinical diagnosis of vegetative state, a 23-year-old girl who suffered a traumatic brain injury 5 months earlier showed signs of consciousness only detectable on fMRI (Owen et al. 2006). She was asked to imagine herself playing tennis and walking

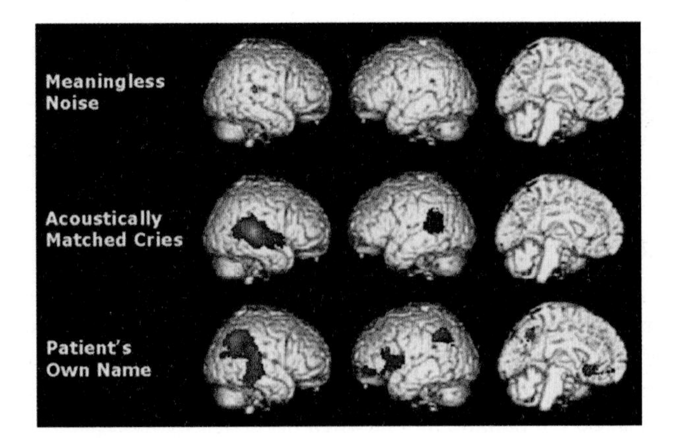

Fig. 2.5 Brain activations during presentation of noise, baby cries, and the patient's own name. Stimuli with emotional valence (baby's cries and names) induce a much more widespread activation than does meaningless noise in the minimally conscious state (taken from Laureys et al. 2004)

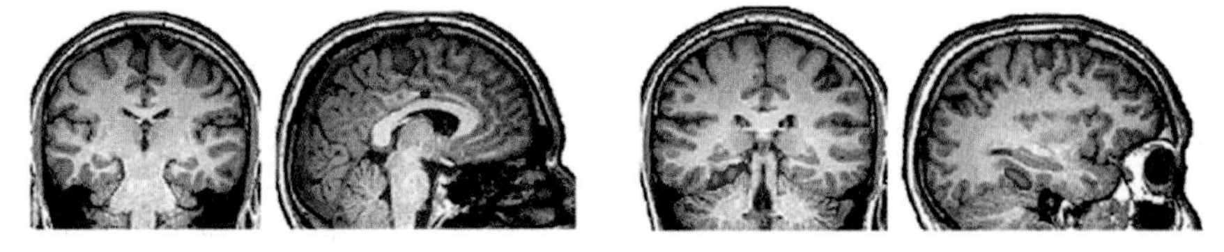

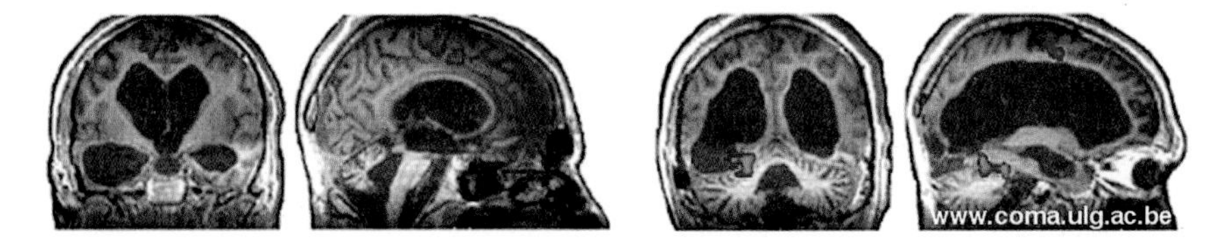

Fig. 2.6 A patient (top two images) who was clinically diagnosed as vegetative showed similar brain activity to a healthy subject (bottom two images) when asked to imaging playing tennis (*left*) or visiting her own house (*right*). A few months after the study, the patient recovered consciousness (taken from Owen et al. 2006)

through her house. The results showed brain activation similar to that of the control subjects for both tasks (Fig. 2.6). This study shows clear evidence of awareness and command-following in the absence of voluntary motor responsiveness. The patient evolved into MCS several weeks later and she was probably in a stage of transition from vegetative to recovery of consciousness at the time of assessment. Finally, another study recently showed that it has been possible for a vegetative patient to communicate through the fMRI. He could answer questions by imaging playing tennis when he wanted to say "yes" and imaging moving into his house when he wanted to respond "no" (Monti and Vanhaudenhuyse et al, 2010).

2.4 Treatment

There is currently no effective standardized treatment for DOC patients. Most of the studies have been conducted under suboptimal settings with methodological and conceptual problems, with the consequence that no strong evidence-based recommendations can be made. However, uncontrolled studies indicate that some rehabilitative procedures can promote the recovery of consciousness, especially in MCS patients. These interventions can be divided into pharmacological and non-pharmacological treatments.

2.4.1 Pharmacologic Treatment

The effect of pharmacological agents on recovery in chronic disorders of consciousness still remains unsatisfactory (Laureys et al. 2006b; Demertzi et al. 2008). Several therapeutic trials have been conducted with post-comatose patients and have led to a marked improvement in their level of consciousness. Zolpidem is a drug originally used in the treatment of insomnia that has occasionally the opposite effect in brain damaged patients. A recent study showed that on 15 patients, only one demonstrated a clinically significant response after the administration of the medication, suggesting a response rate to zolpidem around 7% (Whyte et al. 2009). The effect of Zolpidem was first reported after its use in a 23-year-old man who had been in a vegetative state for more than 3 years following a motor vehicle accident. The patient regained consciousness 15 min after being administered the drug and was able to greet his mother for the first time in 3 years. He was able to sigh, to talk and to communicate with his family (Clauss et al. 2000). But after the effects of the drug wore off, he relapsed, returning to his previous state. Zolpidem has therefore only a temporary effect and lasts for a maximum of a few hours. Temporary improvements have since been observed in stroke and near drowning patients (Clauss and Nel 2004), anoxic brain injury (Cohen and Duong 2008), vegetative (Clauss and Nel 2006) and minimally conscious states (Brefel-Courbon et al. 2007; Shames and Ring 2008). The results varied from a regain of consciousness to an enhancement of motor, verbal and cognitive functions, as well as gestural interaction and arousal. The exact underlying mechanism of the effect of Zolpidem remains unclear (Clauss et al. 2004).

Amantadine is another drug that produces similar effects on VS and MCS patients but its effects last longer (Whyte et al. 2005; Zafonte et al. 2000). It is a dopaminergic agent (also acting on NMDA receptors) initially used against the flu and in the treatment of Parkinson's disease. A recent study involving a chronic anoxic MCS patient showed cognitive improvement after 3 weeks of Amantadine treatment, such as reproducible movement to command (i.e. touching a ball with the feet) and consistent automatic motor responses (i.e. mouth opening when a spoon is approaching). These improvements were associated with an increase in fronto-parietal cortical metabolism which is considered important in consciousness (Schnakers et al. 2008c).

Other pharmacological agents that have been reported as inducing functional recovery are Levodopa, Bromocriptine (Passler and Riggs 2001) Apomorphine (Fridman et al. 2010), and Baclofen (Taira and Hori 2007). Large scale studies on the efficacy of these drugs are still warranted. More specifically, cohort placebo-controlled randomized trials and blinded within-subject crossover designs are needed before reaching any definite conclusions concerning the efficacy of the pharmacological treatment of DOCs.

2.4.2 *Non-pharmacological Treatment*

Deep brain stimulation (DBS) has been proposed as a strategy to improve the functional level of chronic non-communicative patients. This technique consists of implanting an electrode in the brain (more specifically in this case in the thalamus) in order to reactivate a widespread cerebral connectivity mechanism that supports communication and goal-directed behavior. Bilateral DBS of the central thalamus has been performed in a 38-year-old patient who remained in a MCS for 6 years following a traumatic brain injury (Schiff et al. 2007). The electrodes were placed in the intralaminar thalamic nuclei. It has been shown that these thalamic nuclei restore cortical connectivity in the recovery of consciousness after VS (Laureys 2000b, p. 2913). Before applying DBS, the MCS patient failed to recover consistent command following and remained in a non-verbal state without any sign of functional communication. His fMRI, however, showed the preservation of a bihemispheric large-scale cerebral language network, which demonstrates that further recovery was possible (Schiff et al. 2005). During periods in which DBS was on (as compared to periods in which it was off), levels of arousal, motor control and interactive behavior increased considerably. The patient was able to respond consistently to commands and produced intelligible verbalization. The DBS technique can therefore promote a significant functional recovery from severe traumatic brain injury. Nevertheless, replicas of these findings are still needed to validate the technique.

Other non-pharmacologic and non-invasive interventions are the multimodal sensory stimulation techniques which provide frequent sensory input to all five senses in the hope that it will enhance synaptic reinnervation and accelerate neurological recovery (Demertzi et al. 2008; Tolle and Reimer 2003). Sensory stimulation is also intended to prevent sensory deprivation and facilitates coherence between the brain and the body. Sensory regulation is a variant of sensory stimulation that facilitates information processing by adjusting the time exposure and the complexity of the stimuli according to the level of the patient's capacity (Wood et al. 1992). The stimulation sessions are alternated with resting periods in order to increase the ability of the patient to respond during stimulation sessions. Finally, physical and occupational therapy are usually used in rehabilitation centers to prevent complications and enhance recovery. There is uncontrolled evidence that early and increased intervention leads to better outcomes (Oh and Seo 2003; Shiel et al. 2001). The beneficial effects of all these techniques are still debated and are not yet based on evidence.

2.5 Ethical Issues

DOC patients, especially patients in vegetative state, present important ethical and moral issues (Demertzi et al in press). In many countries, it is legally permissible to withdraw life-sustaining treatment once the patient is diagnosed as being in a permanent VS (i.e. with no hope of recovery), if such withdrawal seems likely to be what the patient would have wanted (Jennett 2005). It is therefore recommended to make an advance directive concerning personal wishes in the event of vegetative survival that could legitimately be used by the doctor to withdraw or to sustain the treatment (Demertzi et al 2011).

Three cases have generated considerable debate, positioning pro-life advocates against those defending the right to die with dignity. American Karen Ann Quinlan suffered a cardio-pulmonary arrest in 1975 and became vegetative. Her parents signed the authorization to disconnect the respirator, but the hospital authorities refused because the parents did not have legal custody. A judicial process began and a year later the court gave legal custody to the parents. Karen was disconnected but, against all odds, she continued breathing by herself. She survived in this vegetative state for 9 years until her death in 1985 (Dundon 1978). The case of the American Terri Schiavo is similar but here the parents wished to keep her alive against the wishes of her husband and despite the advice of doctors. After suffering a respiratory insufficiency in 1990, Terri was considered as being in a permanent vegetative state but this diagnosis was criticized by the parents still hoping for a recovery (Cochrane 2006). The Supreme Court of the United States finally rejected the request of her parents to keep her alive and she died in 2005, 13 days after the disconnection of her feeding tube. The most recent case involved the Italian Eluana Englaro, who was left in a vegetative state after a motor vehicle accident in 1992. Her father requested shortly after the accident to have her feeding tube removed but the authorities refused his request. He received the authorization only 17 years later. She finally died in February 2009.

It is also ethically controversial for some whether or not non-communicative patients can be included in clinical trials, as they are unable to provide their agreement. Informed consent is therefore requested from the patient's legal surrogate. The medical community is redefining an ethical framework in order to balance protection for post-comatose patients against the facilitation of research and medical progress (Fins 2003; Fins et al. 2008).

2.6 Conclusion

Defining consciousness as having two components (arousal and awareness) helps us also define the corresponding clinical entities. Coma means lack of consciousness (unarousable unawareness), whereas in the vegetative state arousal is preserved but awareness is absent (arousable unawareness). In the minimally conscious state, arousal is also present but with fluctuating and minimal signs of awareness. Locked-in syndrome has to be differentiated from those disorders of consciousness, as consciousness is intact but voluntary motor control is completely impaired (except for eye movements). In clinical practice, although the Glasgow Coma Scale remains the gold standard for the assessment of comatose patients, the Coma Recovery Scale-Revised is more appropriate in differentiating between vegetative and minimally conscious or locked-in patients. Misdiagnosis is still too frequent in clinical practice despite the introduction of diagnostic criteria. Conscious patients can indeed be diagnosed as vegetative if they have unnoticed paralysis or if voluntary movements are erroneously interpreted as reflexes. Family members of LIS patients are often the first to realize that the patient is conscious (Laureys et al. 2003; Leon-Carrion et al. 2002). Standardized behavioral scales and quantitative individual assessments should therefore be employed repetitively in the clinical routine by trained medical staff, in order to minimize the risk of erroneous diagnosis.

Technological advances in neuroimaging allow us to increase our understanding of the human brain and this knowledge can be exploited in order to develop new diagnostic, prognostic and therapeutic approaches (Laureys and Boly 2008). Studies have shown that the vegetative state is characterized by a functional cortical disconnection syndrome. Only primary cortex can be activated and is disconnected from the higher-order frontoparietal network. In the minimally conscious state, however, the latter areas can be activated especially by emotionally meaningful or noxious stimuli. It has also been shown that neuronal plasticity (e.g. axonal regrowth) may exist, sometimes many months to years after the brain trauma, and this could promote the recovery of consciousness in MCS patients (Schiff et al. 2005; Voss et al. 2006).

DOC patients have rather limited therapeutic options. Basic therapies include life-sustaining therapy (i.e. artificial nutrition and hydration) as well as physical and occupational therapies that are used to prevent complications and enhance recovery (the latter awaits controlled trials). Pharmacologic trials (with Amantadine and Zolpidem) have shown behavioral improvements in some uncontrolled case reports or series of brain injured patients. Deep brain stimulation and multisensory stimulations also showed some positive results but are clearly still in the research domain. Nowadays, therapeutic management lacks large-scale double-blind randomized placebo controlled trials. Much more research and methodical validation are required before accepting or rejecting specific treatments.

Advanced communication techniques based on mental imagery and on cognitive event-related potentials using active paradigms are also currently being investigated. Indeed, brain computer interface (BCI) devices allow brain signals to control external devices without requiring any muscular activity. For example, mental imagery and measurement of the salivary pH can permit a LIS patient to communicate: to say "yes", the patient had to imagine a lemon, which increases salivary pH, whereas to say "no", the patient had to imagine milk, which decreases pH (Wilhelm et al. 2006; Vanhaudenhuyse et al. 2007). Also, it is expected that BCI techniques will be used clinically as a diagnostic tool for differentiating between conscious and unconscious patients (Kubler and Kotchoubey 2007, Sorger 2009), as we have previously seen with the fMRI tennis and spatial mental imagery paradigms.

Neuroimaging studies are moving from the research field to the clinical application of these techniques currently being validated by multi-centric cohort studies. These paraclinical examinations are providing additional information, unavailable through bedside clinical assessments, which is being used to better understand the patient's diagnosis and prognosis. It can be predicted that in the near future, multimodal approaches combining bedside examination, electrophysiology and functional imaging techniques will be routinely employed to assess and treat these challenging patients with disorders of consciousness.

Acknowledgments This research was funded by the Belgian National Funds for Scientific Research (FNRS), the European Commission (CATIA, DISCOS, MINDBRIDGE, DECODER), the James McDonnell Foundation, the Mind Science Foundation, the French Speaking Community Concerted Research Action, the Foundation Médicale Reine Elisabeth and the University and University Hospital of Liège.

References

Alkire MT, Pomfrett CJ, Haier RJ et al (1999) Functional brain imaging during anesthesia in humans: effects of halothane on global and regional cerebral glucose metabolism. Anesthesiology 90(3):701–709

American Congress of Rehabilitation Medicine (1995) Recommendations for use of uniform nomenclature pertinent to patients with severe alterations of consciousness. Arch Phys Med Rehabil 76:205–209

Andrews K, Murphy L, Munday R et al (1996) Misdiagnosis of the vegetative state: retrospective study in a rehabilitation unit. BMJ 313(7048):13–16

Ansell BJ, Keenan JE (1989) The western neuro sensory stimulation profile: a tool for assessing slow-to-recover head-injured patients. Arch Phys Med Rehabil 70(2):104–108

Baars B, Ramsoy T, Laureys S (2003) Brain, conscious experience and the observing self. Trends Neurosci 26:671–675

Bauby J-D (1997) The diving bell and the butterfly (original title: Le scaphandre et le papillon).

Boly M, Faymonville ME, Peigneux P et al (2004) Auditory processing in severely brain injured patients: differences between the minimally conscious state and the persistent vegetative state. Arch Neurol 61(2):233–238

Boly M, Faymonville ME, Peigneux P et al (2005) Cerebral processing of auditory and noxious stimuli in severely brain injured patients: differences between VS and MCS. Neuropsychol Rehabil 15(3–4):283–289

Boly M, Coleman MR, Davis MH et al (2007) When thoughts become action: an fMRI paradigm to study volitional brain activity in non-communicative brain injured patients. Neuroimage 36 (3):979–992

Boly M, Faymonville ME, Schnakers C et al (2008) Perception of pain in the minimally conscious state with PET activation: an observational study. Lancet Neurol 7(11):1013–1020

Boly M, Garrido MI, Gosseries O, Bruno MA, Boveroux P, Schnakers C, Massimini M, Litvak V, Laureys S, Friston K. Preserved feedforward but impaired top-down processes in the vegetative state. Science. 2011 May 13;332(6031):858–62

Born JD (1988a) The Glasgow-Liège Scale. Prognostic value and evaluation of motor response and brain stem reflexes after severe head injury. Acta Neurochir 95:49–52

Born JD (1988b) The Glasgow-Liege Scale. Prognostic value and evolution of motor response and brain stem reflexes after severe head injury. Acta Neurochir Wien 91(1–2):1–11

Brefel-Courbon C, Payoux P, Ory F et al (2007) Clinical and imaging evidence of zolpidem effect in hypoxic encephalopathy. Ann Neurol 62(1):102–105

Bruno M, Bernheim J, Schnakers C et al (2008) Locked-in: don't judge a book by its cover. J Neurol Neurosurg Psychiatry 79:2

Bruno MA, Bernheim J, Ledoux D, Pellas F, Demertzi A, Laureys S (2011) A survey on self-assessed wellbeing in a cohort of chronic locked-in syndrome patients: happy majority, miserable minorityBritish Medical Journal - Open
Cant BR, Hume AL, Judson JA et al (1986) The assessment of severe head injury by short-latency somatosensory and brain-stem auditory evoked potentials. Electroencephalogr Clin Neurophysiol 65(3):188–195
Childs NL, Mercer WN (1996) Misdiagnosing the persistent vegetative state. Misdiagnosis certainly occurs [letter; comment]. BMJ 313(7062):944
Clauss RP, Nel WH (2004) Effect of zolpidem on brain injury and diaschisis as detected by 99mTc HMPAO brain SPECT in humans. Arzneimittelforschung 54(10):641–646
Clauss R, Nel W (2006) Drug induced arousal from the permanent vegetative state. NeuroRehabilitation 21(1):23–28
Clauss RP, Guldenpfennig WM, Nel HW et al (2000) Extraordinary arousal from semi-comatose state on zolpidem. A case report. S Afr Med J 90(1):68–72
Clauss R, Sathekge M, Nel W (2004) Transient improvement of spinocerebellar ataxia with zolpidem. N Engl J Med 351(5):511–512
Cochrane T (2006) Relevance of patient diagnosis of the Terri Schiavo case. Ann Intern Med 144 (4):305–306
Cohen SI, Duong TT (2008) Increased arousal in a patient with anoxic brain injury after administration of zolpidem. Am J Phys Med Rehabil 87(3):229–231
Daltrozzo J, Wioland N, Mutschler V et al (2007) Predicting coma and other low responsive patients outcome using event-related brain potentials: a meta-analysis. Clin Neurophysiol 118 (3):606–614
Damasio AR (1999) The feeling of what happens: body and emotion in the making of consciousness, 1st edn. Harcourt Brace, New York
Demertzi A, Vanhaudenhuyse A, Bruno MA et al (2008) Is there anybody in there? Detecting awareness in disorders of consciousness. Expert Rev Neurother 8(11):1719–1730
Demertzi A, Laureys, S, & Bruno, MA (in press). The ethics in disorders of consciousness. In J. L. Vincent (Ed.), Yearbook of Intensive Care and Emergency Medicine. Berlin: Springer-Verlag
Demertzi, A., Ledoux, D., Bruno, M. A., Vanhaudenhuyse, A., Gosseries, O., Soddu, A., et al. (2011). Attitudes towards end-of-life issues in disorders of consciousness: a European survey. J Neurol, 258(6), 1058–1065
Deuschl G, Eisen A (1999) Recommendations for the practice of clinical neurophysiology: guidelines of the international federation of clinical neurophysiology. Elsevier, Amsterdam
Di HB, Yu SM, Weng XC et al (2007) Cerebral activation to patients' own name uttered by a familiar voice in the vegetative and minimally conscious states. Neurology 68(12):895–899
Dundon SJ (1978) Karen Quinlan and the freedom of the dying. J Value Inq 12(4):280–291
Fang S, Chan H, Chen W (2005) Combination of linear and nonlinear methods on electroencephalogram state recognition. Conf Proc IEEE Eng Med Biol Soc 5:4604–4605
Fins JJ (2003) Constructing an ethical stereotaxy for severe brain injury: balancing risks, benefits and access. Nat Rev Neurosci 4(4):323–327
Fins JJ, Illes J, Bernat JL et al (2008) Neuroimaging and disorders of consciousness: envisioning an ethical research agenda. Am J Bioeth 8(9):3–12
Fischer C, Luaute J, Adeleine P et al (2004) Predictive value of sensory and cognitive evoked potentials for awakening from coma. Neurology 63(4):669–673
Fridman EA, Krimchansky BZ, Bonetto M, Galperin T, Gamzu ER, Leiguarda RC, Zafonte R. Continuous subcutaneous apomorphine for severe disorders of consciousness after traumatic brain injury. Brain Inj. 2010;24(4):636–41
Giacino JT, Ashwal S, Childs N et al (2002) The minimally conscious state: definition and diagnostic criteria. Neurology 58(3):349–353
Giacino JT, Kalmar K, Whyte J (2004) The JFK coma recovery scale-revised: measurement characteristics and diagnostic utility. Arch Phys Med Rehabil 85(12):2020–2029

Gill-Thwaites H (1997) The sensory modality assessment rehabilitation technique: a tool for assessment and treatment of patients with severe brain injury in a vegetative state. Brain Inj 11(10):723–734
Gosseries O, Bruno MA, Vanhaudenhuyse A et al (2009) Consciousness in the locked-in syndrome. In: Laureys S, Tononi G (eds) The neurology of consciousness: cognitive neuroscience and neuropathology. Academic Press Elsevier, Amsterdam
Guérit J-M, Mauguière F, Plouin P (2002) Guide pratique de neurophysiologie clinique. Recommandations de la Fédération internationale de neurophysiologie clinique. Elsevier, Amsterdam
Jennett B (2005) 30 years of the vegetative state: clinical, ethical and legal problems. In: Laureys S (ed) The boundaries of consciousness: neurobiology and neuropathology. Elsevier, Amsterdam
Jennett B, Plum F (1972) Persistent vegetative state after brain damage. A syndrome in search of a name. Lancet 1(7753):734–737
Kane NM, Curry SH, Rowlands CA et al (1996) Event-related potentials–neurophysiological tools for predicting emergence and early outcome from traumatic coma. Intensive Care Med 22 (1):39–46
Kubler A, Kotchoubey B (2007) Brain-computer interfaces in the continuum of consciousness. Curr Opin Neurol 20(6):643–649
Laureys S (2005) Science and society: death, unconsciousness and the brain. Nat Rev Neurosci 6(11):899–909
Laureys S, Boly M (2008) The changing spectrum of coma. Nat Clin Pract Neurol 4 (10):544–546
Laureys S, Fins JJ (2008) Are we equal in death? Avoiding diagnostic error in brain death. Neurology 70(4):e14–e15
Laureys S, Lemaire C, Maquet P et al (1999) Cerebral metabolism during vegetative state and after recovery to consciousness. J Neurol Neurosurg Psychiatry 67:121
Laureys S, Faymonville ME, Moonen G et al (2000a) PET scanning and neuronal loss in acute vegetative state. Lancet 355:1825–1826
Laureys S, Faymonville ME, Luxen A et al (2000b) Restoration of thalamocortical connectivity after recovery from persistent vegetative state. Lancet 355(9217):1790–1791
Laureys S, Faymonville ME, Lamy M (2000c) Cerebral function in vegetative state studied by positron emission tomography. In: Vincent JL (ed) 2000 yearbook of intensive care and emergency medicine. Springer, Berlin
Laureys S, Faymonville ME, Degueldre C et al (2000d) Auditory processing in the vegetative state. Brain 123(Pt 8):1589–1601
Laureys S, Majerus S, Moonen G (2002a) Assessing consciousness in critically ill patients. In: Vincent JL (ed) 2002 Yearbook of intensive care and emergency medicine. Springer, Heidelberg
Laureys S, Faymonville ME, Peigneux P et al (2002b) Cortical processing of noxious somatosensory stimuli in the persistent vegetative state. Neuroimage 17(2):732–741
Laureys S, van Eeckhout P, Ferring M et al (2003) Brain function in acute and chronic locked-in syndrome. Presented at the 9th annual meeting of the organisation for human brain mapping (OHBM), 18–22 June 2003, NY, USA, NeuroImage CD ROM Volume 19, Issue 2, Supplement 1
Laureys S, Owen AM, Schiff ND (2004a) Brain function in coma, vegetative state, and related disorders. Lancet Neurol 3(9):537–546
Laureys S, Perrin F, Faymonville ME et al (2004b) Cerebral processing in the minimally conscious state. Neurology 63(5):916–918
Laureys S, Pellas F, Van Eeckhout P et al (2005a) The locked-in syndrome: what is it like to be conscious but paralyzed and voiceless? Prog Brain Res 150:495–511
Laureys S, Perrin F, Schnakers C et al (2005b) Residual cognitive function in comatose, vegetative and minimally conscious states. Curr Opin Neurol 18(6):726–733
Laureys S, Boly M, Maquet P (2006a) Tracking the recovery of consciousness from coma. J Clin Invest 116(7):1823–1825

Laureys S, Celesia G, Cohadon F, Lavrijsen J, Léon-Carrrion J, Sannita WG, Sazbon L, Schmutzhard E, von Wild KR, Zeman A, Dolce G Unresponsive wakefulness syndrome: a new name for the vegetative state or apallic syndrome and the European Task Force on Disorders of Consciousness BMC Medicine (2010) 8:68

Laureys S, Giacino JT, Schiff ND et al (2006b) How should functional imaging of patients with disorders of consciousness contribute to their clinical rehabilitation needs? Curr Opin Neurol 19(6):520–527

Leon-Carrion J, van Eeckhout P, Dominguez-Morales Mdel R et al (2002) The locked-in syndrome: a syndrome looking for a therapy. Brain Inj 16(7):571–582

Levy DE, Sidtis JJ, Rottenberg DA et al (1987) Differences in cerebral blood flow and glucose utilization in vegetative versus locked-in patients. Ann Neurol 22(6):673–682

Lin JS (2000) Brain structures and mechanisms involved in the control of cortical activation and wakefulness, with emphasis on the posterior hypothalamus and histaminergic neurons. Sleep Med Rev 4(5):471–503

Majerus S, Van der Linden M (2000) Wessex head injury matrix and Glasgow/Glasgow-Liège Coma Scale: a validation and comparison study. Neuropsychol Rehabil 10(2):167–184

Maquet P, Degueldre C, Delfiore G et al (1997) Functional neuroanatomy of human slow wave sleep. J Neurosci 17(8):2807–2812

Monti MM, Vanhaudenhuyse A, Coleman MR, Boly M, Pickard JD, Tshibanda L, Owen AM, Laureys S. Willful modulation of brain activity in disorders of consciousness. N Engl J Med. 2010 Feb 18;362(7):579–89. Epub 2010 Feb 3

Naatanen R, Alho K (1997) Mismatch negativity: the measure for central sound representation accuracy. Audiol Neurootol 2(5):341–353

Naccache L, Puybasset L, Gaillard R et al (2005) Auditory mismatch negativity is a good predictor of awakening in comatose patients: a fast and reliable procedure. Clin Neurophysiol 116(4):988–989

Nieuwenhuijs D, Coleman EL, Douglas NJ et al (2002) Bispectral index values and spectral edge frequency at different stages of physiologic sleep. Anesth Analg 94(1):125–129, Table of contents

Oh H, Seo W (2003) Sensory stimulation programme to improve recovery in comatose patients. J Clin Nurs 12(3):394–404

Owen AM, Coleman MR, Boly M et al (2006) Detecting awareness in the vegetative state. Science 313(5792):1402

Passler MA, Riggs RV (2001) Positive outcomes in traumatic brain injury-vegetative state: patients treated with bromocriptine. Arch Phys Med Rehabil 82(3):311–315

Perrin F, Schnakers C, Schabus M et al (2006) Brain response to one's own name in vegetative state, minimally conscious state, and locked-in syndrome. Arch Neurol 63(4):562–569

Plum F, Posner JB (1983) The diagnosis of stupor and coma, 3rd edn. FA Davis, Philadelphia

Qin P, Di H, Yan X et al (2008) Mismatch negativity to the patient's own name in chronic disorders of consciousness. Neurosci Lett 448(1):24–28

Rappaport M (2000) The coma/near coma scale. The Center for Outcome Measurement in Brain Injury. Accessed 20 May 2009 from: http://www.tbims.org/combi/cnc

Schiff ND (2008) Central thalamic contributions to arousal regulation and neurological disorders of consciousness. Ann NY Acad Sci 1129:105–118

Schiff ND, Rodriguez-Moreno D, Kamal A et al (2005) fMRI reveals large-scale network activation in minimally conscious patients. Neurology 64(3):514–523

Schiff ND, Giacino JT, Kalmar K et al (2007) Behavioural improvements with thalamic stimulation after severe traumatic brain injury. Nature 448(7153):600–603

Schnakers C, Majerus S, Laureys S (2004) Diagnosis and investigation of altered states of consciousness. Reanimation 13:368–375

Schnakers C, Majerus S, Laureys S et al (2005) Neuropsychological testing in chronic locked-in syndrome. Psyche, abstracts from the 8th conference of the association for the scientific study of consciousness (ASSC8), 11(1), University of Antwerp, Belgium, 26–28 June 2004

Schnakers C, Majerus S, Laureys S (2005b) Bispectral analysis of electroencephalogram signals during recovery from coma: preliminary findings. Neuropsychol Rehabil 15(3–4):381–388

Schnakers C, Giacino J, Kalmar K et al (2006) Does the FOUR score correctly diagnose the vegetative and minimally conscious states? Ann Neurol 60(6):744–745

Schnakers C, Majerus S, Giacino J, Vanhaudenhuyse A, Bruno MA, Boly M, Moonen G, Damas P, Lambermont B, Lamy M, Damas F, Ventura M, Laureys S (2008) A French validation study of the Coma Recovery Scale-Revised (CRS-R). Brain Inj. Sep;22(10):786–92

Schnakers C, Ledoux D, Majerus S et al (2008a) Diagnostic and prognostic use of bispectral index in coma, vegetative state and related disorders. Brain Inj 22(12):926–931

Schnakers C, Perrin F, Schabus M et al (2008b) Voluntary brain processing in disorders of consciousness. Neurology 71(20):1614–1620

Schnakers C, Hustinx R, Vandewalle G et al (2008c) Measuring the effect of amantadine in chronic anoxic minimally conscious state. J Neurol Neurosurg Psychiatry 79(2):225–227

Schnakers C, Perrin F, Schabus M et al (2009) Detecting consciousness in a total locked-in syndrome: an active event-related paradigm. Neurocase 15(4):271–277

Sellers E, Kübler A, Donchin E (2006) Brain-computer interface research at the University of South Florida Cognitive Psychophysiology Laboratory: the P300. IEEE Trans Neural Syst Rehabil Eng 14(2):221–224

Shames JL, Ring H (2008) Transient reversal of anoxic brain injury-related minimally conscious state after zolpidem administration: a case report. Arch Phys Med Rehabil 89(2):386–388

Shiel A, Horn SA, Wilson BA et al (2000) The Wessex head injury matrix (WHIM) main scale: a preliminary report on a scale to assess and monitor patient recovery after severe head injury. Clin Rehabil 14(4):408–416

Shiel A, Burn JP, Henry D et al (2001) The effects of increased rehabilitation therapy after brain injury: results of a prospective controlled trial. Clin Rehabil 15(5):501–514

Sorger B, Dahmen B, Reithler J, Gosseries O, Maudoux A, Laureys S, Goebel R (2009) Another kind of 'BOLD Response:' answering multiple-choice questions via online decoded single-trial brain signals Progress in Brain Research, 177 (2009) 275–292, vs/sorger_PBR_coma_science_2009.pdf

Struys M, Versichelen L, Mortier E et al (1998) Comparison of spontaneous frontal EMG, EEG power spectrum and bispectral index to monitor propofol drug effect and emergence. Acta Anaesthesiol Scand 42(6):628–636

Sutton S, Braren M, Zubin J et al (1965) Evoked-potential correlates of stimulus uncertainty. Science 150(700):1187–1188

Taira T, Hori T (2007) Intrathecal baclofen in the treatment of post-stroke central pain, dystonia, and persistent vegetative state. Acta Neurochir Suppl 97(Pt 1):227–229

Teasdale G, Jennett B (1974) Assessment of coma and impaired consciousness. A practical scale. Lancet 2(7872):81–84

The Multi-Society Task Force on PVS (1994) Medical aspects of the persistent vegetative state (1). N Engl J Med 330(21):1499–1508

The Quality Standards Subcommittee of the American Academy of Neurology (1995) Practice parameters for determining brain death in adults (summary statement). Neurology 45 (5):1012–1014

Tolle P, Reimer M (2003) Do we need stimulation programs as a part of nursing care for patients in "persistent vegetative state"? A conceptual analysis. Axone 25(2):20–26

van der Stelt O, van Boxtel GJ (2008) Auditory P300 and mismatch negativity in comatose states. Clin Neurophysiol 119(10):2172–2174

Vanhaudenhuyse A, Bruno MA, Bredart S et al (2007) The challenge of disentangling reportability and phenomenal consciousness in post-comatose states. Brain Behav Sci 30:529–530

Vanhaudenhuyse A, Giacino J, Schnakers C et al (2008a) Blink to visual threat does not herald consciousness in the vegetative state. Neurology 71(17):1374–1375

Vanhaudenhuyse A, Schnakers C, Bredart S et al (2008b) Assessment of visual pursuit in post-comatose states: use a mirror. J Neurol Neurosurg Psychiatry 79(2):223
Vanhaudenhuyse A, Laureys S, Perrin F (2008c) Cognitive event-related potentials in comatose and post-comatose states. Neurocrit Care 8(2):262–270
Vanhaudenhuyse A, Boly M, Laureys S (2009) Vegetative state. Scholarpedia 4:4163
Voss HU, Uluc AM, Dyke JP et al (2006) Possible axonal regrowth in late recovery from the minimally conscious state. J Clin Invest 116(7):2005–2011
Wade DT (1996) Misdiagnosing the persistent vegetative state. Persistent vegetative state should not be diagnosed until 12 months from onset of coma [letter; comment]. BMJ 313(7062):943–944
Whyte J, DiPasquale MC (1995) Assessment of vision and visual attention in minimally responsive brain injured patients. Arch Phys Med Rehabil 76(9):804–810
Whyte J, DiPasquale MC, Vaccaro M (1999) Assessment of command-following in minimally conscious brain injured patients. Arch Phys Med Rehabil 80(6):653–660
Whyte J, Katz D, Long D et al (2005) Predictors of outcome in prolonged posttraumatic disorders of consciousness and assessment of medication effects: a multicenter study. Arch Phys Med Rehabil 86(3):453–462
Whyte J, Myers R Incidence of clinically significant responses to zolpidem among patients with disorders of consciousness: a preliminary placebo controlled trial. Am J Phys Med Rehabil. 2009 May;88(5):410–8
Wijdicks EF (2001) The diagnosis of brain death. N Engl J Med 344(16):1215–1221
Wijdicks E (2006) Minimally conscious state vs. persistent vegetative state: the case of Terry (Wallis) vs. the case of Terri (Schiavo). Mayo Clin Proc 81(9):1155–1158
Wijdicks EF, Bamlet WR, Maramattom BV et al (2005) Validation of a new coma scale: the FOUR score. Ann Neurol 58(4):585–593
Wilhelm B, Jordan M, Birbaumer N (2006) Communication in locked-in syndrome: effects of imagery on salivary pH. Neurology 67(3):534–535
Wood RL, Winkowski TB, Miller JL et al (1992) Evaluating sensory regulation as a method to improve awareness in patients with altered states of consciousness: a pilot study. Brain Inj 6(5):411–418
Young GB (2000) The EEG in coma. J Clin Neurophysiol 17(5):473–485
Zafonte RD, Lexell J, Cullen N (2000) Possible applications for dopaminergic agents following traumatic brain injury: part 1. J Head Trauma Rehabil 15(5):1179–1182
Zandbergen EG, de Haan RJ, Stoutenbeek CP et al (1998) Systematic review of early prediction of poor outcome in anoxic-ischaemic coma. Lancet 352(9143):1808–1812
Zeman A (2001) Consciousness. Brain 124(Pt 7):1263–1289

Chapter 3
Codons of Consciousness: Neurological Characteristics of Ordinary and Pathological States of Consciousness

Gerard A. Kennedy

Abstract It may eventually be possible to identify completely the temporal sequences of electrical microstates that underlie consciousness. From these "codons of consciousness" a DNA-like mathematical model of normal sequencing of the "atoms of thought" could be constructed and matched to subjective experience and behaviour. This would allow the prediction of thought and behaviour from the building blocks of consciousness. Having an intimate knowledge of the sequencing of units of consciousness may allow abnormal patterns of behaviour to be identified even in early childhood and appropriate corrective treatments to be developed. The development of this research may result in treatments for the main pathological problems that trouble all human kind.

3.1 Introduction

In the last few decades the study of the biological basis of consciousness has attracted renewed attention, mainly due to major advances in both neuroanatomical and neurocognitive imaging techniques. The everyday ordinary experience of consciousness reveals only limited information about what may underlie these experiences. In addition, ordinary everyday consciousness and altered states of consciousness have been difficult to study because brain states constantly change and are difficult to associate accurately with external events and subjective reports. Consciousness has been defined in terms of various attributes such as self-reflection, attention, memory, perception and arousal. These attributes can be ordered in a functional hierarchy with the frontal lobes of the brain essential for higher attributes. More recent theories about the neurological correlates of consciousness have tended to focus on the function of frontal cortices in the expression of higher-level attributes. The prefrontal cortex makes up about half the frontal lobes in

G.A. Kennedy (✉)
School of Psychology, Victoria University, Melbourne, VIC, Australia
e-mail: Gerard.Kennedy@vu.edu.au

D. Cvetkovic and I. Cosic (eds.), *States of Consciousness*, The Frontiers Collection,
DOI 10.1007/978-3-642-18047-7_3, © Springer-Verlag Berlin Heidelberg 2011

humans and plays a major role in integrating perceptual information, formulating plans and strategies for appropriate behaviours and instructing the motor cortices in the execution of computed plans of action. The prefrontal cortex is not a single unit and is functionally divided into ventromedial (i.e., lower middle) and dorsolateral (i.e., upper outer sides) apects. This chapter examines some of the studies that have attempted to understand and explain the phenomena of consciousness in terms of their underlying neurobiological basis.

3.2 Ordinary Everyday States of Consciousness

In the course of our everyday activities we experience natural changes in levels of alertness and vigilance as we go through cycles of wakefulness, drowsiness and sleep. In this section naturally occurring states of consciousness are examined. These states include; daydreaming, drowsiness or sleepiness, sleep and dreaming, and also less common states, such as hypnagogic (dream-like hallucinations), sexually induced and near-death states.

3.2.1 Daydreaming

Daydreaming is a subjective experience that usually occurs under conditions of low external stimulation. Daydreaming also includes circumstances where unsolicited thoughts intrude during mental and/or physical activities. Daydreaming is qualitatively different from our usual cognitions focussing on reality and from dreams that occur during sleep. Very few neurophysiological studies have explored daydreaming. However, it has been shown that there are electroencephalogram (EEG) power-spectral signs of lowered attention prior to daydreaming. Lehmann et al. (1995) found that during daydreaming, some EEG power spectral features were correlated with cognition styles. Analysis of 120 EEG power-spectral variables and 20 cognition/emotion variables resulted in four significantly different, independent pairs of canonical (correlated) variables. The first pair had prominent 2–6 Hz and 13–15 Hz EEG power with reality-remote, sudden undirected ideas of low recall quality; the second pair had lowered 10–13 Hz and 15–25 Hz power, with sudden undirected ideas, but with good recall and visual imagery without emotion. These two pairs belong to the hypnagogic family, whereas the remaining pairs were of the awake type. Hypnagogic events are vivid, dream-like states that are experienced as real. The third pair had 10–11 Hz and 19–30 Hz power, with goal-oriented, concatenated thoughts related to present and future and little emotion. Finally, the fourth pair showed a power profile that was the inverse of the second pair, with reality and future oriented thoughts coupled with positive emotion.

Wackermann et al. (2002b) found that EEG power-spectral characteristics during daydreaming conditions were close to characteristics in ganzfeld conditions (sensory homogenization), but were different from those observed during sleep onset. Studies have examined relatively long EEG epochs of 16 s and 30 s, but the basic units of cognitive processes have been shown to fall in the millisecond range (approximately 100 ms). The microstate analysis of EEG data allows a resolution in this time range and has shown that brain electrical states are quasi-stable for fractions of seconds, but rapidly reorganize into another state (Vaitl et al. 2005). This occurs so that brain activity can be parsed into sequences of quasi-homogeneous temporal segments (microstates). Lehmann et al. (1998) showed that different classes of thoughts under daydreaming conditions belong to different classes of microstates with durations of about 120 ms. Lehmann et al. (1998) have suggested that the continuous stream of consciousness is actually discontinuous and consists of sequences of concatenated, psychophysiological building blocks or "atoms of thought" that follow each other in fractions of a second and whose functional significance is identifiable as classes of subjective experiences. They concluded that microstate analyses are required to identify the psychophysiological building blocks of daydreaming and also to discover the rules (syntax) of concatenations of thoughts. The "atoms of thought" or microstates could be thought of in the same way as we think of the elements that make up the genetic codons. Thus, it may be possible to identify particular sequences of microstates and the syntax rules that lead to the formation of actual "codons of consciousness".

3.2.2 Drowsiness and Sleepiness

The changes in alertness and vigilance that occur during the daily circadian cycle of sleep and wakefulness do not usually result in altered states of consciousness. However, alterations in consciousness may occur with only mild levels of sleep deprivation. The phenomena observed include amnesic and/or automatic behaviour, and brief periods of sleep onset (microsleeps). In these states individuals may experience subjective feelings of narrowed attention, reduced volitional control and motivation, decreased memory and impaired cognition. The awareness of time may be reduced, with some studies suggesting it may be overestimated.

The behavioural correlates of sleepiness and drowsiness include performance decrements on psychomotor tasks, such as response omissions and longer reaction times in stimulus detection and discrimination tasks. The physiological correlates reported during periods of reduced vigilance and drowsiness include; changes in EEG, electrooculogram (EOG) and measures of autonomic function. Changes include shifts in EEG spectra to slower frequencies; reduced latency/amplitude of event related potentials (P300); slow eye movements, disappearance of saccades and reduced blinks; and increased pupil diameter variability. The established objective EEG-based vigilance measures include mostly EEG spectral measures and indices. Global brain state descriptors based on multichannel EEG have also been used.

The assessment and classification of states of drowsiness is difficult due to variation in accepted concepts like the activation continuum (i.e., a gradient of consciousness), wakefulness, alertness, vigilance, and sleepiness and drowsiness. The study of states of drowsiness may allow hypnagogic states (dream-like states that occur at sleep onset) and/or hypnopompic states (dream-like states that occur at sleep offset) to be better understood as states of consciousness. Research in this domain has examined the underlying brain electrical activity employing conventional macro analyses that have been used for about 50 years in the area of sleep research. The limitations of these techniques have become apparent more recently when researchers have tried to characterize microsleeps (very brief sleep periods) in an attempt to prevent road accidents. Future research needs to focus more on the microstates (i.e., detailed electrical output analyses of much briefer intervals) that underlie these states consciousness. This may allow each phenomenon to be characterized in terms of the microstate sequences that are peculiar to each. If a more microstate analyses are carried out, the codons of consciousness that underlie these states may be characterised. This may allow us to identify such states and correlate cognitive and behavioural events more accurately to the sequences of microelectric states.

3.2.3 Sleep and Dreaming

Aserinsky and Kleitman (1953) first reported REM sleep and since that time many studies have used physiological recordings to determine sleep stages and awaken subjects to obtain subjective reports. This methodology has allowed the different types of dream states to be described. Dreams occur during all sleep stages, but dreams from REM and non-REM states can be discriminated. REM dreams usually have the highest rates of complete recall and are more extended with a highly narrative structure.

Typical dreams have a narrative structure and consist of story lines that range from realistic to complete fantasy. During dreams parallel cognitive processing and metacognition are reduced and the present moment becomes the most salient. During dreams there is also a virtual sense of reality and a predominance of ego involvement. These two features are the only continuous features of dreaming with the other features being are more or less phasic events.

Hobson and McCarley's (1977) activation-synthesis model suggests that the brainstem contains a dream-state generator that periodically activates subcortical and cortical structures during REM stage sleep. The brainstem-initiated neural activation is most apparent in the visual cortex, the motor cortex, the basal ganglia, and various limbic system structures, particularly the amygdala. The pattern of activation during REM appears to correspond closely to dream content and matches the neural activity of the same behaviours during waking consciousness.

Positron emission tomography (PET) studies have also shown significant deactivations of a large area of the dorsolateral prefrontal lobes. The pattern of deactivation suggests that REM sleep is characterised by a state of general brain activation with a specific exclusion of the executive system. Dreams are devoid of prefrontally dependent cognitions and self-reflection is absent, time is distorted with past present and future freely exchanged, and volitional control is greatly attenuated. Abstract thinking, active decision making, and consistent logic are reduced, and there is not much evidence of focused attention. The capacity for semantic and episodic retrieval of specific memories, which relies heavily on the dorsolateral areas of the prefrontal cortex, is also greatly reduced. In addition, the extent to which a dream is bizarre is related the extent of the prefrontal hypofunction. It is evident that the principal difference between dream cognitions and waking consciousness is due to the deactivation of the dorsolateral aspects of the prefrontal cortex.

Lucid dreaming may be useful for the study of consciousness. During lucid dreaming a person becomes aware that the transpiring events are part of a dream. Lucid dreaming has been defined as an increased level of self-reflective awareness that is usually absent or greatly reduced in other dreaming.

The neurophysiological basis of dreams has been extensively investigated. The scanning hypothesis, suggested that the incidence and directional changes of eye movements during REM sleep were related to the subjects viewing pattern and some early results were promising, but these results could not be replicated. A review of dream research concluded that the results of psychophysiological studies were generally weak and unreliable. Subsequent research has focused more on locating the mechanisms that might generate dreams. Solms (1997) found that patients with bilateral lesions in the occipitotemporal region had no visual dreams and that posterior cortical or bilateral frontal lesions were associated with a total cessation of dreaming. Solms suggested that the forebrain area was a common path to dreaming. Murri et al. (1984) found that subjects with unilateral posterior lesions showed poor dream recall, but that anterior lesions did not produce this deficit. It has been proposed that dreaming results from the excitation of forebrain circuits due to neural impulses arising in the states of consciousness ending activation systems of the brain stem and basal forebrain. However, imaging techniques such as PET employed during REM sleep have not supported this contention. Braun et al. (1998) found that during REM sleep dreaming there were strong activations of extrastriate visual cortices, attenuation of primary visual cortices, activation of limbic and paralimbic regions, and attenuation of frontal association areas. No complete model has yet been proposed that can account for the results from the various different studies of the psychophysiological underpinnings of REM dreaming.

Psychophysiological studies of REM dreaming have been limited by the fact that they still depend on subjects' retrospective reports of dream content. It has been has suggested that this process is impaired because subjects are in different functional states during dreaming and subsequent recall. In addition, as with other waking memorial processes, the reporting of dreams is influenced by interpretive processes, making it very difficult to match physiological activity with psychological events. There needs to be further development of methods that better enable dream researchers to associate neurophysiological events reliably with psychological events. The techniques for describing sleep states are more than 50 years old and have probably outlived their usefulness. The temporal pace of events is too great for the conventional techniques to provide sufficient focus to describe fully the events underlying the ebb and flow of consciousness. Future research in this area could focus more on identifying the microstates that underlie each stage of normal sleep and dreaming. This could help to identify the normal sequencing of microstates associated with each stage of sleep, including dreaming (cf. Lehmann et al. 1998). If the microstate sequences of sleep can be characterized, this would allow us to better distinguish accurately poor sleep from good sleep, and also add vital information to the possible construction of a complete codon of consciousness.

3.2.4 Sleep-Related Hallucinations

Hypnagogic and hypnopompic hallucinations are transient states of decreased wakefulness characterized by short episodes of dreamlike sensory experience that occur at sleep onset and at the end of sleep retrospectively. Hypnagogic-like states can also occur during the daytime when individuals show periods of reduced wakefulness. Hori et al. (1994) described hypnagogic hallucinations as unique periods that cannot be categorized as either waking or sleeping. They also noted that they have unique behavioural, electrophysiological, and subjective characteristics. The subjective experience during hypnagogic states usually consists of both visual and auditory imagery. There appears to be a greater awareness of the real situation during hypnagogic states than during REM dreams. There is little data on behavioural correlates, but leg and/or arm jerks (sleep starts) associated with illusionary body movements have been reported. It has been noted that physiological correlates included drop-offs in alpha EEG activity during the imagery state. Hypnagogic hallucinations are by definition associated with sleep onset [i.e., sleep stage 1 according to Rechtschaffen and Kales (1968)]. However, they may also occur during the presleep period of alpha EEG. Kuhlo and Lehmann (1964) have studied hypnagogic states and EEG correlates both in pre-sleep periods of drowsiness and at sleep onset. They found that spontaneous, transient, fragmentary nonemotional visual and auditory impressions of varying complexity were reported. They reported that these were mainly experienced as unreal and were associated with flattened or decelerated alpha and/or slow theta EEG activity. They suggested that there may be a gradual progression from hypnagogic hallucinations to fragmentary dreams. Wackermann et al. (2000; 2002b) compared hypnagogic phenomena with perceptual phenomena that were observed with reduced sensory input and conceptualized a broader class of "hypnagoid" (i.e., vivid dream-like sensory hallucinations) phenomena and suggested that true hypnagogic hallucinations (primarily visual) are a special case. The conceptualization of these sleep-related states in terms of the underlying sequences of brain microstates would probably show that theses states are cyclic and trap the person between true sleep and wakefulness. Further research to characterize the microstate sequences and hence the particular structure of the codons of consciousness underlying these states would demonstrate how these states differ from sleep or wakefulness consciousness codons.

3.2.5 Exercise Induced States

The most well known exercise-induced altered state of consciousness is the so-called runner's high. Runner's high has been variously described as follows: a state of pure happiness, elation, feelings of unity with one's self and/or nature, endless peacefulness, timelessness, inner harmony, boundless energy, as well as reduction of pain sensations.

Exercise has some well-established beneficial effects on mood states, particularly stress, anxiety and depression. Based on PET studies that have demonstrated hypometabolism in the dorsolateral prefrontal cortex in patients with depression, it has been suggested that this state may be due to hypometabolism in dorsolateral prefrontal cortex. It has been suggested that the widespread activation of motor and sensory systems during exercise comes at the expense of reduced activity in the higher cognitive centres of the prefrontal cortex. In addition, limbic system structures, such as the amygdala, that are not directly required for exercise might similarly show depressed activity. EEG recordings of exercising individuals seem to be consistent with this hypothesis. Exercise is associated with alpha-enhancement, particularly in the frontal cortex. The increase in alpha-activity during and after exercise has been interpreted as indicative of decreased brain activation. Single cell recordings in exercising cats has also provided support for decreased activation in prefrontal regions. It is important to determine the codons of consciousness that underlie intense exercise because it may give us further insight into the brain mechanisms associated with responses such as the fight-flight response, which is one the most basic behavioural sequences and hence consciousness sequences common to all higher organisms.

3.2.6 Sexually Induced States

Several studies have described changes in the electrical activity of the brain that occur during sexual arousal. Mosovich and Tallaferro (1954) reported rapid, low-voltage, activity during early stages of sexual arousal that was followed by slow, high-voltage, paroxysmal (seizure-like) activity during orgasm. Heath (1972) recorded subcortical and cortical EEG in two patients via implanted electrodes and found slow wave and spike activity with interspersed fast activity occurring mainly in the septal region during orgasm. Cohen et al. (1976) showed that alpha waves in both hemispheres during baseline give way to high-amplitude 4 Hz activity over the right parietal lobe during orgasm. Tucker and Dawson (1984) have confirmed this interhemispheric asymmetry and found less alpha power and higher coherence in central and posterior regions of the right compared with the left hemisphere during sexual arousal. Flor-Henry (1980) observed that the orgasmic response was triggered by discharges in the right hemisphere in epileptic patients. Graber et al. (1985) found only a small decrease in alpha activity during masturbation and subsequent ejaculation. The lateralized right hemispheric activation during orgasm has been confirmed with single positron emission computed tomography. It has been shown that there was increased regional cerebral blood flow in the right prefrontal cortex during orgasm.

Neurosurgery and brain lesions following head injury have revealed the involvement of frontal and temporal areas in the inhibitory control of sexual behaviour and septal nuclei in the control of sexual arousal. Men presented with visual sexual stimuli in the PET scanner showed a significant bilateral increase of regional cerebral blood flow in the claustrum and putamen that was correlated with reported sexual arousal. The nucleus accumbens and the rostral part of the anterior cingulate gyrus were also found to be activated. The nucleus accumbens is part of the dopamine incentive system and is involved in the sexual response. The magnitude of activation of the in the nucleus accumbens has been correlated with penile tumescence. Stimulation of this region leads to erection in monkeys. Similarly, it has been found that there is activation in the rostral region of the anterior cingulate gyrus during sexual arousal. However, activity in temporal lobes decreased during an increase of sexual arousal.

In summary, subcortical paroxysmal (seizure-like) and right hemispheric high-amplitude slow activity appear to be related to the partial loss of consciousness during orgasm. Together with sexual arousal and orgasm, lateralized right hemispheric activation tends to occur. Changes in EEG suggest that orgasm is a specific state of consciousness that is not comparable to any other psychophysiological states. Sexual behaviour in human being, unlike many other species, is highly rewarded at the brain level. If we are able to characterize the microstates underlying this behaviour, we will have important basic information about the structure of this part of the consciousness codon. Therefore, further research examining the microstates that occur during sexual activity would be useful in that it would improve understanding of this important aspect of consciousness.

3.2.7 Near-Death Experiences

Near-death experiences often include the following elements (1) feelings of complete peacefulness and wellbeing; (2) a separation of consciousness from the body (out of body experiences); (3) traveling down a dark tunnel; (4) a very bright light often associated with strong feelings of unconditional love; and (5) images of heaven sometimes associated with religious images or figures, or beings of light appear and initiate returning to the body.

Some of the other frequently reported experiences are hearing beautiful music with a slowing down of time and a speeding up of thinking with a complete review of life experiences. Reports of these types of experiences occur in about 10–50% of all near-death situations regardless of subjects' sex, age and profession. The situation in which near-death experiences occur appears to have little or no effect on the occurrence and features of near-death experiences. This suggests that whatever is happening may involve particular neurophysiological processes in the brain. A number of hypotheses have been suggested to explain the neurophysiological processes (cerebral anoxia – lack of oxygen in the brain, depletion of neurotransmitters, release of endorphins, and disinhibition of the brain) and structures (limbic system, septohippocampal formation, temporal lobes, visual cortex) that may be involved in the generation of the near-death experiences. However, given that no experiments can be performed to induce real near-death experiences, neurophysiological models are mostly based on studies that produce altered states of consciousness that are similar to those observed during near-death experiences. These studies use things like acceleration to reduce cerebral blood flow so that near loss of consciousness occurs, the administration of drugs such as ketamine or electromagnetic stimulation of the temporal lobes. However, none of these methods produce states that have all the features of near-death experiences. In particular, it is difficult to ethically produce states in which the psychological load is anywhere near similar to that likely to be experienced in a near-death situation.

The evidence about states of consciousness that occur spontaneously supports the view that they are mediated by changes in cortical activity and arousal levels. Altered states of consciousness are experienced subjectively as dreamlike, illusionary and/or hallucinatory and in most cases are qualitatively different from normal alert, waking and vigilant state perceptions. The changes that occur during altered states of consciousness are transient and disappear rapidly when the central arousal system returns to normal levels of function either via volitional control, changes in circadian rhythms (sleep-wake cycle) or after resuscitation occurs (near death). It would not be difficult to investigate the microstates that accompany death and near death experiences as many people are in situations where they could be monitored. These particular states may inform us about what codons of consciousness remain active until our last moments. This may give us some indication about the hierarchy of importance of the codons of consciousness. Do some remain until the very last moment while others are lost, or is there a general degrading of all codons of consciousness as we near death?

3.3 Pathologically Induced States of Consciousness

There are many degenerative, developmental, and organic brain diseases that are accompanied by alterations of consciousness (e.g., Alzheimer's disease, Parkinson's disease, and various types of dementia). These types of changes in consciousness have not been investigated systematically with respect to the interaction between subjective experiences and specific brain function or malfunction.

3.3.1 Psychotic Disorders

Alterations of consciousness that occur during episodes of psychosis have come under greater scrutiny. These alterations include; hallucinations, delusions, cognitive disintegration and splitting of psychic function. The major hypothesis is that alterations of consciousness are caused by faulty connectivity or by reduced or inhibited interactions between different brain regions. For example, evidence shows that patients with metachromatic leukodystrophy (frontal white matter disorder), show cognitive symptoms that include; thought disorder, hallucinations and delusions. The disconnection hypothesis has focused research on the nature of interhemispheric connection in schizophrenia (split-brain model). However, in the absence of any evidence of structural disconnection, this idea has evolved into one of disordered functional interhemispheric connectivity.

It has been hypothesised that temporally disconnected and abnormal patterns of oscillatory activity contribute to abnormal mental events and that this involves integrative thalamocortical circuits. Gamma oscillations (circa 40 Hz) have been implicated in the binding together of regions subserving conscious perception. It has been suggested that activity of the specific and nonspecific thalamocortical systems underpins conscious experience. Researchers have conceptualized the specific nuclei as providing the content of experience and the altered states of consciousness and the nonspecific thalamic system as providing level of alertness and context. Thus, hallucinations could be interpreted as a state of hyperattentiveness to intrinsic self-generated activity in the absence of appropriate sensory input. Supporting this idea is the coincidence between hallucinations and gamma activity that has been demonstrated in a patient with pseudosomatic hallucinations who was later diagnosed with schizophrenia. Neuromagnetic data also supports this theory. Tononi and Edelman (2000) have shown that conscious experience is dependent on numerous neuronal groups that represent differentiated states distributed in the thalamocortical system. Via corticothalamic and cortico–cortico reentrant interactions, these rapidly bind together into an integrated neural process or a functional cluster. In addition, with conscious experience the neuromagnetic response can become stronger and the cluster more widespread to include frontal, parietal, temporal, and occipital cortices. This is accompanied by increased coherence between distant brain regions, and is characterized by strong and rapid interactions between groups of neurons. Tononi and Edelman studied PET scans from medicated schizophrenic patients during the performance of simple cognitive tasks. They observed that there were similarities in the topography of cluster boundaries, but that functional interactions within the clusters differentiated patients from control subjects. They noted that a coherent network may be disrupted by multiple pathophysiological mechanisms and that many of these have been examined in schizophrenia in past research.

It has been suggested that integrative circuits of the basal ganglia, thalamus, and frontal cortex may be involved in schizophrenia. Thus, schizophrenia has been modeled as a disorder of integration between the sensory systems of consciousness and the motor systems of thought. It has been hypothesised that there are innate cognitive pattern generators that are similar to innate motor pattern generators, which implicates the involvement of the basal ganglia and the frontal cortex in the planning of motor acts and the planning and sequencing of cognitive processes. Similarly, hallucinations, delusions, and disorganized thought have been hypothesized to appear due to the failure of the ability to distinguish external from internal activity, and failure to distinguish self from other. This, in turn, leads to fragmentation of the senses of self and will and results in a subsequent distortion of the boundaries of self.

The diffuse altered states of consciousness ending thalamic systems have also been implicated in the hemispheric activational imbalances found in both schizophrenia and schizotypy. Schizotypy is a conditions that includes some of the same features as schizophrenia, but the severity is much less with sufferers typically showing eccentricity, magical thinking and aloofness. These disorders have been associated with an activated syndrome, possibly underpinned by a left hemispheric activational preference, and a withdrawn syndrome having the opposite lateralized imbalance. Neuroleptic drugs can modify and reverse cognitive and electrophysiological lateral asymmetries. Developmental origins for dispositional imbalances have been suggested. In human infants there are spontaneous asymmetries in gesture and emotion that are evident soon after birth and these may influence the approach–withdrawal balance in social encounters. In schizotypy, factors in the second decade of life may also produce atypical neural connectivity. It has been speculated that early and late pubertal onset are associated with tendencies toward unreality experiences. These effects might be caused by individual differences in regressive events such as synaptic pruning ended by the onset of puberty.

In summary, interruptions ranging from minor asynchrony to complete uncoupling between the conjunctions of specific and nonspecific thalamic systems, and also, in turn, between the content and context of consciousness, may cause many aspects of anomalous processing in schizophrenia and schizotypy. These anomalies include disturbances of sensory processing, sensory gating, and magnocellular functions; of perceptual aberrations, of hallucinations, and with attributions of schizophrenia to a waking dream; of dysregulation of orienting, arousal, alertness, and attention; of mismatches between ongoing and past experiences that may lead to erroneous and delusional thinking; and to truncated microstates (Vaitl et al. 2005). Mental disorders such as schizophrenia are poorly understood and treated. Thus, the opportunity to examine the microstates that underlie different aspects of mental disorders may enhance our knowledge and lead to improved treatments. The microstates underlying the various different states known to occur in schizophrenia should be studied so that codons of consciousness can be constructed that characterize the condition(s). This may help us to understand what goes wrong in the brain in this condition and may also suggest new treatments. New treatments may include modifying the microstates via external or internal computerized microelectrical systems. In the future it may be that diseases such as schizophrenia could be modified or completely cured by an implanted microchip (cf. the cochlear implant).

3.3.2 Coma and Vegetative State

Coma and vegetative states are viewed as decrease or a complete loss of consciousness. However, recent data obtained with PET and evoked response potentials (ERP) suggests that these conditions can also be viewed as involving the fragmentation of the field of awareness, whose single modules may continue to work independently of other modules. This may explain why very simple stimuli may not be the best means to probe the function of fragmented modules and that the optimal level of task complexity may need to be ascertained.

Coma has been is defined as a complete or nearly complete loss of all basic functions of consciousness. This includes; vigilance, mental contents or experience and selective attention. The cause of coma is a dysfunction of brain stem structures regulating sleep and wakefulness. Primary cortical damage may be present, but is not necessary for coma.

Patients who survive coma often pass into another condition referred to as the vegetative state. In contrast to coma, most brain stem functions are preserved in this condition, but all cortical functions are assumed to be lost. Thus, vigilance is preserved, and patients have close to normal sleep–wakefulness cycles. However, mental contents and selective attention appear to be absent. Subcortical responses are usually preserved and often even enhanced, but responses mediated by the cortex are lacking.

The EEG in coma usually shows severe slowing of the dominant rhythms. EEG coma variations with dominant theta or even alpha rhythms have also been described. However, these relatively fast rhythms do not indicate a higher level of vigilance or a better patient prognosis. PET studies show a depression of brain metabolism with a level comparable to that observed during barbiturate anesthesia. The EEG in the vegetative state is usually moderately slowed (4–6 Hz) and dominant delta-rhythm or extremely flat EEG are rarely observed. The decreased level of metabolic activity in subcortical structures in the vegetative state is much more than is observed during sleep, but less severe than that seen in coma patients. However, cortical metabolism may even be more strongly suppressed than in coma.

Early evoked potentials to simple stimuli in both coma and vegetative state can vary from normal through different degrees of suppression up to complete disappearance. The degree of intactness of the evoked potentials serves as a predictor of outcome in coma patients. However, this does not hold for patients in the vegetative state. Late components (100–500 ms post stimulus) or ERP to complex stimulus material such as semantic or syntactic mismatch are often absent in both coma and vegetative states. It has been shown that even in completely non-responsive states, 15–20% of patients show an N400 component to semantically incorrect endings or a P300 component to simple auditory target stimuli. However, absence of evoked brain responses to simple automatic stimuli does not predict absence of evoked brain responses to complex stimuli. Thus, the presence of late components in ERP indicates intact cognitive processing of the presented material in cortical areas, but does not prove conscious experience and control. The subjective state of these patients can only be ascertained in cases where consciousness is regained or if paralysed patients learn to communicate with a brain–computer interface. Deep brain stimulation with electrodes implanted in the nonspecific thalamic nuclei or other parts of the distributed consciousness system have restored to some degree communicative abilities and conscious experience of previously unconscious cognitive processing. Observations about altered states of consciousness in those states from ERP, EEG, or other brain-imaging methods require validation via behavioural, psychological, and/or clinical criteria. Most patients recovering from coma or vegetative states have no memory of the period of unconsciousness. Therefore, it is very difficult to make valid conclusions about the altered states of consciousness in coma, vegetative state, or locked-in-syndrome. If we are able to completely characterize the normal codons of consciousness this may allow us to correct or reconnect microstates in people who have brain damage. These corrections and/or reconnections would most likely involve the implantation of microchips to form a link in the chain of consciousness where it has been identified as deficient.

3.3.3 Epilepsy

Cortical seizures show patterns of paroxysmal activity that are consistent and are a good example of the close connection between neuroelectric (pathological) changes and conscious experience. The location, extension, and intensity of the neurographic signs are correlated with the quality and intensity of the psychological event before, during, and after seizures. The common neurophysiological process underlying different types of epileptic seizures is the hypersynchronization of large areas of the brain. Loss of consciousness occurs if large enough areas of critical cortical tissue are hypersynchronized. This causes the interruption of normal functioning of the neuronal areas involved or leads to deactivations of structures governing consciousness and attention. Partial seizures affect only localized brain areas and provide demonstrations of the modularity of consciousness and the underlying brain mechanisms.

Patients may experience motor activity, sensory symptoms, or cognitive, emotional or autonomic alterations depending on where the paroxysmal neuroelectric discharge occurs in the brain. Visual and auditory hallucinations are particularly frequent after discharges of the memory structures of the medial temporal lobe and the connected hippocampal and cortical regions. Stored memories are excited together with emotional responses in a structured or seemingly chaotic fashion. Out-of-body experiences and autoscopy (observing one's own body in extrapersonal space) are believed to be due to a paroxysmal dysfunction (spasms of seizure-like activity) of the temporoparietal junction in a state of partially and briefly impaired consciousness. Particular seizures of the medial temporal lobe involving the underlying hippocampal and other limbic structures (e.g., amygdala) lead to characteristic and well described altered states of consciousness such as dreamy states and/or distortions of time sense, religious experiences and altered affect.

The multiple pathological processes can cause various different altered states of consciousness. This is probably caused by the impairment of brain functions at different functional levels that appear to be organized in a hierarchical manner. Loss of consciousness during coma is a consequence of severely affected brain stem functions, but in vegetative states brain stem functions are preserved and most cortical functions are lost. At a higher cortical level, paroxysmal neuroelectric discharges (epileptic seizures) result in experiences of altered consciousness. The observation of changing levels of symptoms and the accompanying physiological changes indicates that the normal stream of consciousness is dependent on integrated neural processes and functional clusters subserved by coherent neuronal networks (e.g., via corticothalamic and cortico–cortico reentrant interactions). If these integrative networks are interrupted, altered states of consciousness are likely to occur. The findings on disease-induced altered states of consciousness support the notion of functional and neurobiological modularity. However, the dynamics of altered conscious processes and the content and context of consciousness require a distinct functional organization of integrative neuronal circuits whose nature is still far from being understood. The concept of a codon of consciousness may mean that similarly to people with other forms of brain damage we may be able to correct epilepsy with microchip implants. This idea is not new in this area as various electrical stimulators are already in service. However, these devices are operating at a very course level. Future microchips would be very precise and parse the codons of consciousness so that electrical anomalies were prevented from occurring or errors were corrected.

3.4 Physiologically Induced States of Consciousness

3.4.1 Respiratory Manipulations

Many breathing manipulations are based on meditational practices. These procedures require subjects to focus on breathing and allow slow and shallow respiration to emerge. In meditative techniques, the manipulation of breathing may be paired with other activities such as chanting, counting, or eye-fixation. Deep breathing consists of taking a deep breath, retaining the breath, and exhaling slowly. The method of slow diaphragmatic breathing and paced respiration is also a procedure that alone or in combination with other meditative techniques leads to the reduction of tension in body musculature. The mechanisms of these breathing techniques are based on increases in pCO_2, which results in hypercapnia. Holding one's breath for five or more seconds, and shallow, slow, breathing, results in a hypoventilated state. These methods lead to mild hypercapnia, a slowed heart rate, dilation of the peripheral vasculature, stimulation of gastric secretion, depressed cortical activity and feelings of sleepiness. This process tends to occur naturally during transitions from wakefulness to drowsiness, during hypnagogic, and during normal sleep states.

Hypercapnia effects can also be caused by breathing in a CO_2 enriched environment. Meduna (1950) reported that hypercapnia can cause typical near-death experiences, such as bodily detachment and the perception of being drawn toward a bright light. Both these phenomena were associated with power increases in EEG low-frequency bands. However, a study by Terekhin (1996) was unable to confirm these findings. The differing results may be due to variability in both the duration of breathing the CO_2 enriched air mixture and/or the strength of the mixture itself.

Breathing is also typically altered in other practices, such as ritual dancing and holotropic breathing. Involuntary hyperventilation is often accompanied by physical exertion and sustained emotional tension and mental effort. Hyperventilation is also involved in panic attacks with clinical symptoms including: dyspnea, vertigo, palpitations, chest pain, numbness or tingling, depersonalization and fear of losing control. It has been shown that healthy subjects who are required to hyperventilate experience syncopes. Arrhythmic mycloni occurred in 90% of 42 syncopal episodes. Visual (e.g., colored patches, bright lights, gray haze) and auditory (e.g., roaring noises, screaming) hallucinations were reported by 60% of subjects. Subjects described a state of impaired external awareness, disorientation, weightlessness, detachment and loss of voluntary motor control.

Forced respiration during hyperventilation produces marked physiological effects via alteration of pH and depletion of CO_2. This results in acute or chronic respiratory alkalosis (hypocapnia). The cerebral circulation is highly sensitive to respiratory alkalosis, which develops within the first 15–20 s of hyperventilation. Pronounced hypocapnia (Pa CO_2/22 mm Hg or less) affects regional and local cerebral hemodynamics, circulation, and oxygen supply. Hyperventilation-induced changes in EEG (i.e., hypersynchronisation) were found to be identical to the hypoxia-induced changes, such as arteriole vasospasm, ischemic foci, and redistribution of the blood flow between various brain regions. In both situations, fainting, obnubilation, depersonalization, and similar forms of altered states of consciousness may occur.

There is another respiratory procedure that is part of yoga (Pranayama) breathing exercises. This process involves unilateral nostril breathing. Stancak et al. (1991) studied young subjects during bilateral and 15-min periods of unilateral nostril breathing. In addition to cardiovascular changes (e.g., increased respiratory sinus arrhythmia), lowered peak power of beta-2 EEG activity in the frontal lobes was observed during unilateral nostril breathing. This indicated a relaxation-specific effect, similar to that observed by Jacobs et al. (1996) in their study. A homolateral relationship between the nostril airflow and EEG theta activity was also observed and it was attributed to a lateralized modulation of the subcortical generators of the EEG theta band.

These states may reveal something about the primacy of the codons of consciousness analogous to that discussed in relation to near-death experiences in Sect. 3.2.7. Research in this area may show the structure and type of the codons of consciousness that are preserved until consciousness is finally lost.

3.4.2 Environmentally Induced States of Consciousness

Having to function under extreme environmental conditions can result in altered states of consciousness. Brugger et al. (1999) investigated altered states of consciousness retrospectively via structured interviews about the effects of exposure to altitudes above 8,500 m without supplementary oxygen in eight world-class climbers. The climbers were also examined using neuropsychological tests and electroencephalography and magnetic resonance imaging (MRI) within a week following the interview. All except one of the climbers reported distortions of the body schema to be the main somesthetic experience followed by visual and auditory hallucinations. Illusions and hallucinations were reported more frequently at altitudes above 6,000 m. Hallucinations and illusions were not related to brain abnormalities and were most likely due to severe hypoxia, social isolation and acute stress conditions experienced during climbing.

Studies on the negative effects of extremely high altitudes on cognitive, emotional and behavioural functioning indicate that visual and auditory hallucinations are probably caused by severe hypoxia and acute mountain sickness. Similar reports of illusions, hallucinations, cognitive dysfunction and negative emotional states exist for professional and experienced scuba divers. However, little research is available about the frequency, types and situations associated with the experience of altered states of consciousness, and/or different cognitive and behavioural dysfunctions. There have been a few studies in simulated hyperbaric conditions. These studies have reported deficits in attention, memory, vision, audition and sensation. A few studies found little evidence of such changes in cognitive or perceptual processes. However, other researchers have suggested that altered states of consciousness are more than likely to be due to narcotic properties of gases in breathing mixtures inhaled by divers while in deep water for long periods.

The experience of staying over winter in the Artic or Antartic has given researchers the opportunity to study the effects of extreme climates on psychological and physiological adaptation, performance and well-being. From studies of cognitive and behavioural dsyfunctions and emotional disorders, it has been suggested that most altered states of consciousness are related to various types of illness and/or disease that can occur (e.g., shining, cottage fever, infectious diseases, brain injuries) or with acute stress following interpersonal conflicts between group members.

There is some data suggesting that prolonged exposure to microgravity can affect cognitive functioning and motor behaviour in negative ways. However, no systematic studies are available indicating that hallucinations, sensory illusions or other types of altered states of consciousness are common in those exposed to space travel. The microstates underlying some of these phenomenon may reveal the codons of consciousness that occur when environmental conditions are very challenging. More research is needed to study the brain electrical activity of subjects under such conditions.

3.4.3 *Fasting and Starvation*

Extreme fasting and starvation result in physiological changes that have been associated with cognitive, social, emotional, attentional and behavioural changes. In subjects undergoing extreme fasting and in patients with anorexia, altered states of consciousness and are evident in terms of attentional biases for food-related stimuli, disturbed body image, illusions and/or hallucinations. Studies using a modified version of the Stroop test found that extreme dieters and patients with anorexia show increased reaction times to food-related words not observed in normal or fasting subjects. In addition, patients with anorexia, but not dieters or controls responded with increased reaction times to words related to body image.

Starvation in patients with anorexia has been shown to be associated with decreased regional blood flow in the right hemisphere that normalised after weight gain. It was also shown that early and late components of evoked potentials to somatosensory stimulation were decreased and that spectral power of EEG background activity was reduced in the alpha-1 band in strong dieters and patients with anorexia. There was also decreased theta EEG power over the right parietal cortex during haptic exploration tasks that persisted after the subjects regained weight. There is some evidence for an attentional bias to food in subjects on extreme diets and this is correlated with regional changes in cerebral blood flow and EEG spectral power. The microstates underlying starvation reveal similar information about the primacy of the codons of consciousness that occur under near death conditions.

3.4.4 *Sensory Deprivation, Overload and Homogenization*

Sensory deprivation involves minimizing external sensory input. Research on restricted environmental stimulation has usually been conducted by housing subjects in isolation chambers where sensory stimulation is minimal. An example of this type of chamber is the flotation tank in which visual and auditory stimulation is restricted and tactile stimulation is also reduced. In addition to sensory restriction by isolation, flotation-based minimal stimulation results in the illusion of low gravity and a sensation of floating.

Restricted environmental stimulation results in a number of physiological changes including reductions in plasma levels of epinephrine, of norepinephrine and of stress hormones. Restriction of sensory input has also been shown to be associated with an increase of beta-endorphin levels in the peripheral blood that are not observed during control relaxation conditions.

Some other effects of restricted stimulation have been demonstrated for memory functions, creativity, perception and signal detection, social cognition, and readiness to change one's attitudes on social phenomena, and inducing increased motivation to change critical and maladaptive behaviour patterns. Suedfeld and Eich (1995) observed that autobiographical life episodes were retrieved more intensely and recalled more pleasantly after participation in a restricted stimulation treatment in comparison with a normal rest condition. In addition, when exposed to recognition memory tests for words and unfamiliar faces prior to and following restricted stimulation, subjects showed better performance following a control condition. Memory performance showed a significantly greater enhancement of right hemispheric processing after flotation in comparison to that after a control condition.

Sensory hallucinations do not represent a common phenomenon of restricted stimulation. In a study by Schulman et al. (1967), only 2% of subjects reported perceptions of an outside stimulus that was not physically present, but 74% reported the experience of visual perceptions that appeared under their control. Zuckerman et al. (1969) found that in most cases of sensory minimization, hallucinatory perceptions are transient and impersonal in quality and mostly nondynamic and nonpsychopathological in nature. The hallucinations are mostly visual and composed of simple features, such as sudden flashes, colors, or changes of hues. Hallucinatory perceptions occur most frequently during states of medium to high physiological arousal as opposed to lower states of arousal. Studies have shown that illusions and hallucinations are possibly due to the instructions given to subjects pre-exposure rather than phenomena that occur spontaneously.

Confinement for 72 h did not affect subjects' estimation of time spent in an isolation chamber. Incorrect estimation of elapsed time (over or under) was associated with periods of higher behavioural activity, whereas during periods of fewer time estimation errors, behavioural activity levels were lower (Sugimoto et al. 1968). When groups were examined prior to and following restricted environmental stimulation with and without flotation, using a tactile object-discrimination task carried out with each hand separately while blindfolded and a recognition memory test for words and unfamiliar faces, the flotation group showed a significantly greater enhancement of right hemispheric processing after flotation than was found when retesting the controls. The direction of the lateral imbalance was similar to that observed following hypnosis.

The extent to which changes in perception during restricted environmental stimulation relate to different neural activities in the brain has not been systematically established. However, EEG studies have shown changes of alpha, beta, and theta-band activities in general. There are few systematic studies of the cerebral electrical concomitants of visual, auditory, and other sensory changes, illusions, and hallucinations. Altered sates of consciousness have only been observed in subjects who have had a number of prior experiences with such phenomena due to prior exposure to meditation or consumption of psychedelic drugs or who were instructed that such phenomena might arise. Therefore, whether or not altered states of consciousness are a common consequence of sensory deprivation has not been conclusively determined by any valid research.

Sensory overload has been used as a method of inducing altered states of consciousness. Vollenweider and Geyer (2001) have proposed a cortico–striato–thalamo–cortical loop model of psychosensory processing. They suggested that a deficient thalamic filter, which had been operationalised in animal models by using the prepulse inhibition paradigm, results in sensory overload of the cortex. Knowledge about the relationship between sensory overload and brain function and consciousness related functions is mostly based on clinical research on schizophrenia, intensive care unit syndrome, autism, and trials of hallucinogenic drugs.

In patients with schizophrenia, the ability to inhibit the processing of irrelevant stimuli from the environment is impaired. It has been hypothesized to cause sensory overload in schizophrenic patients and lead to disorganization and thought disorders. There is a cluster of psychiatric symptoms that are unique to intensive care unit environments that may be caused by sensory overload or monotony. Sleep deprivation has also been discussed as a possible cause of sensory overload in intensive care units. McGuire et al. (2000) suggested that intensive care unit psychoses are caused mainly by organic stressors. Autism has been suggested to result from the brain's inability to handle sensory overload. It has been speculated that this results from a reduced capacity for sensory filtering because of the underlying and as yet unidentified neurological disorder. Brauchli et al. (1995) have shown these effects in healthy subjects who were exposed to very high levels of sensory stimulation. However, they were unable correlate the effects with cerebral electrophysiological activity.

Ganzfeld (sensory homogenization) is the term to describe an unstructured, perfectly homogeneous perceptual, usually visual field. A longer exposure (from minutes to tens of minutes) to visual ganzfeld (a homogeneous red light field) and auditory ganzfeld (monotone noise, e.g., "white" or "pink" noise) may induce an altered state of consciousness characterized by episodes of imagery ranging from simple sensory impressions to hallucinatory, and dreamlike experiences. Visual imagery is the most commonly reported experience with perceptions involving other sensory modalities reported less frequently. Ganzfeld procedures are similar to sensory deprivation procedures, but differ in that the level physical sensory stimulation is maintained at medium to high levels. The subjective experience of ganzfeld imagery is similar to that reported for hypnagogic imagery. It has been suggested ganzfeld procedures induce a hypnagogic state. Due to environmental restrictions, subjects under ganzfeld procedures can show little overt behaviour, but can respond verbally. Subjects' verbal expression is often less organized than it is during the ordinary waking-state reports. This has been suggested to be due the dreamlike nature of ganzfeld experience and not to any impairment of cognitive function. It was assumed that ganzfeld induces a state favorable to "extrasensory perception," which would then constitute a special class of altered consciousness. However, the experimental findings remain controversial. Time awareness is little affected by ganzfeld; with verbal estimates of elapsed time intervals (5–15 min) somewhat underestimated in comparison to the waking state.

Studies comparing physiological correlates of ganzfeld, sleep onset, and relaxed waking state, showed EEG spectra in ganzfeld to be more similar to wakefulness than to the sleep onset state. Wackermann et al. (2000, 2002a) found there was slightly accelerated alpha activity and there were no EEG markers indicating decreased vigilance. Therefore, it does not appear that ganzfeld imagery has a hypnagogic basis. In a follow-up study, EEG correlates of vivid imagery were studied with selected "high responders" to ganzfeld. Intraindividual comparisons against no-imagery EEG revealed mostly triphasic courses, consisting of: (1) an initial alpha increase; (2) a burst of accelerated alpha-2 activity (10–12 Hz); and (3) a deceleration and reduction of alpha rhythm and an undulating increase of beta-2 (18–21 Hz) and beta-3 (21–30 Hz) power, indicating a transition from relaxation to outward directed and fluctuating attention and finally preparation for the mentation report (Wackermann et al. 2002a).

3.4.5 Rhythm-Induced States of Consciousness

Rhythmic procedures such as dancing and drumming have been used to induce altered states of consciousness since ancient times and are still common today. These activities can cause a trancelike state of detachment that is usually characterized by a "narrowing of awareness" and stereotyped movements or behaviours that are reported as being beyond the subject's volitional control. In the case of drumming or other rhythmic music and dancing, the rhythmic body movements become synchronized with the beat and seem to occur almost automatically. The subject's awareness of self and surrounding can be markedly reduced as they become absorbed in the experience.

There have been few investigations of these common phenomena. However, Neher (1961, 1962) did study the effects of monotonous drumming on EEG output He found that drum beats (3–8 Hz) can induce EEG waves of the same frequency ("auditory driving"), and speculated that this phenomenon may be responsible for trance-like states. The importance of the role of social stimulation and settings was raised by Rouget (1980). Maurer et al. (1977) found that relaxation and similar shamanic-type experiences (e.g., dissociation from the body, tunnel experiences) were evoked by monotonous drumming in subject who were medium and highly hypnotizable.

In addition to rhythmic auditory stimulation, social settings and personality traits (e.g., absorption), a fourth factor, rhythmic body movements during drumming and dancing, appears to play a major part in the induction of trance states. The rhythmic movement of the body is accompanied by shifts in body fluids (e.g., blood) and respiration tends to synchronize with movements inducing heart rate oscillations known as respiratory sinus arrhythmia. Thus, rhythmic movements may cause respiratory/cardiovascular synchronization with increased blood pressure oscillations that simulate the carotid baroreceptors. The effects of baroreceptor stimulation are not confined to a slowing of the heart rate as they are also known to reduce cortical arousal and excitability. Barostimulation is known to result in increased pain thresholds, increased theta activity, and reduced muscular reflexes which are also typical features of trance states. Studies have been performed using body tilt tables to investigate whether cardiovascular oscillations induced by body movements contribute to trance induction. The subjects were rhythmically tilted between 6° head-down and 12° head-up. Tilting movements were triggered by respiration that was paced at 125 Hz. The respiratory oscillations of the heart rate were amplified in one condition and in the other they were damped. In the condition with increased heart rate variability, power in the EEG theta band was increased and the cross-spectral power between pupil oscillations and respiration was reduced, indicating cortical inhibition. Subjects reported drowsiness, disorientation and hallucinatory experiences. Subjects showed moderate scores on the Phenomenology of Consciousness Inventory (Pekala 1991). It was also reported that subjects responding differently to the two tilting conditions had significantly higher scores on Tellegen's Absorption Scale in comparison to subject who did not respond (Tellegen and Atkinson 1974). The findings of this study were replicated and baroflex sensitivity correlated significantly with the absorption score. These results strongly suggest a relationship between the cardiovascular responsiveness and the absorption personality trait, which is, in turn, known to be positively correlated with hypnotic suggestibility. Similarly to the above section, the codons of consciousness that occur under these conditions are probably very different from those of normal waking consciousness. Therefore, research that is able to characterize the micro-states would be essential for understanding these states.

3.4.6 Meditation

Instructions on how to meditate range from more receptive "wide focus" techniques to those requiring intensive concentration on external objects, imaginings, or parts of the body. The transcendental meditation technique, introduced to the West by Maharishi Mahesh Yogi (1996), is a yogic "mantra" meditation that involves the silent repetition of a syllable in a passive and effortless manner. The procedure is standardized and has been more widely studied than other less known methods of meditation. The subjective experience that occurs during transcendental meditation has been described as a state of blissful mental quiescence, with the absence of thoughts, but yet with consciousness remaining and has been referred to as a state of "transcendental consciousness". Transcendental consciousness is often associated with short periods of cessation of breathing. The breath period, magnitude of heart rate deceleration and skin conductance level have been found to be greater during periods of transcendental consciousness. These observations have suggested that autonomic activity reflects a transition of awareness from active thinking processes to a silent yet alert state of transcendental consciousness.

Observed changes in EEG output during meditation, such as an increase in alpha activity, appear to be nonspecific and do not support the notion of a unique state of consciousness. There is also increased EEG coherence found during transcendental meditation, which has been interpreted as reflecting an "ordering of the mind". However, this coherence is not unique to meditational states. More recent research comparing subjects who have undertaken meditation for periods longer than 3 years with those practicing for less than 6 months has shown a difference in slow wave theta activity. Long-term meditators were characterized by increased theta EEG activity over the frontal region. The intensity of the blissful experience correlated with increases in theta power in anterior-frontal and frontal-midline regions. Long-term meditators showed increased theta synchronization between the frontal and the posterior association cortex that reached its highest level in the left prefrontal region. This finding is comparable to studies that have shown EEG coherence during the experience of positive and negative emotions. These studies suggest that theta activity, peaking in the left prefrontal region, is indicative of emotionally positive experiences.

Some studies have been performed to examine different meditation techniques and determine whether there are any characteristic changes in brain activity. For example, Lehmann et al. (2001b) examined a subject who was able to voluntarily enter distinct meditative states by applying different meditation techniques. It was observed that different centres of gravity of gamma EEG activity related to the various meditation techniques (visualization: right posterior; verbalization: left central; dissolution of the self: right anterior). It has been suggested that meditation is voluntary regulation of attention. They trained subjects who had practiced a breathing meditation for many years to meditate while undergoing fMRI. The subjects showed an activation of neural structures involved in attention and control of the autonomic nervous system. They also observed a signal decrease that appeared to be related to slower breathing during the meditation phases. Similarly, some characteristics of meditative states, such as the sense of becoming one with objects of meditation, have been associated with the blocking of a parietal cortical area that represents body position in three-dimensional space as shown by single positron emission computed tomography.

Traditional meditation procedures are beginning to come under closer scrutiny using modern scientific methods of brain imaging. The thorough cognitive analysis of meditation procedures using trained subjects and brain imaging techniques may lead to a better understanding of biopsychological bases of different altered states of consciousness achieved through meditative procedures. Studies that have used various neuroimaging techniques have shown converging evidence of dorsolateral activation in the prefrontal cortex. PET, SPECT and fMRI studies show an overall increase in dorsolateral prefrontal cortex activity. Many studies of meditation have consistently reported alpha wave activity across the frontal lobe during meditation. Thus, the data from EEG and neuroimaging studies appear to be at odds.

3.4.7 Relaxation

There are many techniques that can be used to induce states of relaxation and subsequent altered states of consciousness. Progressive muscular relaxation, autogenic training, biofeedback, and meditative practices are all commonly used techniques to induce relaxation. Cognitive-behavioral models of relaxation suggest the following three factors are common to all forms of relaxation: (1) focusing, the ability to maintain concentration on and return attention to simple stimuli (acoustic or visual) for an extended period of time; (2) passivity, the ability to refrain from goal-directed and analytic thoughts; and (3) receptivity, the ability to tolerate and accept unusual or paradoxical experiences. It has been proposed that most relaxation techniques elicit a general "relaxation response" that consists of physiological changes that are mainly evoked by decreased autonomic nervous system activity (slowing of heart rate, slow and shallow breathing, peripheral vasodilation, reduced oxygen consumption, and decrease in spontaneous skin conductance responses). These effects are produced by reduced sympathoadrenergic reactivity.

Deep relaxation results in changes in EEG activity that indicates reduced cortical arousal. Relaxation techniques that mainly focus on reducing oculomotor activity (i.e. do not involve vivid imagery of overt movements) increase EEG alpha activity. Jacobs et al. (1996) found that naïve subjects undergoing the progressive muscular relaxation training showed a significant reduction in the EEG beta activity that was mainly located in frontal lobe sites. There were no differences in EEG observed at any of 13 other electrode sites between the relaxation and control groups. Peripheral physiological effects usually follow the reduced EEG cortical arousal. Some authors have investigated the engagement of different brain regions during hypnosis-based suggestion of relaxation compared with suggestion of negative emotion, such as anxiety. Relaxation induced stronger activity compared with anxiety in the left superior temporal gyrus (Brodman Area 22). During the two induced emotional states, brain activity shifted to the right hemisphere in the fronto-temporal regions during negative compared with positive emotions (relaxation). The excitatory frequency EEG band of beta-2 (18.5–21.0 Hz) indicated that the strongest relaxation-related effect on location, with the source centre of gravity located less right-sided during relaxation than during anxiety. The differences in lateralized of brain activity in the two conditions were mirrored symmetrically in the two hemispheres. Relaxation showed greatest activity differences in left temporal areas, and negative emotion in right frontal areas.

Neuroimaging studies have suggested that various brain areas and systems are involved when subjects deliberately decrease sympathoadrenergic activity. Critchley et al. (2001) used PET to examine cerebral activity relating to biofeedback-assisted relaxation and modulation of sympathetic activity. The voluntary control of bodily arousal was related to enhanced activity of the anterior cingulate cortex. This brain area is believed to be engaged in the integration of cognitive control strategies and bodily responses. Critchley et al. (2000) demonstrated that these relations were lateralized, with predominantly left cingulate activity associated with the intention to relax and right cingulate activity occurring during task-related states of sympathetic arousal. Characterizing the codons of consciousness that underlie the relaxation response would have many therapeutic applications and the response may be able to be produced more quickly. Understanding this brain response will also assist in understanding how the brain works.

3.4.8 Biofeedback

Studies suggest that biofeedback and instrumental learning of neuronal activity is a promising noninvasive method of manipulating brain activity as an independent variable and observing altered states of consciousness as dependent variables. The main objective of biofeedback processes is a greater awareness and voluntary control over physiological processes. Various methods have been shown to be highly effective in allowing subjects to achieve greater voluntary control over physiological processes. These processes include heart rate, blood pressure, vasomotor responses and temperature, respiratory activity, gastrointestinal reactions and muscular tension. In 1960s and the early 1970s, biofeedback of EEG alpha rhythm was believed to alter states of consciousness. It was shown that subjects could learn to discriminate whether they were in an alpha or in a beta state, and by means of the brain where biofeedback could produce alpha brain waves. Subjects' verbal reports during moderate to high alpha states indicated a state of serenity and happiness. Therefore, some researchers suggested that states of consciousness can also be modified by brain-wave feedback.

More recently, neurofeedback has been developed as a form of biofeedback linked to aspects of brain electrical activity such as frequency, location, or amplitude of specific EEG activity. Neurofeedback aims to alter electrocortical processes associated with cortical excitability, arousal, and central control of motor performance. EEG components are fed back to subjects by acoustic or visual signals that can be shaped with operant conditioning strategies or cognitive manipulations. The modulation of cortical activity can be achieved via a training process involving real-time representation of EEG parameters paired with positive reinforcement to facilitate successful operant conditioning of desired responses (Egner and Gruzelier 2001). A number of authors have demonstrated operant conditioning and self-regulation of various EEG parameters in animals and humans in both experimental and clinical settings. Sterman and Shouse (1980) showed that epileptic seizure frequency can be reduced by the conditioned enhancement of a low beta-band 12–15 Hz EEG rhythm, sensorimotor rhythm (SMR or u-rhythm), over sensorimotor cortex. These findings are interpreted as improved response inhibition by SMR feedback training. Sterman (1996) suggested that SMR training may lead to increased inhibitory activity of thalamic nuclei interacting with somatosensory and sensorimotor cortex. Similarly, Egner and Gruzelier (2001, 2004) have suggested that inhibitory activity may explain the decrease of performance errors in subjects during sustained attention tasks deficit, as for example in subjects with attention-deficit/hyperactivity disorder. Learning the ability to increase relative and absolute SMR (12–15 Hz) amplitude was shown to correlate positively with improvement of attentional performances and reduction of impulsiveness, whereas the opposite was true for the enhancement of beta-1 (15–18 Hz). The relationship between learning to enhance SMR over sensorimotor cortex and reduced impulsiveness supports the idea that inhibitory processes mediate behavioural and cognitive improvements. In addition, the combination of SMR and beta-1 EEG training has provided evidence, with both behavioural and event-related brain potentials (ERP), of being effective in integrating sensory input and counteracting fast motor response tendencies and error-prone behaviour. Similar frequency-specific effects of neuro feedback were found for cognitive processes, such as working memory and attention.

Negative shifts of slow cortical potential (SCP) in EEG changes reflect widespread depolarization of apical dendrites of pyramidal neurons and a decrease of thresholds for paroxysmal activity. It has been shown in healthy subjects that SCP can be brought under volitional control. Therefore, the use of SCP feedback as an adjunctive treatment for drug-refractory epilepsy has been tried. Some studies have shown that patients with epilepsy are able to learn self-control skills after an extensive training of SCP self-regulation, which has resulted in significant decreases in rates of seizure.

The self-regulation of SCP can also be used for brain–computer communication. It may be useful for patients with neurological diseases that lead to total motor paralysis caused by amyotrophic lateral sclerosis (ALS) or brain stem infarct ("locked-in" patients). Brain–computer interfaces provide a system for communication with a cursor on a screen that is moved (vertically or horizontally) by changes in self-regulated SCP in both directions. Birbaumer and his group have developed a thought translation device and a training procedure to enable patients with ALS, after extensive training, to communicate verbally with other people without any voluntary muscle control (Kubler et al. 2001).

There is a growing catalogue of applications of neurofeedback that have shown that patients have benefited from learning to control EEG parameters. For example, Gruzelier et al. (1999) trained schizophrenic patients to reduce their interhemispheric imbalance via SCP feedback. Future research could use more sophisticated microelectrical implants to correct the codons of consciousness in people with these brain pathologies. Implanted microlectrical devices could be used to correct and parse the flow of codons of consciousness on a permanent, real time, basis.

3.4.9 Hypnosis

Hypnosis has been defined in terms of distinctive psychological processes and also in terms of ordinary processes that underlie everyday behaviour. Hypnosis is a procedure that may or may not involve altered states of consciousness, and the latter is not necessary for most hypnotic phenomena. Hypnosis may be induced by many different procedures. For example, progressive relaxation instructions, focusing on a swinging pendulum, staring at an object or focusing on a sound have all been reported as successful induction procedures. The induction procedure is usually followed by a series of direct or indirect suggestions for changes in experience, behaviour and cognition. Suggestions may be given to behave in ways that are experienced as involuntary, to be unable to behave in ways that are normally voluntary, or to experience changes in belief, perception, or memory that are at odds with reality. Hypnosis states can be induced by others or may also be self-induced. Hypnotic phenomena have also been shown to be produced by suggestion in the complete absence of any formal induction procedure. There a many useful clinical applications of hypnosis, including pain control and changing behaviours (e.g., cessation of smoking), and hypnosis can speed up psychotherapeutic interventions for various conditions (e.g., anxiety conditions) and also improve the immune response.

The effectiveness of hypnotic suggestion in producing changes in experience (i.e., hypnotic susceptibility) shows a high level of inter-subject variability, but appears highly stable within individual subjects. Recent theories of hypnosis stress the importance of subjects carrying out self-directed cognitive strategies, response expectancies, or alterations in central executive control as processes mediating response to hypnotic suggestion. Research has focused on the assessment of hypnotic susceptibility and on behavioural responses that occur during hypnosis. However, the core phenomena appear to lie at the level of the subjects experience. Thus, very similar or even identical behavioral responses often result from markedly different experiences and cognitive strategies produced in individual subjects by very similar hypnotic suggestion.

The neurophysiological basis of hypnosis has been the subject of many studies over the past 50 years. Studies have reported on multiple changes in many different physiological parameters (for reviews, see Gruzelier 1998). The majority of studies of hypnosis have employed evoked potential and/or frequency analysis of EEG data. For instance, a number of studies investigated electrophysiological characteristics of hypnotic susceptibility. These studies reported various differences between high- and low-hypnotically susceptible subjects in different EEG bands. For example, higher power in the theta frequency band in some studies, higher gamma power over the right hemisphere, and higher global dimensional complexity all were reported for high-susceptible subjects. To date no set of neurophysiological measures has been able to reliably categorize data that has been obtained from hypnotized or non-hypnotized subjects while there is a large body of data available, methodological differences make the findings difficult to interpret. One of the main problems has arisen because many different hypnosis procedures have been used.

Gruzelier has consistently supported the idea that hypnosis is a form of frontal inhibition. Well replicated neuropsychological findings (Gruzelier and Warren 1993) show impaired letter fluency (left frontal), but not category fluency (left temporal) performance during hypnosis for subjects with high, but not low in hypnotic susceptibility. This suggests that for highly hypnotizable subjects, hypnosis is associated with inhibition of the left dorsolateral prefrontal cortex (Gruzelier 1998). Selective influences within the cingulate have also been suggested based on evidence of the maintenance of the error-related negativity that was associated with an abolition of the positivity wave in highly hypnotizable subjects during hypnosis (Kaiser et al. 1997). Some studies have investigated hypnotic analgesia to test the hypothesis that part of the phenomena occurring under hypnosis may be explained by dissociation between functional subunits organizing conscious behaviour. For example, Croft et al. (2002) analyzed EEG component frequencies in the period following painful electrical stimulation of the right hand in a control condition during hypnosis, and after the hypnotic suggestion of analgesia. Prefrontal gamma EEG activity localized in the anterior cingulate cortex predicted the intensity of subjects' pain ratings in the control condition. This relationship remained unchanged by hypnosis for the subjects with low susceptibility, but was abolished in highly hypnotizable subjects following hypnosis.

Lehmann et al. (2001a) examined intracerebral source locations during hypnotically suggested arm levitation versus willful initiation of arm raising in highly hypnotizable subjects. They observed sources of delta and theta EEG activity more posteriorly and of alpha and beta-1 activity more anteriorly in the hypnotic arm levitation condition, suggesting the co-occurrence of electrophysiological characteristics of lowered vigilance and increased attention.

Friederich et al. (2001) compared the effects of hypnotic analgesia and distraction on processing of noxious laser heat stimuli in highly susceptible subjects. They found significantly reduced pain reports during both hypnotic analgesia and distraction in comparison with the control condition. However, there were significantly smaller amplitudes of the late laser-evoked brain potential components during distraction in comparison to hypnotic analgesia conditions. In addition, coherence analysis of neural oscillations between different areas of the brain suggested a significant decrease of coherence within the gamma band between somatosensory and frontal sites of the brain in hypnotized subjects in comparison to control subjects. The loss of coherence between somatosensory and frontal brain areas during hypnotic analgesia was hypothesized to reflect a breakdown of functional connectivity between the brain areas involved in the analysis of the somatosensory aspects of the noxious input and areas organizing the emotional and behavioural responses to pain. Therefore, hypnotic analgesia and distraction appear to involve different brain mechanisms. There is probably an early attentional filter for distraction, but a dissociation between early sensory and later higher order processing of noxious input for hypnotic analgesia. These results support neodisossociation theories of hypnosis and hypnotic pain control and are consistent with recent studies demonstrating a dissociation between processing of somatosensory and affective information during hypnotic analgesia by both behavioural and regional cerebral blood flow data.

The results of more recent studies suggest that hypnosis affects integrative functions of the brain and may cause an alteration and/or a breakdown of the communication between subunits within the brain that are responsible for the formation of conscious experience. Recordings from intracranial electrodes in epileptic patients have disclosed the importance of changes during hypnosis in limbic structures including the amygdala and hippocampus. The amygdala has been shown to exert mainly excitatory influences on electrodermal orienting activity, whereas the inhibitory action of the hippocampus facilitates the habituation of the orienting response with stimulus repetition. Neuroanatomically, the influence of hypnosis on electrodermal orienting and habituation was compatible with the evidence from De Benedittis and Sironi (1988) of functional inhibition of the amygdala and activation of the hippocampus by hypnosis. The identification of the codons of consciousness that underlie hypnosis may be useful in that it will allow this therapeutic state to be more reliably reproduced. It would also allow us to determine how the codons underlying hypnosis differs from those underlying normal waking consciousness and thus determine what sequences of microstates contribute to this unique state. This may assist us to better understand how the brain works under different conditions. It may also be useful because in hypnotic states people are more susceptible to suggestions about how they might change their behaviour and attitudes. Microimplants could be developed to assist in reprogramming the thoughts and behaviours of serious criminal offenders (e.g., repeat sex offenders). On a less dramatic scale ordinary people may find technologies could be used to reprogram psychological issues such as anxiety, depression or other more minor issues such as self-esteem, assertiveness, or shyness.

3.5 Summary

Given the extensive range of altered states of consciousness it is a difficult task to understand any common neurobiological basis that they might have. However, attempts have been made to construct lists of basic dimensions or characteristics of altered states of consciousness. Farthing (1992) suggested that 14 dimensions could be used to describe an extensive list of psychological functions and experiences. Pekala (1991) proposed up to 26 dimensions. More recently, Vaitl et al. (2005) have proposed a four-dimensional descriptive system that allows various states of consciousness to be described. Their system includes the following four dimensions:

1. Activation refers to the preparedness of an organism to interact with the physical and/or social environment. Activation is being alert, awake, responsive, and ready to act and react. This dimension spans high arousal, excitement, agitation and low arousal, relaxation, and inertia. Levels of activation below a certain threshold are not compatible with awareness and usually result in sleep.
2. Awareness span refers to changes of content available to attention and conscious processing. Awareness span ranges from narrow, focused attention directed at a singular content to broader, awareness including many things a in a single moment. This variability of awareness span is accessible to the subject and may be reported upon reflection.
3. Self-awareness refers to the other pole of the bipolar self-world structure of human experience. In a reflective attitude, all experience is "mine", that is, related to the subject's self. However, in the flow of immediate experience, the degree of self-reference may largely vary from "forgetting oneself" in absorption or exaltation to an intensified feeling of one's unique being: "I, here and now". Variability of self-awareness may also be accessed subjectively and reported upon reflection.
4. Sensory dynamics consist of various changes in the sensory and perceptual component of subjective experience. As states of consciousness vary, sensation can be reduced (higher thresholds, anesthesia) or enhanced (lower thresholds, hyperesthesia). Some altered states of consciousness are characterized by sensations and perceptions even without an adequate physical stimulus (e.g., synesthesia, dreams, and hallucinations). Sensation components are partly accessible to physiological measurement, but changes in richness, vividness, structure, and contents of the perceptual component are only accessible via subjects reports.

These four dimensions are well anchored in everyday experience and no special or altered states of consciousness need to be referred to in order to define them. The unusual characteristics of altered states of consciousness, often considered to have unique and irreducible features, can be translated into the four-dimensional system. Vaitl et al. (2005) give the example of the "oceanic experience" which results from the expansion of the awareness span with a simultaneous reduction of self-awareness. They suggest that other dimensions might also be useful to describe specific features of altered states of consciousness. However, in the interests of parsimony four dimensions appear to be sufficient to account for all known states of consciousness.

The activation dimension is the closest in terms of physiological description to non-specific activating systems in the brain, to autonomic nervous systems, and to the associated changes that occur in bodily processes. The other three dimensions are probably related to central integrative and regulatory functions of the brain's subsystems (awareness span and self-awareness) or to specific sensory systems and higher level binding and regulatory subsystems (sensory dynamics).

Altered states of consciousness are known to be associated with alterations of brain systems that regulate consciousness, arousal, and attention. These systems have been characterized as non-specific, causing widespread excitatory changes in the cortex. However, specific subsystems serving circumscribed functions in the regulation of attention, wakefulness, and sleep have been identified (Dehaene 2001).

The limited capacity control system (LCCS) becomes active only when processing of new, complex, or vitally important information and voluntary decisions occur. Therefore, the appearance of consciousness at a physiological level consists of an increase in cortical complexity (Tononi and Edelman 1998) and widespread reduction in cortical excitation thresholds regulated by the brain systems. The subcortically mediated increase in excitation of forebrain structure requires a certain minimal anatomical extension and duration. Reentrant pathways particularly between thalamic and cortex produce a reverberation of neuronal activity over a particular time period that is usually longer than 100 ms necessary for maintaining the mutual information exchange (Baars 1988). Altered states of consciousness may result from deviations of inter- and intraregional neuronal information exchange of the LCCS and the connected brain modules. For example, dreamy states may result from excitation of posterior sensory cortical association areas without simultaneous prefrontal activation of working memory, which may result in hallucinatory experiences that are typical of active dreaming.

Altered states of consciousness induced by decreased sensory input, homogenization, sensory overload, or strong rhythmic patterns of input are associated with a large reduction in the naturally occurring variability in sensory input. The reduction of sensory input appears to be an important element in inducing altered states of consciousness.

Illusions or hallucinations are more difficult to explain. Daydreaming and ganzfeld hallucinations are associated with simultaneous awareness of the imaginary nature of experience. Hallucinations during dreaming and hypnosis can be recognized retrospectively. However, schizophrenic patients believe that hallucinationed experiences are real. Brain-imaging studies of patients with schizophrenia during auditory hallucinations indicate activation of a medial temporal lobe-basal ganglia thalamic circuit in the absence of any external stimulation. Hypnotically induced hallucinations are characterized by the dissociation of frontal and parietal lobe connectivity.

Studies of how the brain binds together different features of internal or external stimuli into meaningful representations has suggested that binding may be organised by synchronous neural activation in groups of cells that have specialised functional properties. Thus, stimuli with low complexity may be represented by only a few such cell assemblies with restricted topographical distribution, whereas stimuli composed of many complex features may be represented by larger cell assemblies with widespread topographical organization. It was Hebb (1949) who first suggested these types of functional cell assemblies. More recently, studies have shown that each cell assembly can be characterized by its own high-frequency oscillations. Activities within the gamma band have become the most prominent one and have been shown to be critical for attention, learning and memory, language and meaningful motor behaviours. The breakdown of connectivity between large groups of cell assembles has been suggested to be the basis of hallucinatory and illusory states that occur in patients with schizophrenia, hypnotic states, and loss of consciousness induced by anesthesia. The breakdown of coherent oscillations results in groups of cell assemblies becoming functionally independent units and appears to be the basis for major disturbances in stimulus representations and other cognitive and behavioural functions.

The state of consciousness changes spontaneously over time, during development from birth to childhood to adulthood, during the circadian cycle, and over much shorter periods. These changes suggest a hierarchy of states of consciousness in terms of their characteristic duration and stability. The sleep–wake cycle follows the circadian periodicity and minor fluctuations of vigilance occur on a time scale of hours or minutes, whereas Wundt's "fluctuations of attention" occur in seconds range. The time scale of hierarchically embedded states of consciousness also extends into the subsecond range as shown by spatial analyses of the electrical fields of the brain.

Multichannel recordings of spontaneous or evoked electrical fields of the brain can be converted into a series of instantaneous scalp field configurations. These momentary maps of electrical activity show landscapes with different topographies. Brain field map series exhibit brief epochs of quasi-stable topographies, separated by rapid, discontinuous changes. If it is accepted that the brain at each moment in time is in a particular state and that different spatial configurations of the brain's electrical field must have been generated by different sets of active neurons, it must be concluded that different brain field configurations indicate different states. The brief brain states that occur in the subsecond range are called microstates to distinguish them from global brain macrostates with durations ranging from seconds to minutes. The average microstate durations of spontaneous EEG that have been identified by various segmentation procedures occur within the range of 60–150 ms (Lehmann et al. 1987; Pascual-Marqui et al. 1995; Strik and Lehmann 1993).

Microstates can be classified by their field topographies and compared between groups or states in terms of their mean duration and/or of the percentage of time covered. Topographies of spontaneously occurring microstates in normal individuals have different mentation modes, suggesting that microstates might be building blocks for consciousness or, metaphorically, "atoms of thought" (Lehmann et al. 1998). Microstates differ during maturation (Koenig et al. 2002); differ in awake, drowsy, and REM states (Cantero et al. 1999); show greater variability in schizophrenia (Merrin et al. 1990); and show shortened duration in untreated first-episode schizophrenics (Koenig et al. 1999) and chronic schizophrenics (Strelets et al. 2003).

The mean duration and frequency of occurrence provides a first-order description of microstates. Higher order analyses may focus on the rules (syntax) of concatenations of the atoms of thought. Analyses of transitions between microstates of different classes have shown that transition probabilities are class dependent and revealed asymmetries of transition probabilities between classes (Wackermann et al. 1993). This suggests that sequences of microstates are not entirely random, but that they follow certain probabilistic syntactic rules. Following the general interpretation of microstates as building blocks of consciousness, alterations in consciousness may reflect changes in microstate sequencing rules. Microstate syntax in untreated first-episode schizophrenics has shown clear alteration, with some tendency to reversed sequencing among certain lower classes.

Methods of nonlinear dynamics have also been used to study the electrical activity of the brain and the neural basis of states of consciousness (Pritchard 1997). Nonlinear complexity measures have usually been derived from one-dimensional times series captured from single scalp loci, but these methods can be applied to the entire brain. Various linear measures of covariance complexity of multidimensional brain electrical data that are less computation intensive and easier to interpret have also been suggested (Palus et al. 1990; Wakermann 1996).

Global measures of brain functional states are useful for studying consciousness because it involves the higher level, integrative functions of the brain. The global complexity measure, omega provides a single dimension onto which functional states of the whole brain can be mapped. Systematic variations of global omega complexity with sleep stages have been demonstrated. However, brain states related to different states of consciousness may project to the same position on a one-dimensional continuum. Thus, a three-dimensional system of global descriptors – integral power, generalized frequency, and spatial complexity – has been proposed, which allows a clear separation of sleep stages and indicates that at least two dimensions may be required to account for the variety of wakefulness–sleep transitions. These results have demonstrated the applicability of the global brain state description for consciousness-related studies and have also led to their use in exploring other brain alterations. For example, omega complexity as well as global dimensional complexity have been shown to increase in schizophrenic patients in comparison to control subjects (Saito et al. 1998).

Microstate analysis and multichannel EEG complexity analysis both focus on the spatial aspects of brain states. They operate on the same data from sequences of scalp field maps interpreted as a series of momentary electrical brain states or as a trajectory in a multidimensional state space. There is also a one-to-one correspondence between basic notions of the two analytical approaches. Therefore, higher global spatial EEG complexity corresponds to a higher multiplicity of microstates. Thus, both can be interpreted as a higher diversity of actually active brain electrical generators involved in the corresponding state of consciousness. The convergent findings of microstate and complexity analyses in schizophrenia (Koenig et al. 1999; Koukkou et al. 1994; Merrin et al. 1990; Saito et al. 1998; Strelets et al. 2003) support this theoretical construct. Thus, microstate analysis and global complexity analysis appear to provide a unifying framework for assessment and interpretation of brain functional states underlying states of consciousness and their alterations.

Consciousness depends critically on the proper functioning of several brain systems. These systems can be impaired, by damage to brain tissue, neurotransmitter imbalances, hypo- or hypersynchronization, disconnectivity in the firing of neural assemblies, and fluctuations in arousal. Brain functioning may also be modified by alterations in perceptual input as well as by the use of cognitive self-regulation strategies. During the ordinary waking state, subjective reality is created continuously by processes in the brain. Thus, the maintenance of everyday consciousness requires intact brain tissue, metabolic homeostasis, a moderate level of arousal, a balanced interplay of inhibitory and excitatory networks, and stable environmental conditions. If one of these prerequisites for reliable assembly formation is compromised, alterations of consciousness are highly likely to occur. In addition, by applying psychological procedures like hypnosis or meditation, it is possible to voluntarily suppress or restrict the formation of assemblies. This results in hypnotic phenomena or meditative states of sustained absorption.

The dynamics of micro- and macrostates and the existence of local perceptual binding in the brain and quantification of local nonlinear dynamics can be inferred only by direct cortical recording with electocorticograms (ECoG or MEG) (Vaitl et al. 2005). These approaches both allow the analysis of macro potentials, local field potentials, and oscillations undisturbed by the resistance of bone, cerebrospinal fluid and skin, which can act as filters, particularly for the high frequency gamma band and electrical brain activity resulting from sulci (60% of all cortical surfaces) (Vaitl et al. 2005). A millimeter-scale localization of spontaneous and evoked electrocortical activity is required in order to observe the subtle changes during most altered states of consciousness. ECoG recording is confined to rare cases in which normal subdural or cortical placement of electrodes is ethically acceptable, such as in epilepsy screening, recording during deep brain stimulation electrodes for Parkinson's disease, dytonia and chronic pain and brain tumor surgery. Substantial insight into altered states of consciousness has been gained from MEG recordings (Tononi and Edelman 1998). In addition, magnetic resonance-based noninvasive brain analysis, particularly of limbic and other subcortical regions (e.g., thalamus) that are not accessible to EEG, Cog and MEG recordings, when combined with simultaneous recording of EEG will provide completely new insight into the delicate cortico–subcortical interplay during altered states of consciousness (Vaitl et al. 2005). For example, magnetic resonance spectroscopy allows neurochemical analysis of local brain areas during altered states of consciousness. Diffusion tensor imaging may uncover previously unknown anatomical connections between distant brain areas during altered states of consciousness. The inclusion of anatomically confined lesions and disorders with altered states of consciousness symptoms, as an independent variable coupled with observation of the altered states of consciousness as a dependent variable, may allow new insight. For example, reversible interaction of left operculum with transcranial magnetic stimulation or transcranial direct current stimulation may allow the specification of the role of "mirror" neuronal fields for the experience of self-alienation comparable to those states found in autistic disorder spectrum. In addition, operant conditioning (self-regulation and biofeedback) of local electromagnetic phenomena thought to underlie particular altered states of consciousness could be used to test the suggested correlations. With our increasing knowledge of the neural basis of consciousness, the different and sometimes strange and difficult-to-explain phenomena of altered states of consciousness will become increasingly understandable as natural consequences of the working brain operating under various different environmental conditions.

A major possible direction in this research area would include attempting to further describe and characterize the physiological building blocks of thought (or "atoms of thought"), which are microstates that follow each other in a temporal sequence. If it is possible to identify the microstates and the subjective experiences that accompany them, then it may be further possible to construct a DNA-like mathematical model of normal sequencing of the atoms of thought that could be matched to subjective experience and behaviour. This would allow the prediction of behaviour from the building blocks of thought, and likely patterns of behavioural outcomes could also be predicted. For example, for most animals including human beings we can think of four basic behavioural drive states (i.e., feeding, fighting, mating and fleeing). In humans under normal conditions these behavioural states follow a certain order, but in pathological conditions, such as drug addiction/intoxication, mental illness, criminality or war, the behavioural sequences may be abnormal or disrupted. Having an intimate knowledge of the sequencing of thought may allow abnormal patterns of behaviour to be very accurately identified even in early childhood and appropriate corrective treatments to be developed. The development of this research may result in treatments for the main pathological problems that trouble human kind.

References

Aserinsky E, Kleitman N (1953) Regularly occurring periods of eye motility and concomitant phenomena during sleep. Science 118:273–274

Baars BJ (1988) A cognitive theory of consciousness. Cambridge University Press, Cambridge, England

Brauchli P, Michel CM, Zeier H (1995) Electrocortical, autonomic, and subjective responses to rhythmic audio-visual stimulation. Int J Psychophysiol 19:53–66
Braun AR, Balkin TJ, Wesensten NJ, Gwadry F, Carson RE, Varga M, Baldwin P, Belenky G, Herscovitch P (1998) Dissociation patterns of activity in visual cortices and their projections during human rapid eye movement sleep. Science 279:91–95
Brugger P, Regard M, Landis T, Oelz O (1999) Hallucinatory experiences in extreme-altitude climbers. Neuropsychiatry Neuropsychol Behav Neurol 12(1):67–71
Cantero JL, Atienza M, Salas RM, Gomez CM (1999) Brain spatial microstates of human spontaneous alpha activity in relaxed wakefulness. Brain Tomogr 11:257–263
Cohen AS, Rosen RC, Goldstein L (1976) Electroencephalographic laterality changes during human sexual orgasm. Arch Sex Behav 5:189–199
Critchley HD, Corfield DR, Chandler MP, Mathias CJ, Dolan RJ (2000) Cerebral correlates of autonomic cardiovascular arousal: a functional neuroimaging investigation in humans. J Physiol 523:259–270
Critchley HD, Melmed RN, Featherstone E, Mathias CJ, Dolan RJ (2001) Brain activity during biofeedback relaxation: a functional neuroimaging investigation. Brain 124:1003–1012
Croft RJ, Williams JD, Haenschel C, Gruzelier JH (2002) Pain perception, 40 Hz oscillations and hypnotic analgesia. Int J Psychophysiol 46:101–108
De Benedittis G, Sironi VA (1988) Arousal effects of electrical deep brain stimulation in hypnosis. Int J Clin Exp Hypn 36(2):96–106
Dehaene S (ed) (2001) The cognitive neuroscience of consciousness. MIT, Cambridge, MA, pp 1–37
Egner T, Gruzelier JH (2001) Learned self-regulation of EEG frequency components affects attention and event-related brain potentials in humans. Neuroreport 12:4155–4159
Egner T, Gruzelier JH (2004) EEG biofeedback of low beta band components: frequency-specific effects on variables of attention and event-related brain potentials. Clin Neurophysiol 115:131–139
Farthing GW (1992) The psychology of consciousness. Prentice Hall, Englewood Cliffs, NJ
Flor-Henry P (1980) Cerebral aspects of the orgasmic response: normal and deviational. In: Forleo R, Pasini W (eds) Medical sexology. Elsevier, Amsterdam, pp 256–262
Friederich M, Trippe RH, Özcan M, Weiss T, Hecht H, Miltner WHR (2001) Laser evoked potentials to noxious stimulation during hypnotic analgesia and distraction of attention suggest different brain mechanisms of pain control. Psychophysiology 38:768–776
Graber B, Rohrbaugh JW, Newlin DB, Varner JL, Ellingson RJ (1985) EEG during masturbation and ejaculation. Arch Sex Behav 14:491–503
Gruzelier JH (1998) A working model of the neurophysiology of hypnosis: a review of the evidence. Contemp Hypn 15:3–21
Gruzelier JH, Warren K (1993) Neuropsychological evidence of reduction on left frontal tests with hypnosis. Psychol Med 23:93–101
Gruzelier JH, Hardman E, Wild J, Zaman R (1999) Learned control of slow potential interhemispheric asymmetry in schizophrenia. Int J Psychophysiol 34:341–348
Heath RG (1972) Pleasure and brain activity in men. J Nerv Ment Disord 154:3–18
Hebb DO (1949) The organization of behavior: a neuropsychological theory. Wiley, New York
Hobson JA, McCarley R (1977) The brain as a dream-state generator: an activation-synthesis hypothesisi of the dream process. Am J Psychiatry 134:1335–1348
Hori T, Hayashi M, Morikawa T (1994) Topographical EEG changes and hypnogogic experience. In: Ogilvie RD, Harsh JR (eds) Sleep onset: normal and abnormal processes. American Psychological Association, Washington, DC, pp 237–253
Jacobs GD, Benson H, Friedman R (1996) Topographic EEG mapping of the relaxation response. Biofeedback Self Regul 21:121–129
Kaiser J, Barker R, Haenschel C, Baldeweg T, Gruzelier J (1997) Hypnosis and event-related potential correlates of error processing in a Stroop-type paradigm: a test of the frontal hypothesis. Int J Psychophysiol 27:215–222

Koenig T, Lehmann D, Merlo M, Kocki K, Hell D, Koukkou M (1999) A deviant EEG brain microstate in acute, neuroleptic-naïve schizophrenics at rest. Eur Arch Psychiatry Clin Neurosci 249:205–211

Koenig T, Prichep LS, Lehmann D, Valdes-Sosa P, Braeker E, Kleinlogel H, Robert Isenhart R, John FR (2002) Millisecond by millisecond, year by year: normative EEG microstates and developmental stages. Neuroimage 16:41–48

Koukkou M, Lehmann D, Strik WK, Merlo MC (1994) Maps of microstates of spontaneous EEG in never-treated acute schizophrenia. Brain Topogr 6:251–252

Kubler A, Kotchoubey B, Kaiser J, Wolpaw JR, Birbaumer N (2001) Brain-computer communication: unlocking the locked in. Psychol Bull 127:358–375

Kuhlo W, Lehmann D (1964) Das Einschlaferleben und seine neurophysiologische Korrelate [Experience during the onset of sleep and its neurophysiological correlates]. Arch Psychiatr Z Gesamte Neurol 205:687–716

Lehmann D, Ozaki H, Pal I (1987) EEG alpha map series: brain micro-states by space-oriented adaptive segmentation. Electroencephalogr Clin Neurophysiol 67:271–288

Lehmann D, Grass P, Meier B (1995) Spontaneous conscious covert cognition states and brain electric spectral states in canonical correlations. Int J Psychophysiol 19:41–52

Lehmann D, Strik WK, Henggeler B, Koenig T, Koukkou M (1998) Brain electric microstates and momentary conscious mind states as building blocks of spontaneous thinking: I. Visual imagery and abstract thoughts. Int J Psychophysiol 29:1–11

Lehmann D, Faber P, Isotani T, Wohlgemuth P (2001a) Source locations of EEG frequency bands during hypnotic arm levitation: a pilot study. Contemp Hypn 18:120–127

Lehmann D, Faber PL, Achermann P, Jeanmonod D, Gianotti LRR, Pizzagalli D (2001b) Brain sources of EEG gamma frequency during volitionally meditation-induced, altered states of consciousness, and experience of the self. Psychiatry Res Neuroimaging 108:111–121

Maharishi Yogi M (1996) The science of being and the art of living. International SRM, London

Maurer RLS, Kumar VK, Woodside L, Pekala RJ (1977) Phenomenological experience in response to monotonous drumming and hypnotizability. Am J Clin Hypn 40(2):130–145

McGuire BE, Basten C, Ryan C, Gallagher J (2000) Intensive care unit syndrome: a dangerous misnomer. Arch Intern Med 160:906–909

Meduna LJ (1950) The effects of carbon dioxide upon the functions of the brain. In: Meduna LJ (ed) Carbon dioxide therapy. Charles C Thomas, Springfield, IL, pp 23–40

Merrin EL, Meek P, Floyd TC, Callaway E III (1990) Topographic segementation of waking EEG in medication-free schizophrenic patients. Int J Psychophysiol 9:231–236

Mosovich A, Tallaferro A (1954) Studies on EEG and sex function orgasm. Dis Nerv Syst 15:218–220

Murri L, Arena R, Siciliano G, Mazzotta R, Muratorio A (1984) Dream recall in patients with focal cerebral lesions. Arch Neurol 41:183–185

Neher A (1961) Auditory driving observed with scalp electrodes in normal subjects. Electroencephalogr Clin Neurophysiol 13:449–451

Neher A (1962) A physiological explanation of unusual behavior in ceremonies involving drums. Hum Biol 34:151–160

Palus M, Dvorak I, David I (1990) Remarks on spatial and temporal dynamics of EEG. In: Dvorak I, Holden AV (eds) Mathematical approaches to brain functioning diagnostics. Manchester University Press, Manchester, England, pp 369–385

Pascual-Marqui RD, Michel CM, Lehmann D (1995) Segmentation of brain electrical activity into microstates: model estimation and validation. IEEE Trans Biomed Eng 42:658–665

Pekala RJ (1991) Quantifying consciousness: an empirical approach. Plenum, New York

Pritchard WS (1997) A tutorial of the use and benefits of non-linear dynamical analysis of the EEG. In: Angeleri F, Butler S, Giaquinto S, Majkowski J (eds) Analysis of the electrical activity of the brain. Wiley, New York, pp 3–44

Rechtschaffen A, Kales A (1968) A manual of standardized terminology, techniques and scoring system for sleep stages of human subjects. Public Health Service, U.S. Government Printing Office, Washington, DC

Rouget G (1980) La musique et la trance [Music and trance]. Gallimard, Paris
Saito N, Kuginki T, Yagyu T, Kinoshita T, Koenig T, Pascual-Marqui RD, Kochi K, Wackermann J, Lehmann D (1998) Global, regional and local measures of complexity of multichannel EEG in acute, neuroleptic naïve, first-break schizophrenics. Biol Psychiatry 43:794–802
Schulman CA, Richlin M, Weinstein S (1967) Hallucinations and disturbances of affect, cognition, and physical state as a function of sensory deprivation. Percept Mot Skills 25:1001–1024
Solms M (1997) The neuropsychology of dreams. A clinico-anatomical study. Erlbaum, Mahwah, NJ
Stancak A, Hönig J, Wackermann J, Lepicovska V, Dostalek C (1991) Effects of unilateral nostril breathing on respiration, heart rhythm and brain electrical activity. Neurosciences 17:409–417
Sterman MB (1996) Epilepsy and its treatment with EEG biofeedback. Ann Behav Med 8:21–25
Sterman MB, Shouse MN (1980) Quantitative analysis of training, sleep EEG and clinical response to EEG operant conditioning in epileptics. Electroencephalogr Clin Neurophysiol 49:558–576
Strelets V, Faber PL, Gollikova J, Novototsky-Vlasov V, Koenig T, Guinotti LRR, Gruzelier JH, Lehmann D (2003) Chronic schizophrenics with positive symptomatology have shortened EEG microstate durations. Clin Neurophysiol 14:2043–2051
Strik WK, Lehmann D (1993) Data-determined window size and space oriented segmentation of spontaneous EEG map series. Electroencephalogr Clin Neurophysiol 87:169–174
Suedfeld P, Eich E (1995) Autobiographical memory and affect under conditions of reduced environmental stimulation. J Environ Psychol 15:321–326
Sugimoto S, Kida M, Teranishi T, Yamamoto A (1968) Time estimation during prolonged sensory deprivation. Jpn J Aerosp Med Psychol 61(1):1–5
Tellegen A, Atkinson G (1974) Openness to absorbing and self-altering experiences ("absorption"), a trait related to hypnotic susceptibility. J Abnorm Psychol 83:268–277
Terekhin PI (1996) The role of hypercapnia in inducing altered states of consciousness. Hum Physiol 22:730–735
Tononi G, Edelman GM (1998) Consciousness and complexity. Science 282:1846–1851
Tononi G, Edelman GM (2000) Schizophrenia and the mechanisms of conscious integration. Brain Res Interact 31:391–400
Tucker DM, Dawson SL (1984) Asymmetric EEG changes as method actors generated by emotions. Biol Psychol 19:63–75
Vaitl D, Birbaumer N, Gruzelier J, Jamieson GA, Kotchoubey B, Kübler A, Lehmann D, Miltner WH, Ott U, Pütz P, Sammer G, Strauch I, Strehl U, Wackermann J, Weiss T (2005) Psychobiology of altered states of consciousness. Psychol Bull 131(1):98–127
Vollenweider FX, Geyer MA (2001) A system model of altered consciousness: integrating natural and drug-induced psychosis. Brain Res Bull 56:495–507
Wackermann J, Lehmann D, Michel CM, Strik WK (1993) Adaptive segmentation of spontaneous EEG map series into spatially defined microstates. Int J Psychophysiol 14:269–283
Wackermann J, Pütz P, Büchi S, Strauch I, Lehmann D (2000) A comparison of ganzfeld and hypnogogic state in terms of electrophysiological measures and subjective experiences. In: Proceedings of the 43rd P.A. convention, 17–20 Aug 2000, Freiburg, Germany, pp 302–315
Wackermann J, Pütz P, Braeunig M (2002a) EEG correlates of ganzfeld-induced hallucinatory experience. Int J Psychophysiol 45:17
Wackermann J, Pütz P, Büchi S, Strauch I, Lehmann D (2002b) Brain electrical activity and subjective experience during altered states of consciousness: ganzfeld and hypnogogic states. Int J Psychophysiol 46:122–145
Wakermann D (1996) Beyond mapping: estimating complexity of multichannel EEG recordings. Acta Neurobiol Exp 56:197–208
Zuckerman M, Persky H, Link KE (1969) The influence of set and diurnal factors on autonomic responses to sensory deprivation. Psychophysiology 5:612–624

Chapter 4
Dream Consciousness and Sleep Physiology

Michael Schredl and Daniel Erlacher

Abstract This chapter addresses the question as to how dream consciousness is related to the physiology of the sleeping body. By reviewing studies conducted in the field of dream and lucid dream research on REM sleep it can be shown that correlations between dreamed and actual actions can be found for central nervous activity, autonomic responses and time aspects. In future research, methodological problems have to be considered and fMRI studies will make it possible to study the relationship between brain activation and dream content in a detailed way. Studying the body–mind interaction in sleep seems a promising way to understanding dream consciousness in particular and consciousness in general.

4.1 What Is a Dream? Introduction and Definitions

The body–mind interaction has fascinated – and still fascinates – scientists and philosophers alike. The sleeping state seems to be especially well qualified for studying the relationship between the subjective level (mind) and the physiological level (body) because during this state there is a minimum of external input and a maximum of internally generated experiences (dreaming). In addition, studying this relationship is for dream researchers of the utmost importance because it has been questioned whether dreams (see definitions below) are really a recollection of subjective experiences that occurred during sleep. Maury (1861), a French scientist, woke up from a long dream because he was hit by a part of his bed falling on his neck. Since there was a logical progression of events during the dream that

---

M. Schredl (✉)
Central Institute of Mental Health, PO Box 12 21 20, 68072 Mannheim, Germany
e-mail: Michael.Schredl@zi-mannheim.de

D. Erlacher
Institute for Sport and Sport Science, University of Heidelberg, Im Neuenheimer Feld 700, 69120 Heidelberg, Germany
e-mail: Daniel.Erlacher@issw.uni-heidelberg.de

D. Cvetkovic and I. Cosic (eds.), *States of Consciousness*, The Frontiers Collection,
DOI 10.1007/978-3-642-18047-7_4,

culminated in his execution with a guillotine – at that time he woke up because of the external stimuli – he concluded that the dream is generated during the awakening process within seconds. Relating dreamed images to physiological processes that occurred while sleeping can falsify this hypothesis.

The following definition of dreaming is widely accepted:

> A dream or dream report is the recollection of mental activity which has occurred during sleep. (Schredl 1999, p. 12)

It is important to notice that dream consciousness is not directly accessible in comparison to waking consciousness; two boundaries have to be crossed, the sleep/wake transition and time. As mentioned above, the validity of dream reports – whether they really reflect subjective experiences during sleep (dream consciousness) – has been questioned and the studies reviewed in this chapter are like a foundation for the field. Whereas in the beginning of sleep research researchers assumed that dreaming is exclusively limited to REM sleep, the literature review of Nielsen (2000) clearly indicates that dream recall rates from NREM awakenings are quite high (43.0%), even when the mean recall for REM awakenings is higher (81.9%). Wittmann and Schredl (2004) argued that it is very reasonable to assume that the mind never sleeps, e.g., dream consciousness is continuously active during sleep and that not recalling something from a dream is often due to memory impeded by the sleep/wake transition.

Lucid dreaming – a form of dreaming that is very interesting for this line of research – is defined as dreams where the dreamer is aware – while dreaming – that he/she is dreaming (LaBerge 1985). This knowledge about the present state of consciousness allows prolific lucid dreamers to actively alter dream actions. How this dream type can be used to study the body–mind interaction will be illustrated in this chapter.

Sleep research was enormously stimulated by the discovery of REM (rapid eye movement) sleep, published in 1953. Aserinsky and Kleitman (1953) detected that sleep is not a unitary state of reduced activity but consists of different sleep stages. REM sleep was of special interest because the eyes moved rapidly under the closed eyelids. Awakening from this sleep stage often yielded vivid dream reports indicating that the brain is very active during this state. Subsequently, Rechtschaffen and Kales (1968) developed a scoring manual for classifying the different sleep stages. This manual has been recently updated (AASM 2007). The continuous recording of the electroencephalogram (EEG), the electrooculogram (EOG), and the chin electromyogram (EMG) – called standard polysomnography – is the basis for sleep scoring. The polysomnographic recording is divided into 30 s epochs and each epoch is classified into one of five stages: awake, NREM sleep 1 (sleep onset), NREM sleep 2 (normal sleep), slow wave sleep (NREM), and REM sleep. These classifications (960 epochs for 8 h) are then transformed into a sleep profile that gives information about sleep latency, sleep efficiency, percentage of the sleep stages, and other parameters.

4.2 How to Study the Body–Mind in Sleep: Research Strategies

To study the body–mind interaction during sleep, three paradigms have been adopted (see Table 4.1). The first approach is based on the fact that muscle tone is actively inhibited by the REM sleep generation areas in the brain stem. Jouvet and Delorme (1965) showed that after experimentally damaging the brain areas in cats responsible for muscle atonia, the cats showed typical behavior during their REM phases like prey catching or licking the fur. In humans, a specific sleep disorder called REM sleep behaviour disorder (Schenck and Mahowald 1996) is also associated with a partial loss of muscle atonia during REM sleep, so that these patients act out their dreams, and the relationship between dream consciousness and behaviour can be studied directly.

For normal dreaming, the approach is very simple. Physiological recordings, e.g., heart rate, eye movements, respiration rate, and an electromyogram, are measured and made prior to awakening. After awakening, the participant is asked to report the dream as completely as possible in order to match the subjective experience with the physiological parameters recorded during sleep. Almost all studies are restricted to REM sleep due to the high recall rates of intense dreaming. Dream reports are analyzed by judges, blind to the identity of the dreamer, for specific characteristics (e.g., dream emotion, movements) or sorted into different classes (active vs. passive dream content). Then, the physiological data is compared with the psychological data from the ratings.

Third, lucid dreams, in which the dreamer is aware that she/he is dreaming, offer fascinating opportunities for studying the body–mind relationship during sleep because the dreamer can carry out pre-arranged tasks during the dream and mark their beginning and end by distinct eye movements that can be measured electrically (cf. LaBerge 1985). The volitional eye movements during a lucid dream can be differentiated from spontaneous rapid eye movements with low error rate (Fig. 4.1). Whereas the eye movements themselves document the mind–body interaction (eye movements during the dream correspond with the EOG recording), these eye movements can also be used to analyze the physiological recordings, e.g., the dreamer carried out ten squats (verified by the dream report given by the participant upon awakening) within two eye signals and the respiration rate during this time period changed (Fig. 4.2).

The research regarding the body–mind interaction is presented as follows. First, the lesion study paradigm is illustrated by an example. Second, normal dreaming and lucid dreaming studies relating different aspects of dream consciousness and sleep physiology are reviewed, along with dream time and REM sleep duration,

Table 4.1 Research paradgms to study the body–mind interaction during sleep

Strategies
Lesion studies
Normal dreaming
Lucid dreaming

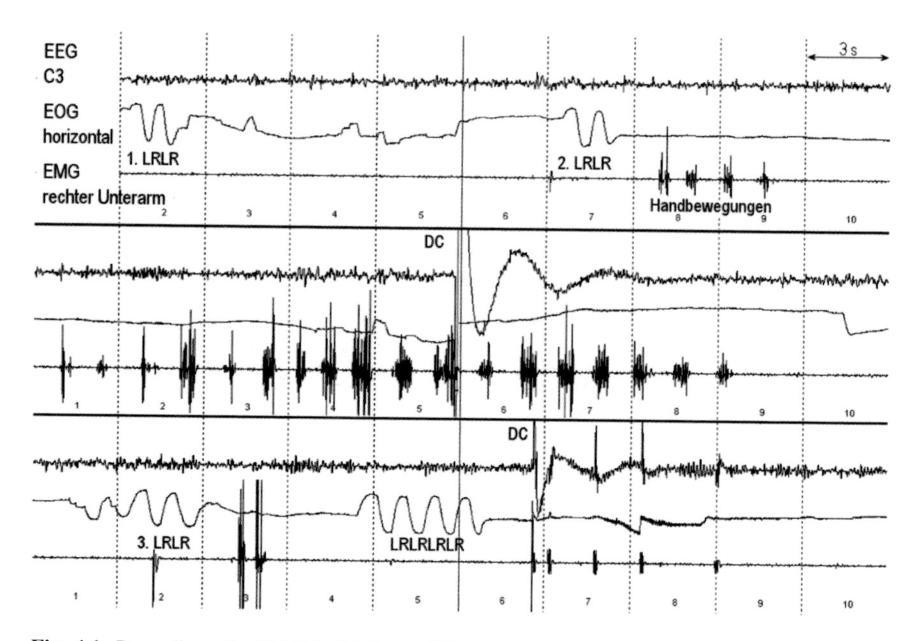

Fig. 4.1 Recording of a REM lucid dream. The task for the participant was to carry out hand clenches with the right hand. In his dream report the subject stated that he continually performed hand clenches between the second block of eye movements (2. LRLR, Left Right Left Right) and the third block (3. LRLR), but the arm EMG amplitude – showing clearly physical hand movements – is not constant throughout this period (from Erlacher 2005)

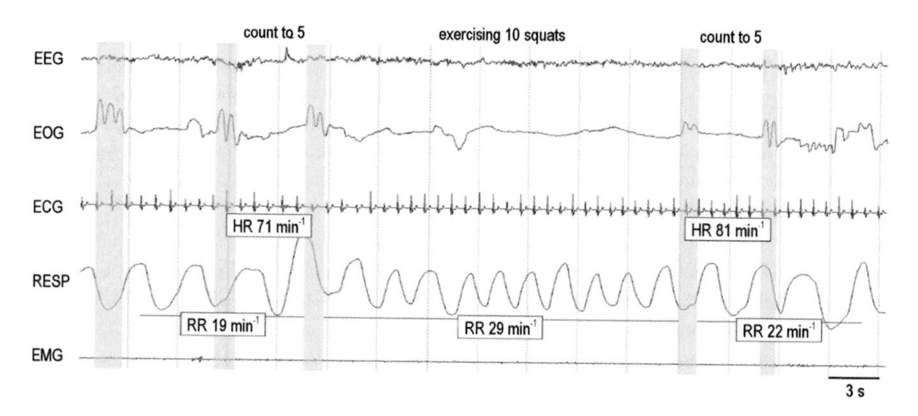

Fig. 4.2 Recording of a lucid dream: Five clear left-right-left-right eye movement signals are shown in the EOG channel. The task for the lucid dreamer was to count from 1 to 5, perform ten squats and finally count again from 1 to 5. The respiration rate and heart rate increase significantly while performing squats in the lucid dreams (from Erlacher and Schredl 2008a)

dreamed gaze shifts and eye movements, brain activation and dream content, dreamed movements and electromyogram, dream content and autonomic parameters (heart rate, sexual excitation).

4.3 Lessons from Patients: Lesions in the Brain Stem

During REM sleep the general muscle tone is actively inhibited by the REM sleep-generating area in the brain stem (e.g. Dement 1974). The brain region responsible for muscle atonia is located in the formatio reticularis, a dense neural network which crosses with several different nuclei from the vertical brain stem. This physiological mechanism seems to serve a biological function in preventing the sleeping person to act out her or his dreams. Patients suffering from REM sleep behaviour disorders have partially lost this muscle atonia during REM sleep (Schenck and Mahowald 1996). The following case vignette illustrates this sleep disorder.

> I was a halfback playing football, and after the quarterback received the ball from the center he lateraled it sideways to me and I'm supposed to go around end and cut back over tackle and – this is very vivid – as I cut back over tackle there is this big 280-pound tackle waiting – so I, according to football rules, was to give him a shoulder and bounce him out of the way, supposedly, and when I came to I was standing in front of our dresser and I had knocked lamps, mirrors, and everything off the dresser, hit my head against the wall and my knee against the dresser. (Man, 67 years old; Schenck et al. 1986, p. 294)

The dream example underlines the direct relationship between dream consciousness and the sleeping body. REM sleep behaviour is a rare disorder and should not be confused with sleepwalking. Sleepwalking occurs during NREM sleep where there is no specific muscle atonia; usually the person does not remember the sleep-walking episodes (Schredl 1999). Comparing these two clinical syndromes, however, raises the question as to why NREM dreaming – which is common – is not accompanied by acting out behaviour despite the lack of muscle atonia during this sleep state. Maybe the parallel between NREM dreaming and daydreaming/imagination is closer than the relationship between REM dreaming and waking imagination because during daydreaming an acting out of the fantasies does not occur. To summarize, the lesion studies show a clear relationship between dreamed activity and physiology in REM sleep, at least in patients with REM sleep disorder.

4.4 How Long Is a Dream? Dream Time and REM Sleep Duration

The famous dream example reported by the French sleep and dream researcher Alfred Maury (1861) raised the question whether the subjectively experienced sense of time matched the real time spent in sleep or REM sleep. Maury reported a long and intense dream about the French revolution which ended with the dreamer in the guillotine and the sleeper waking up with a piece of his wooden bed top having fallen on his neck. Because of the logical line of dream action, Maury (1861) hypothesized that the dream was generated by the arousing stimulus within seconds starting with the end (the guillotine scene) and adding each scene backwards to the beginning. Accounts stemming from other domains, e.g., near-death experiences (Moody 1989), also found that sense of time can be altered considerably (rapidly reviewing one's whole life). The relationship between subjectively estimated time in dreams and real time was first experimentally studied by Dement and Kleitman (1957). The participants were awakened in a random order after either 5 or 15 min of REM sleep. After awakening, participants were asked to estimate whether the elapsed sleep interval was 5 or 15 min. From 111 awakenings, 83% judgments were correct. Furthermore, the elapsed time of the REM period correlated with the length of the dream report (Dement and Kleitman 1957). The latter findings were replicated by Glaubman and Lewin (1977), as well as by Hobson and Stickgold (1995) using an ambulatory measurement unit ("night cap") in a home setting. Jovanovic (1967) reported that awakenings conducted 20 s after the beginning of REM sleep yielded a report of a dream that had just begun.

For lucid dreams, LaBerge (1985) showed in a pilot study that time intervals for counting from one to ten in the dream (by counting from 1,001 to 1,010) are comparable to the time intervals for the same counting during wakefulness. Also the time intervals for two different activities (performing squats and counting) carried out within lucid dreams and wakefulness were quite similar (Erlacher and Schredl 2004). The results for counting replicates the finding of LaBerge's (1985) that time intervals for counting were quite similar in lucid dreams and in wakefulness but performing squats required 44.5% more time in lucid dreams than in the waking state. Unpublished data of the second author confirm that there is on average an overestimation of the duration of motor activities like walking compared to cognitive tasks like counting. The motor tasks within the dream took longer than in waking life, i.e., the hypothesis that dreaming replays of experiences are speeded up can be rejected based on these findings.

4.5 Looking Around: Dreamed Gaze Shifts and Eye Movements

In dream research, the relationship between dreamed eye movements and eye movements measured on the sleeping body by EOG is known as the scanning hypothesis (Ladd 1892). In modern dream research two approaches were applied to test this hypothesis. First, global measures of visual dream activity reported by the dreamer were correlated with the number of eye movements occurring in the EOG recording prior to the awakening. Second, attempts were made to relate the EOG pattern directly to the dreamed gaze shifts.

Dement and Wolpert (1958) scored 105 dream reports as to whether the last sequence was visually active (e.g., looking around) or visually passive (e.g., looking at a distant object). In a similar way, they classified the EOG recordings of the REM period prior to the awakenings. These independently made classifications were correctly matched in 74% of the ratings. Although several subsequent studies replicated this finding, other studies were not able to detect substantial relationships between visual dream content and global measures of eye movement activity (e.g., Firth and Oswald 1975). The mixed results might be explained by methodological issues such as averaging over distinct time periods to derive global scores for visual dream activity and eye movements. The averaging over time might conceal underlying and substantial correlations.

In a study by Roffwarg et al. (1962) the following approach was chosen to enable a direct matching of dream content and eye movements. Shortly after awakening, the subjects were interrogated by an interviewer, who had no access to the EOG recordings of the prior REM period, about the last 10–20 s of the dream experience. This interviewer, who was familiar with relating particular actions to the corresponding eye movement patterns, predicted the eye movements that he would have expected to occur due to the elicited dream report. The following example from Roffwarg et al. (1962, p. 240) gives an impression of this approach.

> The last thing I remember is looking down at a small piece of paper, held at about chest level, trying slowly and haltingly, dwelling on each word, to translate something that looked like three lines of French poetry. It took about 20 or 30 s to do it, probably. I don't remember if I looked up at any time from the paper. As I remember, essentially, I kept my eyes on the paper.

Interrogator's prediction: "There should be relative REM quiescence with the exception of a few spaced leftward glances if the subject has finished a line of reading and return to the next line".

The fact that the interviewer did not predict small eye movements (reading word by word) can be explained by the impossibility of detecting such eye movements or the position of the eyeballs by the commonly used AC amplifiers. The corresponding EOG pattern showed three small but rapid eye movements to the left (22, 8 and 0.5 s prior to the awakening). Overall, 80% of the clearly recalled dream actions corresponded with the EOG recording in this study (Roffwarg et al. 1962). Subsequent research confirmed small but substantial relationships between reported gaze shifts within the dream and recorded eye movements (Herman et al. 1984).

Another interesting approach was adopted by Dement and Kleitman (1957). They awakened their subjects after recording a distinct EOG pattern, for example, solely horizontal or vertical eye movements or the lack of any eye movements. One subject who was awakened after vertical eye movements dreamed of standing at the bottom of a tall cliff and looking up at climbers at various levels. In a dream reported after a REM period with horizontal eye movements the subject was watching two people throwing tomatoes at each other. A quiescent eye movement pattern was associated with driving a car (looking ahead) in the dream.

The results from dream research showed promising results, but it is still not clear how tight this relationship is, i.e., whether all eye movements are related directly to the dream gaze shifts. Additionally, methodological issues have to be kept in mind, e.g., concerning the electrical measurement of eye movements occurring during sleep, and that similar research areas, for example, relating eye movements to waking imagery, encounter the same difficulties in matching subjective reports with actual eye movements. This may be explained by the fact that most eye movements are carried out without consciousness, such as looking at a small object or glancing during a conversation.

In lucid dreams, a strong correlation between deliberately carried out dream gaze shifts (e.g., look left, right, left, right) and the corresponding eye movements of the sleeping body measured by EOG has been demonstrated (Erlacher 2005). Figures 4.1 and 4.2 depict two examples for typical lucid dream signals. LaBerge and Zimbardo (2000) found that dream gaze is more closely related to perception than to imagination. Participants were instructed to carry out a specific tracking protocol where they had to trace a circle with their stretched arm and follow the tip of the finger with their gaze. The participants performed this task during wakefulness, in waking imagination and during lucid dreaming. For all conditions the eye pattern was recorded by EOG. The analysis of the saccadic eye movements could clearly discriminate between dreaming and perception from imagination by showing more saccadic movements for imagination than for perception and dreaming.

Although the findings regarding dreamed gaze shifts and recorded eye movements are not homogeneous, one might safely conclude that at least some of the dreamed eye movements are associated with real eye movements of the sleeping body.

4.6 Which Parts of the Brain Dream? Dream Content and Brain Activation

Several dream studies investigated the question as to whether corresponding brain areas are active during dreaming specific actions in the same way as they are in the waking state. Zadra and Nielsen (1996) analyzed the EEG recordings of dreams with strong negative affects and found higher EEG activity for the central and occipital cortex areas. The authors assumed that this higher activity was caused by the intense pictures (occipital area, primary visual cortex) and the movements in the dream (central area, motor cortex), e.g., running away from unfriendly dream characters. In a single participant, Etevenon and Guillou (1986) found an alpha power decrease over the left central cortical areas in correspondence with dreamed performance of the right hand.

A study by Hong et al. (1996) investigated the relationship between talking and listening in the dream and EEG parameters. The advantage of analyzing these activities is the relatively good localization of the corresponding brain areas (Broca's area for talking and Wernicke's area for listening). The results confirmed the hypothesis that the lower the alpha power on the left sides of those areas corresponding to Broca's and Wernicke's areas was, the more expressive or receptive language was present in the dream.

In lucid dream research, Holzinger et al. (2006) explored differences in EEG recordings between lucid and non-lucid dreams in REM sleep. One finding indicated a higher activation of the left parietal lobe for lucid dreams in comparison to non-lucid dreams. Holzinger et al. (2006) stated that this brain area is considered to be related to semantic understanding and self-awareness and therefore reflects the self awareness of the dreaming person in lucid dreams. The lucid dreamers in a study by LaBerge et al. (1981) were instructed to perform two simple cognitive tasks (counting vs. singing) in their dreams. The analysis of the EEG alpha power showed higher activation of the left hemisphere during counting and higher activation of the right hemisphere during singing, a pattern one would expect also for wakefulness. Erlacher et al. (2003) instructed a single participant to carry out hand clenching either with the right or the left hand or – as a control condition – counting. Those events had to be marked by eye movements allowing the analysis of the EEG alpha power over the motor cortex (C3, CZ and C4) for the left or right hand movements. Results showed that the EEG alpha band over bilateral motor areas decreased while the lucid dreamer executed left or right hand clenching in contrast to dream counting.

In a recent study Strelen (2006) was able to record event-related potentials during lucid dreams by using an odd-ball paradigm with acoustic stimuli. The participants were listening throughout the night to two types of tones (high vs. low tone). The lucid dreamers were instructed to respond to the high tones and signal a single left right eye movement for such an event. The EEG analysis of three lucid dreamers who were able to accomplish this task showed a P300 activation. This evoked potential (P300) can be interpreted as a conscious processing of the acoustic information during the lucid REM dreams – similar to what one would find in wakefulness.

To summarize, a relationship between dream content and brain activity measured by EEG was demonstrable. However, it is not clear how direct this relationship is, i.e., whether all dreamed activities are related directly to specific brain activation; especially in the non-lucid dream studies it is difficult to find correspondences between dream content with measures of brain activation.

From a methodological viewpoint, it would be desirable to use modern imaging techniques like functional magnetic resonance imagery (fMRI) to assess brain activation patterns and relate them to dream content. Due to difficulties related to the setting, e.g., the head of the participant has to be kept in exactly the same position while sleeping, there is loud noise during data acquisition (up to 80 dB), and the strong magnetic fields interfere with the EEG recordings necessary to assess whether the participant is still sleeping, imaging studies are rare (Wehrle et al. 2007). But the problems can be resolved, so one can expect that the existing imagining studies of sleeping persons will be extended to study dream content as well.

4.7 Movement in Dreams: Dream Content and EMG Activity

Muscle tone is inhibited during REM sleep. Despite this general muscle blockade, electrical impulses can be measured in the limb muscles, which might cause small twitches (Dement 1974). Several studies (Dement and Wolpert 1958; Wolpert 1960; Gardner et al. 1975) have shown that dreamed physical activities correspond with the EMG activities of the sleeping body. Wolpert (1960) demonstrated that dreams in which the participants reported a lot of dreamed body movements are associated with an elevated EMG activity in arms and legs. The studies of Grossman et al. (1972) and Gardner et al. (1975) extended the findings from Wolpert (1960) by finding associations between dreamed arm and dreamed leg activity and the corresponding limb EMG recordings. McGuigan and Tanner (1971) have measured chin and lip EMG during REM sleep and demonstrated that dreams with talking were accompanied by elevations of EMG activity in chin and lip muscles. This finding was confirmed by Shimizu and Inoue (1986). Based on research studies investigating the expression of emotions in the waking state, Gerne and Strauch (1985) recorded EMG potentials of corrugator (brow lowered, negative emotions) and of zygomatic (lip corner pulled, positive emotions) movements. In their study, statistically significant relationships between EMG potentials and dream emotion were obtained, but a study by Hofer (1987) failed to replicate them. Overall, the findings indicate that muscle activations found by the EMG recordings correspond to the activity reported by the dreamer during REM sleep.

Several studies with lucid dreamers (Fenwick et al. 1984; Hearne 1983; LaBerge et al. 1981) also demonstrated a close correlation between dreamed limb movements and EMG activities in the corresponding limb. For example, LaBerge et al. (1981) observed that a sequence of left and right fist clenches carried out in a lucid dream resulted in a corresponding sequence of left and right forearm twitches as measured by EMG. Fenwick et al. (1984) showed that for movements carried out in a lucid dream with distal muscle groups (e.g., hand) the EMG activity was higher than for movements that were performed with proximal muscles (e.g., shoulder). In proximal muscles the EMG activity can disappear completely. In several case reports from Erlacher (2005), it was shown that the EMG activity from the forearm of a lucid dreaming person changes during a series of dreamed hand movements (Fig. 4.1). The findings in lucid and non-lucid dreams indicate that physiological mechanisms inhibit most of the activity of the skeletal muscles but residue movements can still be measured.

4.8 Emotions in Dreams: Dream Content and Autonomic Parameters

The studies subsequently reviewed investigated the relationship between dream content and autonomic parameters such as respiratory rate, heart rate, skin resistance or sexual excitation. Psychophysiological studies of nightmares (e.g., Fisher et al. 1970) have shown that severe anxiety dreams are accompanied by elevations in heart rates and respiratory rates. Several correlational investigations (Fahrion 1967; Baust and Engel 1971; Engel 1972; Stegie 1973; Hauri and Van de Castle 1973) have studied the hypothesized association between dreaming of emotional material and physiological arousal. Overall, the results are inhomogeneous, for which Stegie et al. (1975) gave responsibility to two factors. First, the night-time effect was not controlled in all studies, i.e., since physiological arousal as well as dream intensity tend to increase during the course of the night, artificial correlations will be computed in all cases in which awakenings across the night are included. Second, Stegie et al. (1975) pointed out that psychophysiological research has demonstrated what is known as "individual response specificity" in the waking state, i.e., different persons will show different physiological arousal patterns in response to the same emotional stimulus. If the data of several persons were included into the correlational analysis the specificity will markedly reduce the correlation coefficient. In order to avoid this problem, Stegie et al. (1975) elicited dream series (14–28 dreams) of six subjects. Indeed, the results confirm their hypothesis since correlation patterns differed considerably from subject to subject. However, one should consider that 74 correlation coefficients were computed in that study and only 4 of them reached significance ($p < 0.05$) so that this may be explained by chance.

Hobson et al. (1965) have reported that in dreams of healthy persons after awakening following short apneas, specific respiratory themes, e.g. being choked, occurred more often. On the contrary, dreams with respiratory content were seldom found in patients with sleep apnea who often experience long apneas with a considerable reduction of blood oxygen saturation (Schredl 1998a). The following dream example was reported by a patient with sleep apnea.

"During the dream I felt tied up or chained. I saw thick ropes around my arms and legs and was not able to move. I experience the fear of suffocation without being able to cope with the situation. Powerlessness and also resignation came up." Patient with sleep apnea, male, 39 years, respiratory disturbance index (RDI): 68.1 apneas per hour, maximal drop of blood oxygen saturation: 43% (Schredl 1998a, p. 295)

For another physiological variable, penile erections that occur regularly during REM sleep in males, the relationship with dream content was investigated. Fisher et al. (1965) hypothesized an interaction between physiological and psychological factors. Since penile erections occur very often during REM sleep, they assumed an underlying physiological control system. Psychological factors (e.g., erotic dream content or dreamed anxiety) can modulate the degree of the erection. Out of 30 dream reports elicited after REM periods with marked erections, 8 were characterized by erotic themes whereas this topic was not present in any dream report elicited after a REM period without erection (Fisher 1966). In the latter, anxiety was often found. However, the phenomenon of "wet dreams" seems not to be related directly to dream content. Fisher (1966) measured one nocturnal ejaculation of a subject in the sleep laboratory and reported that his dream was not explicitly sexually toned: the dreamer held the hand of his girl friend. Similarly, LaBerge (1985) reported than none of his dreamed ejaculations in lucid dreams was accompanied by a real ejaculation.

In a pilot study, LaBerge et al. (1983) showed a correspondence between subjectively experienced sexual activity during REM lucid dreaming and several autonomic parameters such as respiration rate, skin conduction, vaginal EMG and vaginal pulse amplitude. The parameters increased significantly during experienced lucid dream orgasm, but contrary to expectations, heart rate increased only slightly; the increase was not significant. In another study by LaBerge and Dement (1982), it was shown that lucid dreamers can exert voluntary control over their respiration during lucid REM dreaming. In a study by Erlacher and Schredl (2008a), proficient lucid dreamers were instructed to carry out specific tasks (counting vs. performing squats) while dreaming lucidly. During the night, the heart rate and respiration rate were measured continuously. The results showed an increase of heart rate while performing squats in the lucid dream. The results for respiration rate were less clear but showed the expected changes with higher respiration rates while squatting in the dream. Figure 4.2 shows a recording associated with a succession of squats.

4.9 Body/Mind Interaction in Sleep: Current Direction and Open Problems

The studies reviewed in the present chapter demonstrate clearly an interaction of mind and body during REM sleep, i.e., the two levels (dream consciousness and sleep physiology) are linked. All three approaches (lesion studies, normal dreaming, and lucid dreaming) are valuable tools that help shed light on the correlations between dreamed actions and physiological processes. Overall, the findings are promising, underlining the advantage of studying sleeping persons (low external input, high internal input). The data also showed no one-to-one correlation between dream consciousness and physiology. So it is of importance to differentiate between these two levels and measure each in a reliable and valid fashion. It is not possible to "explain" dreaming by neural activity even though there are associations between physiology and dream content.

Despite the fact that many topics of the body–mind relationship in sleep have been covered, it is important to notice that many findings have not been replicated (as in other psychological research areas). So, current research activities should focus on pursuing the paradigms outlined in this chapter. The mixed results in research with normal dreams, for example, might be explained by the large variability of the dream content (Schredl 1998b). Anyone who wishes to study the relationship between dream content and specific physiological parameters will have to include a large number of participants because the expected correlation coefficients are small. This is of course a problem in sleep laboratory studies because of financial issues. In lucid dream research – where researchers can apply innovative approaches – the samples often consists of only a few participants because it is difficult to recruit many proficient lucid dreamers who are able to control their dreams in the sleep laboratory setting. A prerequisite for strengthening this approach will be the development of techniques to enhance lucid dreaming ability, because existing techniques like carrying out reality checks during the day have only a small effect on lucid dreaming frequency (LaBerge 1985).

The number of open and unaddressed problems regarding the body–mind interaction in sleep is large. In the following, several topics that are worth studying in future are listed.

A major point is that research has mainly focused on REM dreams but not on NREM dreams. As there is no specific muscle atonia present during NREM sleep as there is in REM sleep (to prevent the body from acting out intense dreams), it seems promising to link dreamed actions from NREM dreams to physiological parameters recorded prior to the awakening, which might answer the question of why there is no acting out of dreams during NREM sleep. Keep in mind that sleepwalking is not an acting out of dreams but a disorder of arousal, i.e., the person's brain does not fully wake up but remains in an intermediate state between waking and sleeping. Although the abilities of sleepwalkers are limited they act like awake people – with eyes open and carrying out habitual activities of the waking state. A study by Dane (1986) showed that lucid dream experience also can occur during NREM sleep. To integrate those findings will be a challenging task for future research because it will be difficult to explain why dreamed movements, for example, do not result in body movement (lack of muscle atonia, see above).

One of the most promising future research strategies will be the use of modern imaging techniques such as functional magnet resonance imaging (fMRI). Like the studies carried out with awake participants in the scanner, this approach allows comparison of dream reports after awakening with the brain activation patterns in this period of sleep. This will shed more light on the question as to whether the brain regions that are necessary for carrying out actions or processing incoming information in the waking state are also active while dreaming these actions or processes. Even though modern neuroimaging techniques have their problems, e.g., noise and the difficulty of sleeping with retained head in the scanner, the latest results demonstrate that those problems can be solved (cf. Dang-Vu et al. 2006). Thus, one of the most interesting tasks for future research will be to study lucid dreaming by modern neuroimaging techniques (Hobson 2009).

Another exciting opportunity using lucid dreaming is to find out whether it is possible to communicate with a sleeping person. This is clearly a very interesting question and was addressed in a pilot study carried out by Strelen (2006). He showed that it was possible for participants to "hear" acoustic stimuli from the external "world" within the dream and to discriminate two different tones by carrying out specific eye movement patterns. This is a simple kind of communication between the researcher in the sleep lab and the lucid dreamer in his or her dream world. The researcher uses different tones to communicate whereas the dreamer uses eye movements to signal a reaction. To extend the approach of this pilot study further is a promising research area.

Another open question is whether dream content is related to memory processes. Sleep-dependent memory consolidation has been demonstrated in a large number of studies (Walker 2005) for a variety of declarative (learning word pairs) and procedural (motor learning) tasks. A pilot study carried out by De Koninck et al. (1996) reported a significant correlation ($r = 0.71$) between performance in a reading task wearing goggles inverting the visual field (with training during the previous day) – a task that can be classified as procedural – and the incorporation of task-related elements into subsequent dreams. If future studies confirm this first finding, one will have an answer regarding the function of dreaming, namely that they assist in memory consolidation. Up to now, it is not clear whether these sleep-dependent memory consolidation processes operate solely on a physiological level – like long-term potentiation of synapses – or whether the subjective experience is necessary too. For lucid dreams, this would open up a new venue of training opportunities. Training while dreaming lucidly could enhance performance in athletes – like the mental practice commonly used by professional athletes. Erlacher and Schredl (2008b) have collected several case reports that lucid dream training can be beneficial.

To summarize, it seems very important for future dream research to integrate psychological and physiological approaches. This will shed more light on the body–mind interaction in general and enhance our understanding of dream consciousness and REM sleep in particular.

References

American Academy of Sleep Medicine (2007) The AASM manual for scoring of sleep and associated events: rules, terminology and technical specifications, 1st edn. American Academy of Sleep Medicine, Westchester

Aserinsky E, Kleitman N (1953) Regularly occurring periods of eye motility and concomitant phenomena during sleep. Science 118:273–274

Baust W, Engel RR (1971) The correlation of heart and respiratory frequency in natural sleep of man and their relation to dream content. Electroencephalogr Clin Neurophysiol 30:262–263

Dane J (1986) Non-REM lucid dreaming. Lucidity Lett 5:133–145

Dang-Vu TT, Desseilles M, Albouy G, Darsaud A, Gais S, Rauchs S, Schabus V, Sterpenich V, Vandewalle G, Schwartz S, Maquet P (2006) Dreaming: a neuroimaging view. Schweizer Archivs für Neurologie und Psychiatrie 156:415–425

De Koninck J, Prevost F, Lortie-Lussier M (1996) Vertical inversion of the visual field and REM sleep mentation. J Sleep Res 5:16–20

Dement WC (1974) Some must watch while some must sleep. W. H. Freeman, San Francisco

Dement W, Kleitman N (1957) The relation of eye movements during sleep to dream activity: an objective method for the study of dreaming. J Exp Psychol 53:339–346

Dement W, Wolpert EA (1958) The relation of eye movements, body motility, and external stimuli to dream content. J Exp Psychol 55:543–553

Engel RR (1972) Experimente zur Psychophysiologie des Traumes. Universität Düsseldorf, Mathematische-Naturwissenschaftliche Fakultät, Düsseldorf

Erlacher D (2005) Motorisches Lernen im luziden Traum: Phänomenologische und experimentelle Betrachtungen. University of Heidelberg. http://www.ub.uni-heidelberg.de/archiv/5896

Erlacher D, Schredl M (2004) Required time for motor activities in lucid dreams. Percept Mot Skills 99:1239–1242

Erlacher D, Schredl M (2008a) Cardiovascular responses to dreamed "physical exercise" during REM lucid dreaming. Dreaming 18:112–121

Erlacher D, Schredl M (2008b) Do REM (lucid) dreamed and executed actions share the same neural substrate? Int J Dream Res 1:7–14

Erlacher D, Schredl M, LaBerge S (2003) Motor area activation during dreamed hand clenching: a pilot study on EEG alpha band. Sleep Hypn 5:182–187

Etevenon P, Guillou S (1986) EEG cartography of a night of sleep and dreams. A longitudinal study with provoked awakenings. Neuropsychobiology 16:146–151
Fahrion SL (1967) The relationship of heart rate and dream content in heart-rate responders. Dissertation Abstracts 27(9-B):3307
Fenwick P, Schatzmann M, Worsley A, Adams J, Stone S, Backer A (1984) Lucid dreaming: correspondence between dreamed and actual events in one subject during REM sleep. Biol Psychol 18:243–252
Firth H, Oswald I (1975) Eye movements and visually active dreams. Psychophysiology 12:602–606
Fisher C (1966) Dreaming and sexuality. In: Loewenstein RM, Newman LM, Schur M, Solnit AJ (eds) Psychoanalysis – a general psychology. International Universities Press, New York, pp 537–569
Fisher C, Gross J, Zuch J (1965) Cycle of penile erection synchronous with dreaming (REM) sleep. Arch Gen Psychiatry 12:29–45
Fisher C, Byrne J, Edwards A, Kahn E (1970) A psychophysiological study of night-mares. J Am Psychoanal Assoc 18:747–782
Gardner R, Grossman WI, Roffwarg HP, Weiner H (1975) The relationship of small limb movements during REM sleep to dreamed limb action. Psychosom Med 37:147–159
Gerne M, Strauch I (1985) Psychophysiological indicators of affect pattern and conversational signals during sleep. In: Koella WP, Rüther E, Schulz H (eds) Sleep 1984. Gustav Fischer, Stuttgart, pp 367–369
Glaubman H, Lewin I (1977) REM and dreaming. Percept Mot Skills 44:929–930
Grossman WI, Gardner R, Roffwarg HP, Fekek AF, Beers L, Weiner H (1972) Relation of dreamed to actual movement. Psychophysiology 9:118–119
Hauri P, Van de Castle RL (1973) Psychophysiological parallels in dreams. Psychosom Med 35:297–308
Hearne K (1983) Lucid dream induction. J Ment Imagery 7:19–24
Herman JH, Erman M, Boys R, Preiser L, Taylor ME, Roffwarg HP (1984) Evidence for a directional correspondence between eye movements and dream imagery in REM sleep. Sleep 7:52–63
Hobson JA (2009) The neurobiology of consciousness: lucid dreaming wakes up. Int J Dream Res 2:41–44
Hobson JA, Stickgold R (1995) The conscious state paradigm: a neurocognitive approach to waking, sleeping, and dreaming. In: Gazzaniga MS (ed) The cognitive neurosciences. MIT, Cambridge, pp 1373–1389
Hobson JA, Goldfrank F, Snyder F (1965) Respiration and mental activity in sleep. J Psychiatr Res 3:79–90
Hofer S (1987) Emotionalität im Traum und EMG der Gesichtsmuskeln. Unpublished Lizentiatsarbeit, Universität Zürich
Holzinger B, LaBerge S, Levitan L (2006) Psychophysiological correlates of lucid dreaming. Dreaming 16:88–95
Hong CC-H, Jin Y, Potkin SG, Buchsbaum MS, Wu J, Callaghan GM, Nudleman KL, Gillin JC (1996) Language in dreaming and regional EEG alpha power. Sleep 19:232–235
Jouvet M, Delorme F (1965) Locus coeruleus et sommeil paradoxal. Comptes Rendus des Séances et Memoires de la Societé de Biologie 159:895–899
Jovanovic UJ (1967) Über den Traumbeginn. In: Merz F (ed) Bericht über den 25. Kongress der deutschen Gesellschaft für Psychologie (Münster 1966). Hogrefe, Göttingen, pp 557–562
LaBerge S (1985) Lucid dreaming. Tarcher, Los Angeles
LaBerge S, Dement WC (1982) Voluntary control of respiration during lucid REM dreaming. Sleep Res 11:107
LaBerge S, Zimbardo PG (2000) Smooth tracking eye-movements discriminate both dreaming and perception from imagination. Paper presented at the toward a science of consciousness conference, Tucson, April 10–15

LaBerge S, Nagel LE, Dement WC, Zarcone VP (1981) Lucid dreaming verified by volitional communication during REM sleep. Percept Mot Skills 52:727–732
LaBerge S, Greenleaf W, Kedzierski B (1983) Physiological responses to dreamed sexual activity during REM sleep. Psychophysiology 19:454–455
Ladd GT (1892) Contribution to the psychology of visual dreams. Mind 1:299–304
Maury A (1861) Le sommeil et les rèves. Didier, Paris
McGuigan FJ, Tanner RG (1971) Covert oral behavior during conversational and visual dreams. Psychon Sci 23:263–264
Moody RA (1989) The light beyond. Bantam Books, New York
Nielsen TA (2000) A review of mentation in REM and NREM sleep: "covert" REM sleep as a possible reconcilation of two opposing models. Behav Brain Sci 23:851–866
Rechtschaffen A, Kales A (1968) A manual of standardized terminology, techniques and scoring system for sleep stages of human subjects. US Public Health Service, Washington
Roffwarg HP, Dement WC, Muzio JN, Fischer C (1962) Dream imagery: relationship to rapid eye movements of sleep. Arch Gen Psychiatry 7:235–258
Schenck CH, Mahowald MW (1996) REM sleep parasomnias. Neurol Clin 14:697–720
Schenck CH, Bundlie SR, Ettinger MG, Mahowald MW (1986) Chronic behavioral disorders of human REM sleep: a new category of parasomnia. Sleep 9:293–308
Schredl M (1998a) Träume und Schlafstörungen: Empirische Studie zur Traumerinnerungshäufigkeit und zum Trauminhalt schlafgestörter PatientInnen. Tectum, Marburg
Schredl M (1998b) The stability and variability of dream content. Percept Mot Skills 86:733–734
Schredl M (1999) Die nächtliche Traumwelt: Einführung in die psychologische Traumforschung. Kohlhammer, Stuttgart
Shimizu A, Inoue T (1986) Dreamed speech and speech muscle activity. Psychophysiology 23:210–214
Stegie R (1973) Zur Beziehung zwischen Trauminhalt und der während des Traumes ablaufenden Herz- und Atmungstätigkeit. Universität Düsseldorf, Mathematisch-Naturwissenschaftliche Fakultät, Düsseldorf
Stegie R, Baust W, Engel RR (1975) Psychophysiological correlates in dreams. In: Koella W (ed) Sleep 1974. Karger, Basel, pp 409–412
Strelen J (2006) Akustisch evozierte Potentiale bei luziden Träumen. Unpublished doctoral thesis, University of Mainz
Walker MP (2005) A refined model of sleep and the time course of memory formation. Behav Brain Sci 28:51–64
Wehrle R, Kaufmann C, Wetter TC, Holsboer F, Auer DP, Pollmächer T, Czisch M (2007) Functional microstates within human REM sleep: first evidence from fMRI of a thalamocortical network specific for phasic REM periods. Eur J Neurosci 25:863–871
Wittmann L, Schredl M (2004) Does the mind sleep? An answer to "What is a dream generator?". Sleep Hypn 6:177–178
Wolpert EA (1960) Studies in psychophysiology of dreams: II. An electromyographic study of dreaming. Arch Gen Psychiatry 2:231–241
Zadra AL, Nielsen TA (1996) EEG spectral analyses of REM nightmares and anxiety dreams. Paper presented at the thirteenth international conference of the association for the study of dreams, Berkeley

Chapter 5
Dream Therapy: Correlation of Dream Contents with Encephalographic and Cardiovascular Activations

Agostinho C. da Rosa and João P. Matos Rodrigues

Abstract Sleep and dreaming are overlapping and inseparable phenomena, but they have not often been addressed simultaneously in the scientific sleep research literature. This chapter describes dream research with a focus on objective dream content analysis and on neurocognitive theory analysis. Special emphasis is placed on connecting dream content analysis with current and advanced sleep research methodologies.

This chapter presents some of the traditional and current interventions on dreaming during the periods of REM and NREM sleep, namely the behavioral and physical interventions. A more holistic view is provided through the description of the relationship of REM–NREM sleep and dreaming neurophysiology with the autonomous nervous system.

A survey of current dream therapy usage is discussed in the light of the holistic approach provided, with the aim of showing pathways for future applications.

5.1 Introduction

Sleep and dreaming have been addressed separately as independent phenomena instead of in a holistic perspective. Sleep was studied as part of the circadian waking-sleeping cycle. The introduction of the electroencephalogram, by Berger in 1927, and more recent imaging techniques have boosted the mostly empirical and observational studies to fully fledged scientific research. Multiple approaches have been used, from neurological, physiological and metabolic point of views, looking for biological rhythms and perturbations of different disorders. The task of distinguishing sleep from the other phenomena is now almost complete. However, most research considers sleep as a specific state without much room for cognitive activities, even less for brain control. Dreaming, as a phenomenon that happens only

A.C. da Rosa (✉) • J.P. Matos Rodrigues
Evolutionary Systems and Biomedical Engineering Lab – ISR, Instituto Superior Técnico, Universidade Tecnica de Lisboa, Av Rovisco Pais 1, Torre Norte 6.20, 1049-001 Lisbon, Portugal
e-mail: acrosa@laseeb.org

D. Cvetkovic and I. Cosic (eds.), *States of Consciousness*, The Frontiers Collection,
DOI 10.1007/978-3-642-18047-7_5,

during sleep, has been regarded mostly from a psychological point of view. This gap has only very recently been partially filled with highly controversial theories of dreaming, such as the cognitive approach leading to a more objective and quantitative analysis of dreams distinct from the traditional psychological and subjective analysis, and more recently the proposed neurocognitive theory of dreaming, which adds the neurological aspects of dreams, and especially the REM–NREM controversies over dreaming.

A more medically oriented technique which has developed strongly since the dawn of electro cortical physiology is that of so-called evoked potentials, where electromagnetic brain patterns are recorded and studied during both the awake and sleeping states in response to somatosensory stimuli. Stimuli in afferent pathways and cortical connections have been found to be so consistent that a normative database has been established and used for diagnosis purposes.

Brain control (BC) research attempts to change brain waves or produce specific brain patterns. Two main streams have been pursued intensively, with specific applications in mind. First, brain-computer interface (BCI) work uses specific patterns of brain activity for activating or controlling certain computer mediated devices. Second, neuro-feedback (NF) uses feedback to mediate voluntary control of brain patterns. As we describe in detail in this chapter, specific brain patterns may have strong influences on brain functions, like memory and attention, but more importantly can be used as alternative therapy for many neurological, psychiatric or psychological disorders. The main advantages of this technique are that it is free of side effects and, more interestingly, that the training results are long lasting, with reported durations of more than 2 years after the end of the training sessions.

The focal point of this work is a holistic approach to working memory in its multiple manifestations, from simple sequence rehearsal up to artistic creativity. Many standard and well established tests are available for memory performance, but the quantification of creativity in general is still an open and controversial problem. Our working approach proposes a differential quantification of creativity based on an objective classification of dream contents.

Why do we dream, what is the function of dreams and what they can tell us? Our work covers a very specific aspect of the wider question, which is huge and still mostly unaddressed. The following sections describe relevant aspects of the different contributions, summarize the groundwork for the proposed research, and derive the integrative perspective. The final sections sketches a research proposal for exploring further the integrative approach.

5.2 Neurophysiology of Sleep

5.2.1 EEG During REM and NREM Sleep Stages and Correlations with the ANS

Figure 5.1 shows a hypnogram based on EEG recordings during REM and NREM sleep stages over the course of one night.

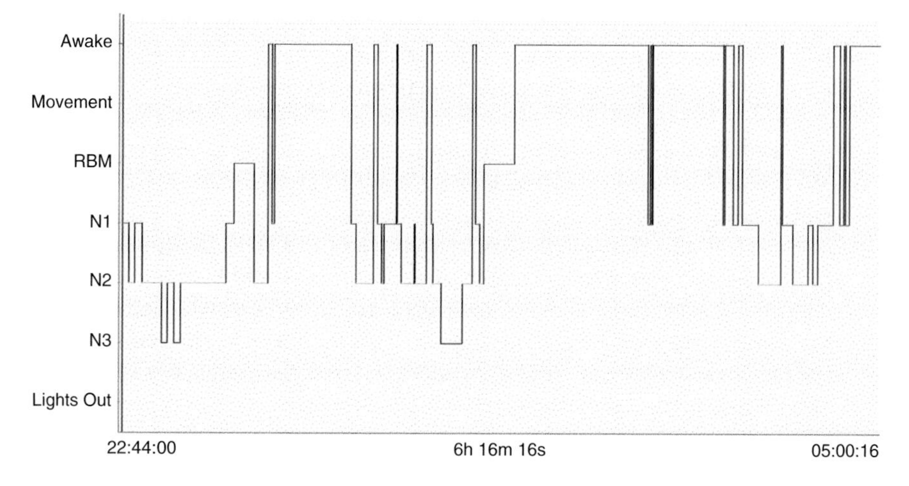

Fig. 5.1 Hypnogram: Epochs for every 30 s throughout one night are classified into stages for REM and three depths of NREM: N1, N2 and N3

Whereas in active wakefulness low amplitude high frequency waves are seen in the EEG, in slow wave sleep there is the opposite – high voltage and low frequency (delta) waves (Coenen 1998). In the waking state, thalamocortical neurons show low synchronization in their firing pattern. Their membranes are depolarized,[1] leading to sustained high activity that allows information to pass easily through the neurons. The correlation dimension is a quantitative estimation of the degrees of freedom of a time series and it is used to distinguish between deterministic chaos and random noise (Grassberger and Procaccia 1983). When used in studies for the analysis of the brain electrical activity, it is usually related positively to the amount of cognitive or mental processing (Lamberts et al. 2000) so, during wakefulness, this value is high. As cells undergo moderate hyperpolarization,[2] the consciousness level drops and afferent information reduces, by synaptic inhibition (Steriade et al. 1993), until slow wave sleep is reached. Here sensory inputs are largely blocked at thalamic level by the rhythmic burst firing of the thalamic network. Although the peripheral sensory organs keep sending impulses resulting from sensory stimuli, as the impulses reach the thalamic network, a stereotyped oscillation is produced that masks the stimuli. During REM sleep, the EEG pattern resembles wakefulness and thalamocortical neurons are also depolarized. The correlation dimension of REM EEG is also higher than for slow wave sleep (SWS) and sometimes can reach waking levels. The fact that this indicator varies during REM is in tune with

[1]When the neuron's membrane potential becomes more positive, the neuron is said to depolarize. This brings the membrane potential closer to the threshold for the cell to produce an action potential, and thus to conduct.

[2]Hyperpolarization is the opposite process to depolarization: the membrane potential becomes more negative, increasing its distance to the necessary threshold for producing an action potential.

findings in evoked potential studies that suggest different amounts of information processing in consecutive episodes of REM (Coenen 1998).

Sleep spindles have been recorded in human subjects during stage 2 NREM and during general SWS in animals. Their frequency ranges from 7 to 14 Hz and they are generated in the thalamus as a result of synaptic interactions (GABAergic inhibition) between neurons of the reticular thalamic nucleus (nRT), thalamocortical cells and cortical pyramidal neurons. Spindles can last from 1 to 3 s and recur every 3–10 s (Steriade et al. 1993). During this time, thalamocortical relay cells are moderately hyperpolarized with membrane potentials around −60 mV.

In late sleep stages, along with further hyperpolarization of thalamocortical cells, spindles reduce and lower frequency/higher voltage thalamocortical oscillations occur as delta (1–4 Hz) and slow oscillation ($<$1 Hz). The higher amplitude of delta oscillations implies that during this activity neurons fire synchronously. Delta oscillations during sleep might be responsible for the reorganization of cortical networks – synaptic connectivity patterns – and the regulation of biochemical activity between single neurons (Steriade et al. 1993).

Ponto-geniculo-occipital (PGO) waves occur during REM, have their origin in the pontine reticular formation and travel to the cortex. These waves are said to be the pacemakers of the thalamus and extended cortical areas as their neurons become depolarized and start firing in a tonic mode similar to wakefulness (Steriade et al. 1993). Meanwhile, neurons from the peripheral motor system are deeply hyperpolarized, resulting in the relaxation of almost every muscle except for those responsible for the movements of the eyes and extremities.

To summarize, during wakefulness there is the highest information flow through relay neurons, the EEG correlation dimension is also the highest and the characteristic EEG pattern is beta activity. When drowsiness starts, the membrane potential of thalamocortical cells drops, the EEG can show predominant alpha activity and spindles (complexity decreases), and the cortical network starts working in an oscillatory mode. During slow wave sleep, membrane potentials drop further along with the EEG complexity and external information processing. Delta activity is predominant in the EEG as neurons fire in burst mode. During REM sleep, neurons fire continuously and EEG complexity is variable, sometimes reaching values similar to those in wakefulness.

Heart rate variability (HRV) is used as a measure of cardiovascular autonomic regulation. More specifically, HRV measures the interaction between the sympathetic and parasympathetic activity in autonomic functioning. This measure has several approaches but for all of them the HRV time series must consist in a sequence of values that represent the interval in time between each R-wave peak in the electrocardiogram (ECG) (Figs. 5.2 and 5.3). Two of these approaches cover the time and the frequency domains. For the time domain, the standard deviation of the time series for each interval is used as a measure of HRV. For the frequency domain, a frequency analysis of the time series is performed (Figs. 5.3 and 5.4). With the frequency analysis, the high frequency (HF) – related to parasympathetic activity – and low frequency (LF) oscillations – sympathetic activity – are observed and their ratio (LF/HF) calculated in order to establish a balance between them (Herbert and Gaudiano 2001). Subjects with poor cardiac health associated with a

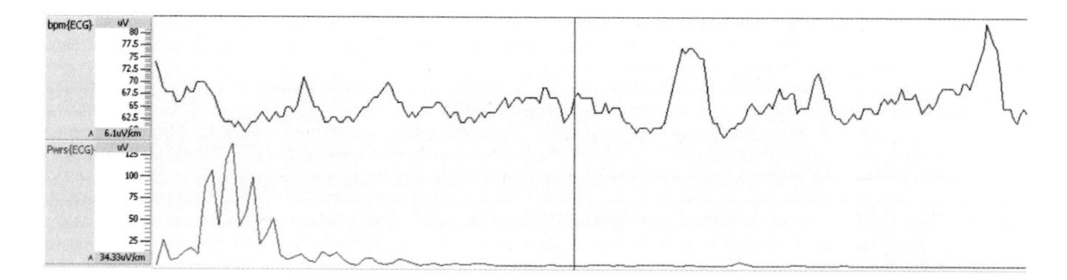

Fig. 5.2 HRV detection: compressed view of 5 min section of a sleep recording. The top trace (*red*) is the ECG (electrocardiogram), the second curve is the R wave interval detection, followed by the corresponding beat per minute curve (bpm), and the bottom curve is the periodogram power spectrum from 0 to 1 Hz (equivalent) of the R–R Intervals

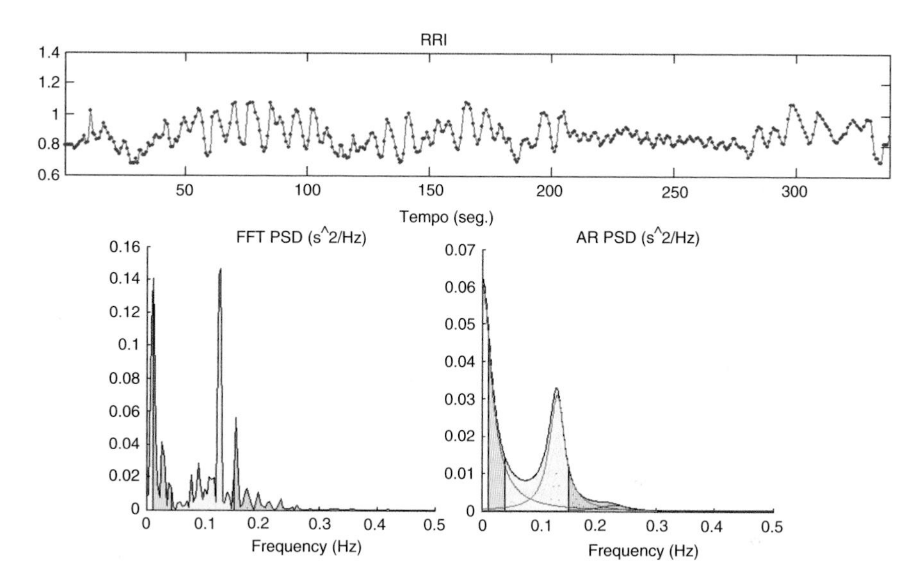

Fig. 5.3 HRV bands calculation: screen view of FFT based and AR model based power spectrum of the RRI. The HRV is divided into VLF, LF and HF bands

slower reaction mechanism show a less erratic HRV time series compared to young and healthy subjects whose heart are more robust to interactive factors. Low HRV can be related to autonomic dysfunction and to an increased risk of future cardiac events (Manis et al. 2007).

The HRV is used to analyze the activity of the autonomic nervous system (ANS) during the different sleep stages. For healthy subjects, the HRV time series shows dominant frequency components with different occurrences during sleep: the HF oscillations are predominant during NREM stages while LF oscillations increase during REM (Ako et al. 2003). In a study realized with seven test subjects, a significant negative correlation was found between delta EEG power and LF oscillations and LF/HF on the first to third NREM period while beta EEG power correlated negatively with HF in the fifth NREM period. Alpha EEG activity is also

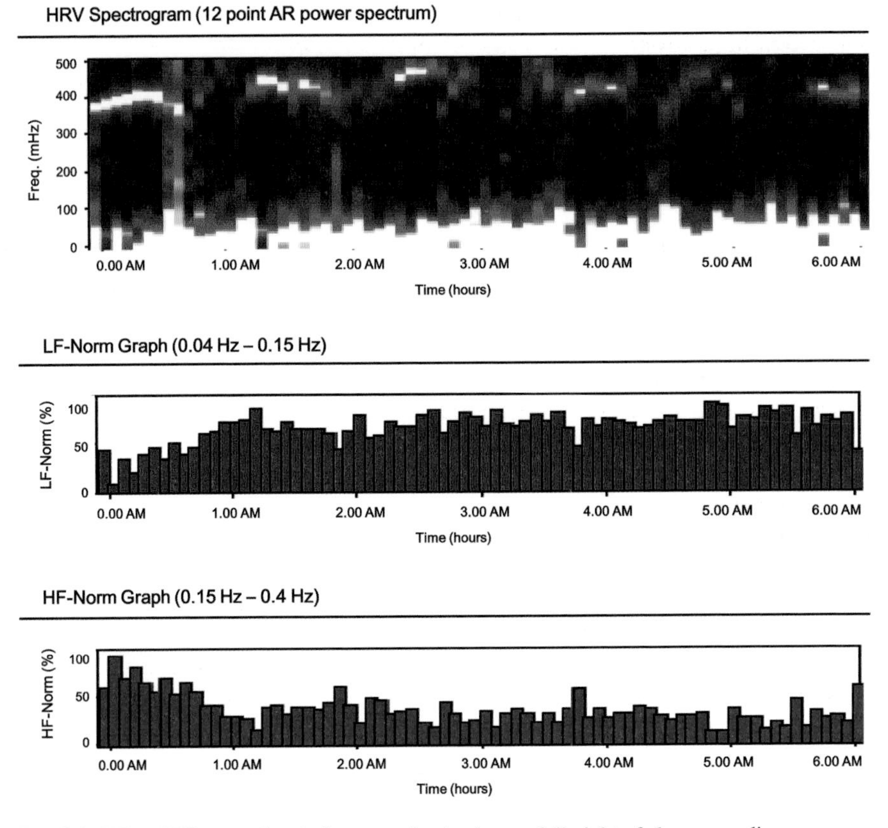

Fig. 5.4 HF and LF power bands for every 5 min along a full night of sleep recording

negatively correlated with heart rate and LF/HF ratio (Ehrhart et al. 2000). Very low frequency oscillations on HRV were found to be closely related to the occurrence of each sleep stage as well as to the SEF90 (spectral edge frequency below which 90% of power of the spectrum resides) (Zhuang et al. 2005). Another contrast between EEG and HRV is that changes in HRV can precede those in the EEG from 10 s to 5 min (Zhuang et al. 2005). The low and high frequencies during one night can be seen in Fig. 5.4.

5.2.2 Neurophysiological Interpretations

The two main probes of brain activity have been functional magnetic resonance imaging (fMRI) and electromagnetic scalp surface recordings. Both have advantages and disadvantages and are mostly complementary, therefore the current research trend is to use both methodologies, but so far the added value of such approaches have been rather limited and very specific. Another drawback of a combined approach is the complexity of the procedure and its cost: fMRI and MEG are far from affordable for routine use. The time resolution of fMRI is also rather limited: only recent high-power machines can achieve near real-time acquisition.

So the method of choice is still the old EEG, where long studies continued over day and night, multiple electrodes and channels, and high sampling frequency recordings are affordable, simple and robust.

The neurophysiology of the EEG is also well known and has been part of routine clinical practice for many years. There are six main rhythms:

- The occipital alpha rhythm (from 8 to 12 Hz) plays a central role in the waking state and has known reactivity to visual processing (attenuation of alpha amplitude) and mental operations, with an increase in the alpha amplitude in a relaxed state or with eyes closed.
- The theta rhythm (from 4 to 8 Hz) is associated with drowsiness and hypnagogic states and occurs mainly during the transition from awake to asleep.
- The delta rhythm (from 0.1 to 4 Hz) is associated with sleep states, in particular of NREM sleep, and gives an indication of sleep depth and long term memorization.
- The sigma rhythm (from 12 to 16 Hz) is also associated with sleep states and is attributed a sensory blocking function.
- The beta rhythm (from 16 to 20 Hz) has been associated with intensive cognitive processing in the brain, both during sleeping and waking states.
- The gamma rhythm is recorded during the waking state with specific procedures, is related to high functional capabilities, and can appear in both sleeping and waking states.

The rhythms are grouped as tonic activity, characterized by slowly varying characteristics and continuously present. By contrast, brief activities that appear suddenly in the EEG trace are known as phasic events and regarded as the system response to external or internal stimuli. In the waking state, phasic events are of low amplitude, with negative signal to noise ratio, and averaging is necessary to extract them. During sleep, the phasic events are usually discernible and easily spotted from the tonic EEG.

The boundaries of the different frequency bands are standardized by averages of the normative population. As a consequence, each individual measure suffers from this standardization. More representative or indicative results would be obtained if it were possible to determine individualized frequency bands. We therefore put the utmost emphasis on the importance of tailoring the analysis by the individual characteristic frequencies in the EEG rhythms.

5.3 Studies of Dreams, Content and Objective Outcomes

5.3.1 Introduction to Dream Theories

There have been several theories and assumptions about the process of dreaming. All these theories need to be reconsidered in view of new findings that appear every year in the field of neuro-imaging, medicine and psychology. To establish a relation with the physiology of sleep and dream content and to determine the role of dreaming in the regulation of the human metabolism and mental states, one needs to consider evidence from dream content analysis, brain functional activation, electroencephalographic data from the various sleep states and wakefulness, the influence of neuromodulators in the sleep states, the limitations and characteristics of dreams in patients with brain lesions and illness, and also in research with animals.

While most of the dream characteristics find explanation, or at least support, in findings in these fields, there are still some doubts about the functions of sleeping and dreaming. There is strong evidence in support of its role in the restitution of the brain, but evidence about how it affects memory, despite its growing volume, is still considered by some as insufficient. Anyway, several theories about sleep and memory exist and most of them have empirical support for some of their claims.

Here we discuss some of these theories along with the other findings from dream content related to functional activation, sleep states and neuromodulators.

5.3.2 Neurocognitive Theories of Dreaming

Dreams have well defined characteristics common to almost any healthy subject. Functional activation evidence from several studies (Dang-Vu et al. 2005) aims to establish a connection between dream characteristics and neurophysiologic aspects. For example, visual imagery during dreams is always present while auditory components are present in 40–60% of dreams and movement and tactile sensations on 15–30%, according to a study by Strauch (Strauch et al. 1996). The activation of temporo-occipital cortices in REM seen in neuro-imaging studies along with evidence that there is a cessation of visual imagery in dreams from some patients with lesions in these cortices (Ako et al. 2003) strongly relates them to these sensory components. Deactivation in the prefrontal areas of the cortex during REM sleep explains dream characteristics like lack of orientation, alteration in time perception and belief that one is awake, as their activation in waking states is responsible for these perceptions (Hobson et al. 1998, 2000).

Dreams in the REM stage are usually presented in a narrative manner although they suffer from some discontinuity with abrupt scene changes. These discontinuities were one of the aspects of dreams that were explained by the activation-synthesis hypothesis of Hobson and McCarley (1977) and were later found in 34% of 200 dream reports analyzed by Rittenhouse et al. (1994). However, these discontinuities, along with bizarre features, were overemphasized and it has been shown that during waking cognition (some experiences were in a darkened room but the subject was still awake) the level of discontinuity and spontaneity in thoughts is not so different from the dreaming experience (Domhoff 2005).

Other similarities between dream and waking cognition are shown by Domhoff (1999a), e.g., the deficits present in the waking thoughts are also present in sleeping mentation and speech patterns in dreams have the same grammatical correctness as in waking life. Domhoff also attributes importance to Calvin Hall's studies with dream content analysis, which showed significant continuity between dream content, waking conceptions and emotional concerns. Dreams that show more aggression are related to people with whom the subject has conflicts in the waking life. Dreams also have components related to recent waking activity in more than half the cases (Fosse et al. 2003). In a study by Maquet et al. (2000), areas involved in learning prior to sleep are reactivated, during REM, in subjects previously trained on motor learning tasks, as opposed to subjects who weren't trained and showed no activation. These findings suggest reprocessing, during REM, of procedural memory acquired during previous waking states. Both amygdala and hippocampus, responsible for emotional conditioning and declarative knowledge in waking cognition (Bechara et al. 1995), are active during REM sleep and can participate in the processing of memory traces during this state (Ako et al. 2003). The amygdala is also responsible for activating the medial prefrontal cortical structures, associated with the highest order regulation of emotions, in REM sleep (Stickgold et al. 2001). Results from a study by Sterpenich et al. (2009) that compared memory consolidation of emotional and neutral pictures between sleep deprived subjects with well slept controls suggested that sleep during the first post-encoding night influences the long-term consolidation of emotional memory. Roffwarg and Walker's work is shown as evidence of this characteristic, where the first dreams in the night include memory fragments from recent experiences and later dreams include fragments increasingly farther back in the past (Paller and Voss 2004).

During REM there is activation in the extra-striate cortex along with a deactivation of the striate cortex (primary visual cortex). During wakefulness, their activity is usually positively correlated and this pattern of functional connectivity suggests that REM sleep allows internal information processing in a closed system disconnected from the external world (via striate cortex) (Ako et al. 2003). Although some information can reach the cortical levels, the thresholds for awakening are lower for relevant stimuli than for irrelevant ones, suggesting a subconscious evaluation during sleep. In REM there is also no coordinated motor behavior (muscular atonia), but in subjects suffering from REM sleep disorders, coordinated behavior is observed that is usually somehow related to the dream narrative provided when woken up (Schenck et al. 1986). In contrast to "acting the dream" is the "controlling the dream" characteristic of lucid dreams. The presence of activity in the alpha band on the sleep EEG during REM can indicate an ongoing lucid dream. This dreaming state is characterized by more realistic content in the dreams as well as an improved consciousness of the self (Domhoff 2001a).

A neurocognitive theory of dreams that takes into account all these findings can be constructed. Domhoff has done much work in this direction by reviewing previous theories, such as activation-synthesis and psychoanalysis, and by proposing new elements for a cognitive theory (Domhoff 1999b, 2001b; 2005).

- Dreaming is a gradual cognitive achievement that correlates with the development of visuospatial abilities. After REM awakenings with children with ages from 5 to 8, only 20% of them reported dream content and it was poor both in content and in visual complexity. The possible inhibiting effect of the laboratory in these dream recalls was shown to be irrelevant by Foulkes, by finding the same results in dreams collected at home (Weisz and Foulkes 1970).
- Dream content and characteristics can suffer from specific neural defects or lesions. The location and nature of each defect can be compared to the changes in dream content.
- Dreams express concepts about self and others.
- It is most likely that dreams are the accidental by-product of sleep and consciousness.
- Dream content is impermeable to pre-sleep and concurrent stimuli that have been used to influence it (Foulkes and Rechtschaffen 1964).
- A new understanding in cognitive linguistics needs to be established for the neurocognition of dreams, in order to understand their frequent figurative aspects – metaphors, resemblance metaphors, metonymies and conceptual blends.
- Some dreams may use the same system of figurative thinking used in waking cognition.

In light of these findings, the activation-synthesis and psychoanalytic theories are found to be incorrect in many aspects, some of them pointed out by Domhoff. First, the existence of dream reports after NREM awakenings invalidates the possibility that dreams only occur during REM. Dream content seems to be much more related to the waking life than to spontaneous image activations, which depend on the area of the brain stimulated, or to the representation of one's unconscious needs, fears or fantasies. Developmental changes between the ages of 3 and 8 in dream content and its subsequent continuity also deviate from what is claimed in both of these theories. Figures 5.5 and 5.6 show the most common characteristics and elements present in the dreaming context.

5.4 Neurophysiological Correlates of Dreaming

5.4.1 REM and NREM Sleep and Dream Collection

The REM sleep state was described in 1953 by Eugene Aserinsky and Nathaniel Kleitman. Along with the rapid eye movement, EEG patterns and autonomic nervous system changes were also observed, which led to the conclusion that these phenomena were all manifestation of a particular level of cortical activity encountered normally during sleep (Aserinsky and Kleitman 1953). For a long period of time, there was a strong belief that dreams only occurred during REM sleep so most of the dreams were collected after this sleep state. In fact, sleep

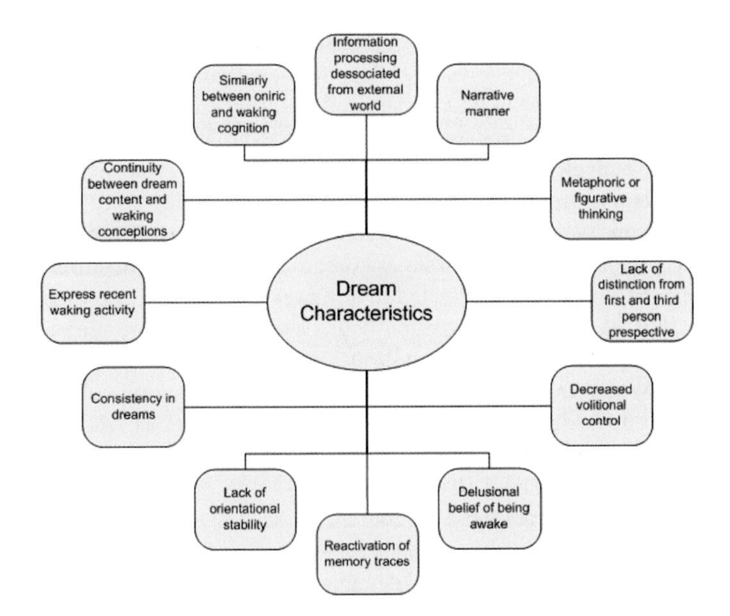

Fig. 5.5 Characteristics of dreams

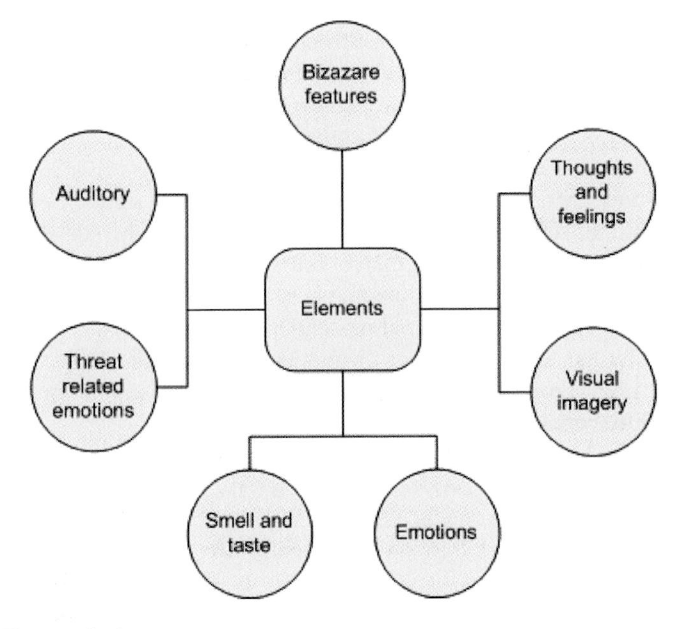

Fig. 5.6 Elements in dreams

mentation also occurs during NREM but, as reported by Nielsen, with a lower recall after waking. In his review, there was an 80% average of REM dream recall rate and a 50% average of NREM dream recall rate (Nielsen 2000).

These dream recalls are reports from dreams collected in various ways. The most efficient and systematic one is in sleep laboratories. Four or five dream narratives can be acquired in a single night by a subject when this subject is awakened during REM periods or NREM periods to maximize the probability of recall (Foulkes 1979). Dream collection can also rely on dream diaries requested by a dream researcher, personal dream journals and dream discussions recorded in psychotherapy (Domhoff and Schneider 1998). A more recent method of dream collection, adopted because of the lack of funding for dream studies, is the Most Recent Dream method. The dream report is collected in an interview fashion. The interviews last 15–20 min and are done in convention halls, waiting rooms and classrooms. The samples are representative if at least 100–125 reports are collected for each age group (Avila-White et al. 1999). Foulkes states that the reports collected by this method do not differ from those collected in sleep laboratories when controls are introduced (Lamberts et al. 2000) but there are different opinions about the reliability of this method.

To analyze the content of these dream reports, several methods are applied, such as asking the dreamer for the meaning of each element in the report (without censoring) or finding metaphoric meanings for the dreaming narrative and searching for repeated themes in series of dreams. For a quantitative analysis to be reliable, it is necessary to define clear categories of dream content that lead to the same results when used by different investigators so that their findings can relate quantitatively to others. This is achieved by content analysis. The scales used can be based on rating each element or can be nominal, reporting just the presence of an element or characteristic. Hall and Van de Castle developed a set of nominal categories that are widely used in content analysis (Domhoff 1999b).These categories include characters, social interactions, activities, striving, misfortunes and good fortunes, emotions, physical surroundings and descriptive elements. Each category is further subdivided into subcategories. For example, social interactions are subdivided into friendly, aggressive and sexual (Manis et al. 2007). These categories allow the use of indicators that consist of the ratios between groups of categories. For example, the "animal percent" is the ratio between characters that are animals by the total number of characters (Manis et al. 2007). Finally, the indicators of one group are compared with the norm to search for statistically significant differences.

5.4.2 *Memory and Learning During Sleep*

Several insights about dream content, especially during REM sleep, have been presented. Memory reactivation is one the characteristics of the dreaming experience. We shall now analyze it in more detail.

Declarative (or explicit) memory is referred as the ability to recall and recognize episodes and complex facts and has two subclasses: the episodic memory, which consists in personal facts and experiences, and semantic memory for impersonal facts (Vertes 2004). According to Paller, after an analysis of several studies (Paller and Voss 2004), this aspect of memory depends on multiple neocortical regions and requires "cross-cortical storage". This means that the memory fragments by themselves are not sufficient to comprise the memory but need to be linked together for it to exist. So the cognitive characteristics of declarative memory probably include a reliance on relational representations (Eichenbaum 2001). Episodic memory, one type of declarative memory, is encoded by the hippocampal networks as sequences of events and places where these traces were created (Paller and Voss 2004). Cross-cortical storage is in tune with other models that suggest a gradual formation of the representation of memories in the cerebral cortex. According to this model, declarative memories exist in a fragile state after initial learning. The hippocampus has a major role in establishing these new connections between separate cortical representations, and these quick links are temporary. After sufficient cross-cortical consolidation, the hippocampal neurons are no longer necessary to unite this set of distinct cortical networks and the declarative memory endures. This process depends on the nature and frequency of memory access from the initial acquisition to the retrieval. These individual memories can change after their initial learning period. So two things can occur when declarative memory is accessed: the fragments of episodic of factual representation become more strongly bound together or they become more strongly linked to new representations of that fact or to related stored information. These connections to stored information provide additional routes for subsequent retrieval and this gives the stored information connections with aspects disjoint in time. The declarative memory network changes with post-acquisition processing, and if this processing provides cross-cortical consolidations these memories can be retrieved in the absence of the hippocampus. The cross-cortical storage links different components of memory together. The memory of an event can include individual links to each item of the event as well as to the context when this learning occurred and emotional experiences created by this event. The conscious experience of remembering declarative memories also depend on such aspects as attention, cognitive control and working memory.

Duda relates two types of amnesiac patients and their memory limitations (Dudai 2004). Amnesiacs who have difficulties recalling recent events but whose remote memory is spared (temporally graded retrograde amnesia for declarative memory) usually show damage in the medial temporal lobe, hippocampus formation and associated cortices. Patients whose amnesia extends over many decades usually show additional damage in the neocortex in the lateral and anterior temporal lobe. This suggests that memory storage is related to the neocortex while the hippocampus formation is related to a time-limited retrieval of intact declarative long-term memory.

Dudai also defines two types of memory consolidation: synaptic consolidation and system consolidation (Dudai 2004). The synaptic consolidation model consists in the consolidation of short-term and labile memories to long-term and stable ones by means of long-term modification of synaptic proteins and in the remodeling and growth of these synaptic connections. The activation of this process occurs in a limited time window during and immediately after the information acquisition and can be disrupted by various types of agents such as inhibitors of protein synthesis. System consolidation explains how older memories are secured by shifting their storage from the hippocampus to the neocortex over a period of weeks.

There are three different views of consolidation of a memory trace, differing in how they account for reconsolidation. The "weak" view states that, when retrieved, a memory trace only has its new parts updated. The existing parts are not changed because consolidation only happens once. The "intermediate" view claims that after retrieval, memory traces become modifiable and labile, and so need to undergo a new process of stabilization. If an appropriate interference occurs, parts of a trace can become corrupted. The "strong" view differs from the intermediate view by including the new information to be processed in the list of parts that can become corrupted.

Procedural memory is a type of non-declarative memory that involves unconscious acquisition and utilization of perceptual and motor skills (Vertes 2004). Despite this difference of role, procedural memory traces are also reconsolidated when accessed.

While it is now evident that declarative memory is accessed during sleep, there still is a void about their consolidation and the role of dreams and sleep stages in this process. Evidence in these areas exists, but it is fragmented and sometimes contradictory. Fortunately, this has stimulated several studies and theories. Because most of the content in dream reports show that sleep mentation is flooded with fragments of recent events and knowledge, Paller assumes that during sleep cross-cortical consolidation of declarative memory traces also occurs (Paller and Voss 2004). He also hypothesizes that these connections between memories can be central for problem solving and that connections between fragments, if they occur on larger scales, can connect episodic memories with behavioral strategies (these behavioral strategies are taken as cognitive fragments based on past experiences and, as they consolidate, become aspects of the subject's personality). A similar view is presented by Rasch and Born where newly acquired memory traces are gradually redistributed, from fast learning regions of the temporal lobe (hippocampus mainly), to neocortical regions by strengthening cortico–cortical connections during off-line periods (Rasch and Born 2007). Repeated reactivation of the new information is essential for this process and it is accompanied by reactivations of related older memory traces. This pattern of reactivation is thought to gradually integrate the new memories to the pre-existing network of long term memories and may produce qualitative changes in the respective memory representation. The authors present several studies that show the influence of slow wave sleep in declarative memory reactivation in animals. During post-learning SWS periods, firing patterns in the hippocampus similar to the ones present during learning were observed and it is suggested that the learned information is repeated during sleep. Furthermore, it was observed that in rats these reactivations in the hippocampus alternate with reactivations in the visual cortex, an aspect that can be seen as the gradual transfer of the replayed information from the hippocampus to the cortical regions (Ji and Wilson 2007). These results led to the replication of the studies in human subjects, where the same reactivation of hippocampal regions that were activated during a spatial navigation task (declarative memory) was observed, along with evidence for a positive relation between the performance of memory retrieval the next day and the amount of activity in the hippocampus during SWS (Peigneux et al. 2004). Still, slow wave sleep is characterized by a reduced ability to induce synaptic long term potentiation (LTP) while the REM stage seems more appropriate for that effect due to the presence of more neurotransmitters and modulators responsible for synaptic plastic changes. Gais and Born attribute importance to the sequence of SWS and REM for memory consolidation as the hippocampal reactivation serves to tag the synapses that are then strengthened in REM (Gais and Born 2004b), or in wakefulness if REM is not possible (Morris 2006).

Other studies by Gais and Born (2004a) suggest that slow wave sleep is responsible for declarative memory consolidation and that this consolidation cannot be achieved without SWS's characteristic low levels of the neurotransmitter acetylcholine in the hippocampus. Still, SWS length does not increase after increased declarative learning and declarative memory performance does not increase with the amount of SWS. According to Gais and Borne, high levels of cholinergic activity in SWS, induced by post-trial administration of physostigmine (cholinesterase inhibitor), completely eliminated the consolidating effect of sleep on hippocampus-dependent declarative memory.[3] These were results from a word pair test compared to a control population that was given a placebo instead of physostigmine and also to a set of experimental and control populations that stayed awake. Hippocampus-independent memory was also tested with a mirror tracing task and no relation was found between subjects administered physostigmine and control subjects. Further studies from the same authors assign acetylcholine the role of a switch between modes of acquisition and consolidation of memory traces, as its low levels potentiate consolidation in SWS and high levels during wakefulness support encoding (Rasch et al. 2006).

Another study showed that it was possible to externally induce reactivation during slow wave sleep by presenting the sleeping subject with cues that were presented during the previous learning period (Rasch et al. 2007) and that this led to increased consolidation. An odor cue was used because it is the least sleep disturbing stimulus. It was observed that this cueing affected neither sleep architecture nor sleep EEG and subjects did not recall the odor treatment during sleep. In the experimental procedure, this stimulus was presented while subjects performed learned object locations in a two-dimensional object location memory task (declarative memory) in the evening before sleep and represented during the first two periods of subsequent SWS. Alternative trials included presenting the odor only during SWS, only during learning and REM and only during learning and waking. The control trial did not include an odor cue in any condition. Improvements in location recall compared to the control trial were only seen in

[3]Activity in neurons that has acetylcholine as its neurotransmitter is called cholinergic activity. Cholinesterase is a group of enzymes responsible for the decrease in the acetylcholine levels. If these enzymes' action is inhibited, in this case by physostigmine, the levels of acetylcholine and the activity in the cholinergic neurons are maintained.

the first experimental procedure accompanied with activation in the hippocampus. To test the odor cueing effect on procedural memory, a finger tapping speed test, with the same procedures as before, was also done but did not show any significant improvements.

Transcranial direct current stimulation (tDCS) also changes the process of consolidation in declarative memory when applied during SWS. Declarative memory consolidation was improved (word pairs) relative to a control group whereas procedural memory was not.

Although memory consolidation is improved, the accuracy of the recalled memories after sleep may not be perfect and false memories can be originated. In a study that used the Deese–Roediger–McDermont (DRM) paradigm for memory recall, an increase was observed in both veridical and false memories in subjects retested after sleep compared to subjects who were retested after the same period of daytime wakefulness (Payne et al. 2009). The DRM paradigm consists on a list of words learning task and the recalled words are interpreted as veridical, false and critical. Veridical words are the words presented on the list, semantically associated words are interpreted as critical and unrelated words are interpreted as false. This study shows that there is an increase in the recall of critical words over learned words after a night of sleep compared to equivalent periods of wakefulness (nevertheless learned words recall also increased compared to equivalent periods of wakefulness). After a daytime nap the recall of critical words was 50% greater than the control recall while there was no significant increase in the recall of veridical words. This at least suggests that sleep enhances recall not only of the exact words but also of the surrounding network of semantically related words, thus pointing to an effect of strengthening and creating connections between memory traces.

Although procedural memory is not improved in the previous experiences, it is a robust and evident finding that it benefits from sleep (Walker and Stickgold 2006). It can be divided into several functional domains, such as perceptual – visual and auditory – and motor skills.

Motor adaptation (like learning to use a computer mouse) and motor sequence learning (like learning a piano scale) are two forms of qualifying motor skills. Some of the following studies presented were reviewed and related to each other by Walker (Walker and Stickgold 2006).

It has been demonstrated that a night of sleep after learning a sequential finger tapping task can produce significant improvements in speed and accuracy while the same amount of wake time provides no significant benefit (Walker et al. 2002). In this study the improvements were positively correlated to stage 2 NREM sleep, but this relation is not consistent between experiments. In other study by Kuriyama et al. the speed of each step in a motor sequence, before and after sleep, was taken into account. It was found that sleep-related improvements were not equal among steps (Kuriyama et al. 2004). In each sequence, some steps were more problematic than others and this was reflected by decreased speed compared to the easier ones. Again, significant improvements were only verified on subjects who slept between learning and testing, and these improvements were greater in those slower steps.

Walker interprets these results as a characteristic of sleep-dependent consolidation where a single memory element, comprised of smaller motor memory units, is selectively improved, giving priority to the more problematic units. This suggests that sleep not only increases the strength of existing memory representations but also does it in a selective manner. The need to sleep in the first night or day following the learning, for improvements in a finger tapping task to be noticed, is also confirmed by Fischer (Fischer et al. 2002). In this study, the improvements seem to be more related to the amount of REM sleep than to stage 2 NREM. This finger-tapping test was different from the others because it consisted of finger-to-thumb movements rather than keyboard typing like the previous. Walker hypothesizes that the REM correlation is due to the novelty in the procedure, because keyboard typing is a simple variant of a well learned task while finger-to-thumb movements introduce new tasks (Walker and Stickgold 2006). Other studies that use selectively sleep-deprived subjects on the first night prior to learning show that retention of a visuomotor adaptation task suffers more with the deprivation of stage 2 NREM and that the learning of a motor-reaching adaptation task was accompanied by a discrete increase of the subsequent SWS activity at the start of the night, where this increase was proportional to the enhancement of the tested motor skill on the next day (Walker and Stickgold 2006).

In other studies, areas involved in learning prior to sleep were seen to be reactivated during REM in subjects previously trained on motor learning tasks while subjects who were not trained showed no activation (Maquet et al. 2000). During REM after procedural task learning, the firing sequence of hippocampal neurons seemed similar to that of the task learning phase (Battaglia et al. 2004). According to Paller, this can serve to strengthen synaptic connections between neurons that process information in a manner relevant to performing the task by mechanisms for circuitry remodeling produced by LTP (Paller and Voss 2004).

Visual and auditory sleep-dependent skill learning is also confirmed by studies of the same nature as the previous ones. Learning a visual or pitch discrimination task also shows significant improved results following a night of sleep compared to the same amount of wake time, where no improvement is shown (Walker and Stickgold 2006).

Daytime naps also seem to improve procedural memory. There is an interesting effect seen in a study that applied a sequential finger tapping task (Walker and Stickgold 2005). Two groups were trained on a task in the morning and tested. One of the groups remained awake until the retesting, later in the day, while the other underwent a midday nap of 60–90 min. When retested, the group whose subjects napped showed 16% improvement whereas those who did not nap showed no significant improvement. The interesting fact is that on the next day, after both groups slept through the night, when retested, the control group showed 24% improvement and the nap subjects showed 7% improvement (improvements compared to the previous retest, not the first). So, on the next day, both groups showed more or less the same improvement. The difference is that the first had 23% improvement split in 16% nap plus 7% night sleep while the second had the whole 24% improvement in night sleep.

Still in the field of procedural memory improvement due to sleep, one characteristic that is sometimes interpreted as a limitation of this learning is that it is specific to the sequence learned and to the limb used (Vertes 2004). Improvements gained in one procedural task are not present on other tasks, and similarly for the limb used.

Benefits from sleep were also found on working memory (Kuriyama et al. 2008). In a test used to determine working memory capacity and response time, subjects were divided into three different groups. Group A trained at 8 a.m. and were retested at 3 p.m. and 10 p.m. without any sleep in between. Group B trained at midday and were retested at 10 p.m., went to sleep and were again retested at 8 a.m. on the next morning. Finally, group C trained at 10 p.m., went to sleep and were retested at 8 a.m. and later at 6 p.m. This different training and retesting hours are important to see in which case working memory benefited most. The results showed that improved working memory results only showed up in retests after sleep and there was no time of day specific to better results (which excludes a circadian influence).

Besides declarative, procedural and short-term memory, there are also other aspects of cognition that seem to benefit from sleep. In a study by Wagner, subjects were required to perform a mathematical operation on a string of digits (Wagner et al. 2004). The time spent on the operations could be significantly reduced by applying a certain rule that was left for the subjects to discover. The insight of this rule was more frequent when the training was followed by a night's sleep than by a sleepless night or the equivalent time awake during the daytime.

Now it seems evident that the role of sleep is more than physiological recuperation or a period of inactivity to keep our primitive ancestors idle during the night, and that dreams can be the manifestation of some of the processes that occur during this period.

Having these findings in mind, it is possible to propose some characteristic relations between sleep, memory and known dream elements and to test new possibilities for their interplay. For example, some environments or logical sequences of episodes during a dreaming episode can be due to the activation of a memory trace and further chain of activation of the linked traces. Because these traces were linked together by some logical or temporal relation, when reactivated, the result also seems logical instead of the random activation proposed by the activation-synthesis theory. This could be one way to create a coherent dreaming scenario.

5.5 Sleep and Dreaming as Targets for Interventions

The main result of our integrative approach is a conceptual and practical framework that enables different analytic perspectives of the brain through the EEG recordings. The aim is to segment the continuous EEG into different functional sections, for restorative processes, memory related functions (storage, consolidation, recall, association, etc.), and cognitive processes (expectation, attention, decision, deduction, etc.), not only during the awake state but especially during sleep.

5.5.1 *Cognitive Processing During Sleep and Dreaming*

The degree of consciousness or cognitive function in the brain during sleep is still a controversial and debated topic. One possible source of answers or new arguments for the discussion is research conducted in accordance with the following hypothesis:

- Cognitive processing in the waking state can be characterized by its EEG characteristics and these are assumed to be the same during dreaming.
- The dream recall immediately after a REM episode is a faithful report of the dream content at least regarding the time sequence of the salient contents.
- The Van Castle dream classification is an adequate tool for objective dream content classification into categories that can be associated with cognitive processes.
- NREM dreaming can be treated in the same way as REM dreaming.

The research protocol is straightforward. Subjects under study are monitored with standard EEG recordings while awake and performing several cognitive tasks and again during sleep, when they are awakened every 10 min during REM episodes and their dream recall narration digitally recorded.

The dream content of each episode is analyzed by the Van Castle technique in order to establish the time sequence of cognitive states. The awake EEG is used to determine the individual alpha frequency (IAF) and the theta frequency (TF) in order to establish the alpha-1, alpha-2 and upper alpha bands, then the lead and significant electrodes and bands are calculated for the different cognitive tasks. This process is repeated for every REM dream episode resulting in sequences of cognitive processing states. A correlation between the sequences obtained by the dream recall or EEG characteristics is calculated in order to establish a plausible causality. A significant individual and/or group positive correlation indicates the presence of cognitive processing during the REM dream episode.

The same research can be applied to NREM sleep stages 2 and 3. If the correlation is still positive then we have strong evidence that the EEG characteristics identified are indeed descriptors of cognitive processing during sleep and the whole night's sleep can be analyzed under this perspective.

A final procedure is the recording of uninterrupted sleep with whole-night dream recall the next morning. The recall is then correlated with the cognitive processing analysis of the whole night EEG. If a positive correlation is found, it might indicate the possibility of estimating the dream content in cognitive processing terms without the need for dream recall.

5.5.2 *Memory Enhancement*

As described above, short-term memory recall has been shown to be positively correlated with alpha frequency band. It has been reported that shorter reaction times and longer recall sequences are positively related to the IAF. The underlying hypothesis includes the assumption that the IAF can be changed by some sort of training procedure, i.e., that the neuro-feedback paradigm is an effective way to teach the subject how to control the brain EEG characteristics. Subjects with surface electrodes recording in specific regions of the scalp EEG are asked to increase in a sustained way and for a controlled period of time a feedback signal extracted from the EEG, which can be the power of a specific band of alpha, the frequency of the IAF, a alpha-theta ratio, or coherence measure between two EEG channel, etc. The objectives are progressively and adaptively increased between sessions. The subject usually learns to control the EEG characteristics after six training sessions and a total of 20–30 sessions are required to obtain the first significant results.

5.5.3 Creativity Enhancement

Creativity is still a much debated concept and many definitions exist. The concept of creativity used here is the production of any new idea or object by new means that are not exactly like any seen or proposed before. The precise definition of creativity is not very important, because it is almost impossible for anyone to create an idea or an object that has never been partially suggested or partially produced before. However defined, the creativity must be confirmed by peers. The basic idea is the production of different and varied ideas from existing knowledge during sleep, mainly by associative combinations of atoms or acquired ideas already present in the brain. The hypothesis is that during sleep the available ideas are combined or associated in a random way. More ideas mean more material for combinations, increasing the chance to produce new ideas or objects. Since these objects are obtained by free combinations, it is rather unlikely that these ideas will result in a useful or important idea. It is also hypothesized that while the subject is awake all the combinations are filtered by a more conscious or rational filter, thus limiting the chain of innovative ideas. By contrast, in REM sleep, there is no filtering, resulting typically in bizarre and mostly impractical ideas. The virtue lies in the middle, a state between being awake and REM sleep, known as light REM, where the brain state is between REM and awake states. The practical implementation of this experiment is to deliver subliminal stimuli to the subject during REM sleep. These subliminal stimuli will lighten the REM sleep toward a waking state, obtaining an intermediate filtering state, hopefully with combinatorial power and at the same time some realistic natural filtering. This setup will promote a more open and well behaved subconscious state.

5.6 Conclusion

During sleep it is possible to distinguish different stages occurring in a cyclic pattern. Each of these stages shows unique characteristics in the brain activity, heart rate and muscle tone. Although the exact reason for the presence of these different stages during sleep is not yet established, several studies assign different roles to each one of them. This brings up the question of their meaning and origin. If the length of sleep is due to our ancestors' primitive need to remain inactive through the night, because of the lack of light, then today, as the problem of illumination no longer applies, its importance may be exaggerated. Is our circadian clock out of date, keeping us from being awake more of the time because of a self-defense mechanism that is no longer necessary? Without the knowledge presented in this chapter, one could easily agree with this observation. It seems that the restorative role of sleep does not act only on the metabolic processes of the body, since the cognitive processes are also influenced. Memory consolidation and reasoning capability are examples of processes that are affected by sleep deprivation or even selective sleep stage deprivation. It is possible that each stage has a more relevant role for each cognitive process, but in the case of memory consolidation the evidence points to the importance of the interplay between stages.

Before the recording of brain activity or heart rate was possible, dreams were the only window to the sleeping process. Today, all the recordings are taken into account and correlated with each other. The results enable us to see that some regions of the brain, responsible for some emotional states for example, are activated during sleep when the corresponding emotional state is present in the dream. Other previously known characteristics of the dreaming scenario are also explained by activation or deactivation of brain regions that are known to be responsible for those characteristics in the waking state. However, these activations are not totally random, since reports of dream episodes show some consistency in their plot. One possible explanation for this consistency is the activation in sequence of several memory traces that are connected to each other. Because these traces were connected during the same time period or in the same context, when reactivated they produce a consistent plot. In other words, everything we know is connected somehow in our brain. During sleep, these fragments of knowledge are activated and linked to each other, producing the dream episode. For example, it is possible that during a dream, if the right associations occur, a solution to a problem that seemed unsolvable is achieved. Such a solution would result from a very improbable event, but if the way associations are made could be induced somehow, this kind of event would be less improbable.

In summary, even if the time one should spend sleeping is debatable and varies between individuals, it is not fair to underestimate its underlying importance.

References

Ako M, Kawara T, Uchida S, Miyazaki S, Nishihara K, Mukai J, Hirao K, Ako J, Okubo Y (2003) Correlation between electroencephalography and heart rate variability during sleep. Psychiatry Clin Neurosci 57:59–65

Aserinsky E, Kleitman N (1953) Regularly occurring periods of eye motility, and concomitant phenomena, during sleep. Science 118:273–274

Avila-White D, Schneider A, Domhoff GW (1999) The most recent dreams of 12–13 year-old boys and girls: a methodological contribution to the study of dream content in teenagers. Dreaming 9:163–171

Battaglia FP, Sutherland GR, McNaughton BL (2004) Hippocampal sharp wave bursts coincide with neocortical "up-state" transitions. Learn Mem 11:697–704

Bechara A, Tranel D, Damasio H, Adolphs R, Rockland C, Damasio AR (1995) Double dissociation of conditioning and declarative knowledge relative to the amygdala and hippocampus in humans. Science 269:1115–1118

Coenen AM (1998) Neuronal phenomena associated with vigilance and consciousness: from cellular mechanisms to electroencephalographic patterns. Conscious Cogn 7:42–53

Dang-Vu TT, Desseilles M, Albouy G, Darsaud A, Gais S, Rauchs G, Schabus M, Sterpenich V, Vandewalle G, Schwartz S, Maquet P (2005) Dreaming: a neuroimaging view. Swiss Arch Neurol Psychiatry 156(8):415–425

Domhoff GW (1999a) Drawing theoretical implications from descriptive empirical findings on dream content. Dreaming 9:201–210

Domhoff GW (1999b) New directions in the study of dream content using the Hall and Van de Castle coding system. Dreaming 9:115–137

Domhoff GW (2001) A new neurocognitive theory of dreams. Dreaming 11:13–33

Domhoff GW (2005) Refocusing the neurocognitive approach to dreams: a critique of the hobson versus solms debate. Dreaming 15:3–20

Domhoff GW, Schneider A (1998) New rationales and methods for quantitative dream research outside the laboratory. Sleep 21:398–404

Dudai Y (2004) The neurobiology of consolidations, or, how stable is the engram? Annu Rev Psychol 55:51–86

Ehrhart J, Toussaint M, Simon C, Gronfier C, Luthringer R, Brandenberger G (2000) Alpha activity and cardiac correlates: three types of relationships during nocturnal sleep. Clin Neurophysiol 111:940–946

Eichenbaum H (2001) The hippocampus and declarative memory: cognitive mechanisms and neural codes. Behav Brain Res 127:199–207

Fischer S, Hallschmid M, Elsner AL, Born J (2002) Sleep forms memory for finger skills. Proc Natl Acad Sci USA 99:11987–11991

Fosse MJ, Fosse R, Hobson JA, Stickgold RJ (2003) Dreaming and episodic memory: a functional dissociation? J Cogn Neurosci 15:1–9

Foulkes D (1979) Home and laboratory dreams: four empirical studies and a conceptual reevaluation. Sleep 2:233–251

Foulkes D, Rechtschaffen A (1964) Presleep determinants of dream content: effect of two films. Percept Mot Skills 19:983–1005

Gais S, Born J (2004a) Low acetylcholine during slow-wave sleep is critical for declarative memory consolidation. Proc Natl Acad Sci USA 101:2140–2144

Gais S, Born J (2004b) Multiple processes strengthen memory during sleep. Psychol Belg 44:105–120

Grassberger P, Procaccia I (1983) Measuring the strangeness of strange attractors. Phys D 9:189–208

Herbert JD, Gaudiano BA (2001) The search for the holy grail: heart rate variability and thought field therapy. J Clin Psychol 57:1207–1214

Hobson JA, McCarley RW (1977) The brain as a dream state generator: an activation-synthesis hypothesis of the dream process. Am J Psychiatry 134:1335–1348

Hobson JA, Stickgold R, Pace-Schott EF (1998) The neuropsychology of REM sleep dreaming. Neuroreport 9:R1–R14

Hobson JA, Pace-Schott EF, Stickgold R (2000) Dreaming and the brain: toward a cognitive neuroscience of conscious states. Behav Brain Sci 23(6):793–842; discussion 904-1121. Review. Erratum in: Behav Brain Sci 2001 Jun;24(3):575. [PMID: 11515143]

Ji DY, Wilson MA (2007) Coordinated memory replay in the visual cortex and hippocampus during sleep. Nat Neurosci 10:100–107

Kuriyama K, Stickgold R, Walker MP (2004) Sleep-dependent learning and motor-skill complexity. Learn Mem 11:705–713

Kuriyama K, Mishima K, Suzuki H, Aritake S, Uchiyama M (2008) Sleep accelerates the improvement in working memory performance. J Neurosci 28:10145–10150

Lamberts J, van den Broek PLC, Bener L, van Egmond J, Dirksen R, Coenen AML (2000) Correlation dimension of the human electroencephalogram corresponds with cognitive load. Neuropsychobiology 41:149–153

Manis G, Nikolopoulos S, Alexandridi A, Davos C (2007) Assessment of the classification capability of prediction and approximation methods for HRV analysis. Comput Biol Med 37:642–654

Maquet P, Laureys S, Peigneux P, Fuchs S, Petiau C, Phillips C et al (2000) Experience-dependent changes in cerebral activation during human REM sleep. Nat Neurosci 3(8):831–836

Morris RGM (2006) Elements of a neurobiological theory of hippocampal function: the role of synaptic plasticity, synaptic tagging and schemas. Eur J Neurosci 23:2829–2846

Nielsen TA (2000) A review of mentation in REM and NREM sleep: "covert" REM sleep as a possible reconciliation of two opposing models. Behav Brain Sci 23:851–866

Paller KA, Voss JL (2004) Memory reactivation and consolidation during sleep. Learn Mem 11:664–670

Payne JD, Schacter DL, Propper RE, Huang LW, Wamsley EJ, Tucker MA, Walker MP, Stickgold R (2009) The role of sleep in false memory formation. Neurobiol Learn Mem 92(3):327–334

Peigneux P, Laureys S, Fuchs S, Collette F, Perrin F, Reggers J, Phillips C, Degueldre C, Del Fiore G, Aerts J, Luxen A, Maquet P (2004) Are spatial memories strengthened in the human hippocampus during slow wave sleep? Neuron 44:535–545

Rasch B, Born J (2007) Maintaining memories by reactivation. Curr Opin Neurobiol 17:698–703

Rasch BH, Born J, Gais S (2006) Combined blockade of cholinergic receptors shifts the brain from stimulus encoding to memory consolidation. J Cogn Neurosci 18:793–802

Rasch B, Buechel C, Gais S, Born J (2007) Odor cues during slow-wave sleep prompt declarative memory consolidation. Science 315:1426–1429

Rittenhouse CD, Stickgold R, Hobson JA (1994) Constraint on the transformation of characters, objects, and settings in dream reports. Conscious Cogn 3:100–113

Schenck CH, Bundlie SR, Ettinger MG, Mahowald MW (1986) Chronic behavioral disorders of human REM-sleep: a new category of parasomnia. Sleep 9:293–308

Steriade M, Mccormick DA, Sejnowski TJ (1993) Thalamocortical oscillations in the sleeping and aroused brain. Science 262:679–685

Sterpenich V, Albouy G, Darsaud A, Schmidt C, Vandewalle G, Vu TTD, Desseilles M, Phillips C, Degueldre C, Balteau E, Collette F, Luxen A, Maquet P (2009) Sleep promotes the neural reorganization of remote emotional memory. J Neurosci 29:5143–5152

Stickgold R, Hobson JA, Fosse R, Fosse M (2001) Sleep, learning, and dreams: off-line memory reprocessing. Science 294:1052–1057

Strauch I, Meier B, Foulkes D (1996) In search of dreams: results of experimental dream research. State University of New York Press, New York

Vertes RP (2004) Memory consolidation in sleep: dream or reality. Neuron 44:135–148

Wagner U, Gais S, Haider H, Verleger R, Born J (2004) Sleep inspires insight. Nature 427: 352–355

Walker MP, Stickgold R (2005) It's practice, with sleep, that makes perfect: implications of sleep-dependent learning and plasticity for skill performance. Clin Sports Med 24:301–317, ix

Walker MP, Stickgold R (2006) Sleep, memory, and plasticity. Annu Rev Psychol 57:139–166

Walker MP, Brakefield T, Morgan A, Hobson JA, Stickgold R (2002) Practice with sleep makes perfect: sleep-dependent motor skill learning. Neuron 35:205–211

Weisz R, Foulkes D (1970) Home and laboratory dreams collected under uniform sampling conditions. Psychophysiology 6:588–596

Zhuang Z, Gao X, Gao S (2005) The relationship of HRV to sleep EEG and sleep rhythm. Int J Neurosci 115:315–327

Chapter 6
The Substrate That Dreams Are Made On: An Evaluation of Current Neurobiological Theories of Dreaming

Janette L. Dawson and Russell Conduit

Abstract Theories regarding ambiguous consciousness states, such as dreaming, often attract questions regarding the scientific status of the experiments on which they are based. Rarely, however is the scientific status of the theory itself scrutinized. There are basic principles of theory construction that can provide a framework for evaluating current neurobiological theories of dreaming. This chapter places particular emphasis on the activation-synthesis (AS) and activation, input and modulation (AIM) models, developed by Hobson and colleagues over the past three decades (Hobson and McCarley, Am J Pschiatry 134:1335–1348, 1977; Hobson et al., Behav Brain Sci 23:793–842, 2000). This theory set was chosen as it can be considered one of the most widely cited and publicized neurobiological theories of dreaming today. Our aim in this chapter is not to criticize this work specifically, but to draw attention to the problems of theory development presently inherent in all dream research. The nature of the assumptions which underlie dream theories, the logic of argument, and the validity of methodologies used in collecting the empirical evidence, are scrutinized according to principles of theory construction and validity. We argue that modern theories of dreaming, whilst evolving ad hoc modifications in the face of new and sometimes anomalous evidence, are essentially unfalsifiable, and by definition do not qualify as scientific theories. However, as methodologies and technologies improve, particularly in the areas of sleep stage recording and neuroimaging, a new paradigm for the neurobiology of dreaming may emerge.

J.L. Dawson (✉) • R. Conduit
School of Psychology, Psychiatry, and Psychological Medicine, Faculty of Medicine, Nursing and Health Sciences, Monash University, Clayton, VIC 3800, Australia
e-mail: Janette.Dawson@med.monash.edu.au; Russell.Conduit@med.monash.edu.au

D. Cvetkovic and I. Cosic (eds.), *States of Consciousness*, The Frontiers Collection,
DOI 10.1007/978-3-642-18047-7_6,

6.1 Introduction

Neuroscience is "the collective sciences of brain and behaviour" (Rose 1998, p. 1). For both neuroscientists and philosophers, an intriguing question concerns the degree to which behaviour can be accounted for by neurobiological mechanisms. This question has stimulated much debate, particularly with reference to the problem of consciousness. The philosopher David Chalmers divided the problem of consciousness into "hard" and "easy" components. The "easy" components are those that are susceptible to explanation in terms of, say, computational or neurophysiological mechanisms, depending on the level of explanation required. The "hard" problem of consciousness concerns its experiential aspects – how consciousness emerges from neural substrates (Chalmers 1996). As yet, neuroscientists are still struggling with the easy problem, that is, the identification of the neural correlates of phenomenal experience and their possible underlying causal mechanisms.

Dreaming has been regarded by some as a special state of consciousness that may shed light on consciousness overall. Since the discovery in the 1950s of rapid eye movement (REM) sleep, which seemed to be a clear-cut neurophysiological correlate of the phenomenological experience of dreaming, much research has taken place to investigate the dimensions of this apparent correlation, with several theories subsequently put forward to explain the underlying mechanism. One of the most prominent researchers in the field has been J.A. Hobson. In collaboration with various colleagues, Hobson has proposed an influential and evolving neurobiological theory of dreaming, initially presented as the activation-synthesis (AS) hypothesis (Hobson and McCarley 1977) and later modified as the activation, input and modulation (AIM) model (Lim et al. 2007). This theory has prompted extensive research and reflection by others (e.g. Lim et al. 2007), and much supporting evidence for the AS/AIM dream theory has been cited by these authors (e.g. Hobson et al. 2000). This theory has also been widely disseminated in introductory psychology textbooks (Squier and Domhoff 1998).

Nevertheless, the AS/AIM theory has also been controversial. Key assumptions and evidence cited in its support have been disputed by numerous researchers (Antrobus 2000a, b; Solms 1997, 2000), who have advanced alternative accounts of the mechanisms of dreaming. Clearly, not all researchers in the field regard the AS/AIM theory as paradigmatic; and, if it was ever so regarded, there are those who believe that a paradigm shift is now warranted (Domhoff 2000).

Our aim in this chapter is to evaluate the AS/AIM neurobiological theory of dreaming. This particular work was not chosen based on its relative weakness as a dream theory. To the contrary, it was selected as it can be considered one of the most widely cited and recognized neurobiological dream theories today. Hence, our aim is not to criticize this work, but to draw attention to the problems of empirical investigation and theory development presently inherent in dream research, using this long standing and well known theory as an example. Although a vast amount of empirical evidence has been cited both for and against this theory, our focus in this chapter is not to demonstrate whether or not this evidence indicates support for its claims. Rather, non-empirical criteria for theory evaluation are outlined and subsequently utilised to assess AS/AIM as a theory. Evaluation will therefore proceed via analysis of the structure of the theory itself. We shall emphasise the nature of the assumptions which underlie the theory; the logic of its argument; the validity of methodologies used in its testing; and how the empirical evidence cited in its support is collected, analysed and interpreted. We shall propose that the AS/AIM theory, whilst evolving somewhat makeshift or ad hoc modifications in the face of new and sometimes anomalous evidence, still relies to some degree on early, inaccurate assumptions, and is still pervaded by an early conflation of causation with correlation. Along with these issues, methodological limitations inherent in all current investigations of dreams constrain the project of testing both theoretical assumptions and assertions. However, as methodologies and technologies improve, particularly in the areas of sleep stage recording and neuroimaging, finer-grained analysis of data and detection of correlations between variables will become possible, and new hypotheses will be generated to account for them. From this process of scientific bootstrapping, a new paradigm for the neurobiology of dreaming may possibly emerge.

6.2 Scientific Theories

6.2.1 The Nature of Scientific Theories

A scientific theory may be defined as an explanation that accounts for the covariance between two phenomena. This may consist of a proposed mechanism whereby one event causes another, which accounts for a contiguous covariation (Koslowski 1996). Theories may also account for the covariation of two phenomena by positing an underlying third variable. Thus, a neuroscientist might identify covariations between physiological variables, or between physiological and phenomenological variables. However, in searching for underlying mechanisms, distinguishing causation from correlation may be difficult, due to the reciprocal interactions between the brain and behaviour and the limitations in the methodologies used to observe the variables of interest.

Philosophers of science have argued that statements about observations of variables can only be made within the context of some sort of theory, even if low-level; that is, that all observations are already both theory-dependent and theory-laden (Chalmers 1982). In science, the methodology used to manipulate and/or collect observations about variables of interest will also be influenced, if not entailed, by the theoretical framework itself and the nature of the covariations to be thus investigated. The methodologies can be effective only to the extent that the theories about the phenomenon are more or less veracious. Additionally, only those observations pertinent to the theory will be recorded, so that if the theory is misleading, other potentially relevant factors may be passed over (Koslowski 1996). Meaningful arrangement and interpretation of findings will be theory-dependent also, for even if the findings are anomalous, they will be interpreted, initially at least, in relation to the original theory.

Theories, methodologies and observations are also interdependent (Koslowski 1996; Laudan 1984). The nature of observations gathered may suggest improvements to methodological instruments and practices. Empirical discoveries and new technologies may shape new methodological attitudes, enabling more rigorous testing of theories through more precise observations. Observations consistent with the theory will strengthen the theory's claim to veracity. Anomalous observations may lead to the articulation or modification of the original theory. Articulation includes revising, clarifying or refining a hypothesis in answer to new data, both congruent and incongruent; it may also involve identifying the limits of an explanation by specifying the situations to which the posited mechanism does or does not apply, or by specifying a variant on the mechanism under differing conditions (Kuhn 1996). Changes to a theory may suggest the use of other methodological instruments and practices in order to test new hypotheses generated by the modified theory. New sets of observations will then feed back at the theoretical and methodological levels, a process that can be described as scientific bootstrapping (Koslowski 1996).

As Koslowski argues, science may be said to progress by this process of bootstrapping when theory, method and observations enhance each other over time (Koslowski 1996). Thus, if a working hypothesis derived from a theory is supported by the observations collected, thereby supporting the existence of the posited mechanism, then this mechanism may suggest likely additional covariations. These in turn help to refine the working hypothesis about how the mechanism may work, and further generating predictions about the conditions in which various covariations will occur. Over time, an increasingly comprehensive account, or theory, of such covariations will thereby evolve (Popper 1968).

Of course, the opposite case may also occur. That is, when an inaccurate theory constrains the methodologies and potential observations, so that relevant factors are missed or dismissed as irrelevant because they do not fit into a theory's conceptual field, or when inadequate methodologies limit the usefulness of any observations made, the correction or disconfirmation of the original theory is hampered. This may set limits to a theory's generalisability; more seriously, it may constrain the growth of a whole scientific field if the theory is influential (Popper 1968).

6.2.2 Evaluating Scientific Theories

A scientific theory may be evaluated according to various criteria. Initially, one can assess the structure of the theory as a whole in order to determine its success as an *argument*. Thus, is the logic of the theory valid? Does it follow principles of inference if causal mechanisms are posited? Is the inference of causation from purely correlational data avoided? Are the assumptions of the theory reasonable, justified and accurate in themselves? Do the methodologies used in testing the theory conform to generally accepted scientific standards, such as the use of reliable and valid instruments?

A theory may also be assessed in regard to how problematic data has been managed. Truly anomalous findings that *falsify* the theory must lead to outright rejection of a theory (Popper 1968). But it may be justified to question or reject the *data* outright, especially when the theory has already gathered a layer of supporting findings, if its explanatory mechanism seems plausible and persuasively explains the data that inspired the original theory. Also, theories may not be precise propositions that can be clearly *falsified* (Popper 1968), but may be working hypotheses, so that it is appropriate to modify, or articulate, rather than reject when confronted with anomalous data (Kuhn 1996). Such decisions concerning the rejection or revision of (and the nature of any revisions to) a theory are of themselves highly theory-dependent. The appropriateness of such modifications can be judged by whether they are theoretically motivated or makeshift ad hoc modifications. Modifications are theoretically based rather than ad hoc when anomalous data is accounted for by plausible mechanisms that fall within the basic explanatory framework of the original theory. An example might be when a pattern in the anomalous data refines the understanding of the original mechanism by indicating another mechanism or variable that constrains it (Koslowski 1996). An ad hoc modification might consist of the invocation of another mechanism to account only for the anomalous data, so that the theory becomes potentially unfalsifiable (Popper 1968). Theoretically motivated modifications are seen as appropriate because they represent a strategy for fine-tuning or amplifying understanding of relevant causal factor(s), part of the process of scientific bootstrapping (Koslowski 1996). A theory, or sequence of theory-articulations, may also be evaluated as to the progress made relative to the set of aims it was devised to achieve, or as to progress made toward realising or achieving a desirable goal state (Laudan 1984).

The criteria discussed above are those used in the remainder of this chapter in order to evaluate the AS/AIM neurobiological theory of dreaming. A brief summary of this theory and its development follows.

6.3 The Evolution of the AS and AIM Neurobiological Theories of Dreaming

6.3.1 Historical Background

Rapid eye movement (REM) sleep was first described scientifically by Aserinski and Kleitman (1953). Periods of rapid, jerky and simultaneous eye movements were observed over the night in sleeping participants, accompanied by a distinctive low-voltage electroencephalogram (EEG) and a high respiratory rate compared to periods of ocular quiescence. When woken during these periods of eye movement, participants reported dreams containing visual content in 20 of 27 awakenings. There was no dream recall in 19 of 23 awakenings from non-REM (NREM) sleep (Aserinski and Kleitman 1953). Dement and Kleitman (1957) then used a larger sample and found 152 instances of dream recall and 39 of non-recall from REM sleep and 11 dream recalls and 149 non-recalls from NREM sleep. The next year, Dement described sleep similar to human REM sleep in adult cats, with a low voltage, fast rhythm EEG, and rapid eye movements (Dement 1958). Jouvet and Jouvet observed EEG, behavioural, structural and functional similarities between the stage of sleep with rapid eye movements in humans and the same stage in cats, and concluded that they depend upon the same neural structures in a study published in 1963. The authors stated that they believed that "the very surprising similarity between this stage of sleep in man and cat makes possible the correlation between a structural analysis of cat (*sic*) and the subjective data obtained in man" (Jouvet and Jouvet 1963).

6.3.2 The Activation-Synthesis Model of Dreaming

In the activation-synthesis (AS) model of dreaming developed by Hobson and McCarley (1977), REM sleep was assumed to be the physiological substrate of dreaming. REM sleep was observed to occur in both humans and cats, and, because they (then) had no evidence of significant differences between human and cat brains, they felt justified in using the cat for their "study of the brain as a dream process generator, whether or not cats dream" (Hobson & McCarley 1977, p. 1338). The AS model was principally based on microelectrode recordings of ponto-geniculate-occipital (PGO) waves in cats, noted to take place primarily in REM sleep. Hobson and McCarley observed eye muscle activity occurring several milliseconds before PGO waves registered at the occipital cortex, arguing therefore that eye movements and pontine activity preceded REM cortical activity, rather than being a product of it. Thus, visual cortical events were determined by events in the oculomotor brainstem, with PGO waves transmitting information from the pons to the occipital cortex. Since PGO waves were related to the activation of these cortical visual processing areas, it was inferred that the PGO pathways also conveyed pseudosensory visual information to the forebrain, which "may be making the best of a bad job in producing even partially coherent imagery from the relatively noisy signals sent up to it from the brainstem" (Hobson & McCarley 1977, p.1347). Speculating that access to memory is facilitated during dreaming sleep, "best fits to the relative inchoate and incomplete data provided by the primary stimuli are called up from memory... The brain, in the dreaming sleep state, is thus likened to a computer searching its addresses for key words" (Hobson & McCarley 1977, p.1347), and then synthesising the dream content from these disparate elements.

More recent neuroimaging and lesion data has been used to enrich the detail of the AS model in regard to the neuroanatomical bases of cortical synthesis (Hobson et al. 2000). Such neurobiological dream research supports the principle of brain-mind isomorphism, where dream activity is presumed to have underlying brain activity. This research has provided analysis of the "forms of dreams which might be expected to have their roots traced to isomorphic forms of brain activity" (Hobson et al. 2000, pp. 822–823). Lesion study data and neuroimaging studies identifying areas believed to be involved in dreaming and their putative functions in waking are used to account for particular dream features. For instance, the deactivation of the prefrontal cortex (believed crucial to executive functioning) (Maquet and Phillips 1998; Maquet 2000) and REM-associated increase in activation of the limbic structures (associated with emotion) (Braun et al. 1997) "greatly enrich and inform the integrated picture of REM sleep dreaming as emotion-driven cognition with deficient memory, volition and analytic thinking" (Hobson et al. 2000, pp. 808–809).

6.3.3 The AIM State Space Model of the Brain/Mind

The AS model of dreaming has been updated and extended, transmuting into the AIM state space model of the brain-mind isomorphism, wherein a conscious state may be understood as a point in a three-dimensional state space, with three axes, A, I, and M (Hobson et al. 2000). The level of brain activation, A, may be defined as the mean firing frequency of brain stem neurons, reflected by levels of high and low frequency in the EEG. The high levels of A in REM sleep are "a correlate of the mind's ability to access and manipulate significant amounts of stored information from the brain during dream synthesis." Input source, I, is a measure of how much sensory data being processed originates from external or internal sources (for example, internally generated REM sleep events such as PGO waves replace blockaded external sensory input, activating sensory and affective centres, which then prime the cortex for dream construction via stimulation of the visual cortex by PGO waves during REM sleep). In REM sleep, this can be estimated from the density of eye movements, because this is postulated to be a reflection of brainstem PGO and motor generator activity. Neuromodulation, M, relates to the levels of the brainstem aminergic neurotransmitters noradrenaline and serotonin, which are released by cholinergic PGO burst cells in the brainstem. The AIM model posits that moving to the cholinergic state in REM alters the mnemonic capacity of the brain-mind and reduces the reliability of cortical circuits, increasing the likelihood of bizarre temporal sequences and associations accepted as reality when dreaming. Other neurotransmitters such as dopamine and histamine may also determine the mode of information processing in various states of consciousness (Hobson et al. 2000).

The AIM model has been used to describe waking, REM and NREM sleep states as follows:

If *A*, the electrical activation is high, the brain will be in waking state or REM sleep.
If *I*, the input-output gating is high, the brain will be in REM sleep. If *I* is low, the brain will be in waking.
If *M*, the aminergic to cholinergic modulatory ratio is high, the brain will be in waking state. If *M* is low the brain will be in REM sleep.
If all these three functions are at intermediate levels, the brain will be in NREM sleep (Hobson 1997, p. 393).

This updated AS/AIM theory retains most of the earlier assumptions of the original AS model in regard to dreaming: that the brainstem generates both REM sleep and dreaming; that REM sleep is the chief sleep correlate of dreaming; and PGO waves are still seen both as reflections of REM brain activity and as causal to dreams due to their stimulation of the cortex (Hobson et al. 2000).

An exhaustive list of studies has been cited as providing convergent evidence for the AS/AIM theory (Hobson et al. 2000). Most physiological evidence is derived from studies using animal models, particularly the cat (Hobson et al. 1998). Dream evidence has been derived from self-reports of participants in studies rated for phenomenological variables according to a variety of scales (Hobson and Stickgold 1995). Correlation of dream occurrence has been made with sleep stages, usually determined using the sleep scoring criteria developed in 1968 by Rechtschaffen and Kales. Other convergent evidence has been derived from the findings of lesion studies and recent neuroimaging research (Hobson et al. 2000).

6.4 Evaluation of AS/AIM Neurobiological Theories of Dreaming

6.4.1 The Structure of the AS Model of Dreaming

The argument of the original AS theory ran as follows:

1. The brainstem activity that results in REM sleep in cats also results in accompanying PGO waves that may reflect concurrent processes occurring within the brain
2. Cats and humans share similar characteristics in REM sleep
3. Human dreaming occurs only during REM sleep.

   Therefore:

4. The brainstem activity that results in REM sleep in humans also results in PGO waves, which may be the reflection of concurrent processes realised in the human brain as the dream.

The first and second premise of this argument still underlies the updated version of the AS/AIM dream theory. The first premise does not include consideration of other putative roles for PGO waves in sleep (e.g. alerting: Bowker and Morrison

1976; Sanford et al. 2001). The second premise relies upon the validity of animal models of human neurobiology, and on the validity of mapping from the neurobiology of the animal model onto corresponding human neurobiology and phenomenology research. These issues are discussed in detail in the following sections.

6.4.2 Assumptions Underlying the AS/AIM Models of Dreaming

6.4.2.1 Alternative Hypotheses for the Role of PGO Waves

There are other hypotheses concerning the possible role of PGO waves. For example, in both awake and sleeping cats, PGO waves can be elicited by presenting loud tones, suggesting that these waves are related to an alerting response (Bowker and Morrison 1976; Sanford et al. 2001). Therefore, PGO waves, whether occurring in animals or in humans during REM sleep, indeed may not be reflections of dream processes at all, but merely reflect the processes related to REM sleep (Solms 2000).

6.4.2.2 The Use of Animal Models in Sleep and Dream Research

Theories about human physiology can be neither confirmed nor countered by studies using animal models; only studies using humans can ultimately test such theories. Animal models can only be *suggestive* of hypotheses that may be applicable to humans (Sarter and Bruno 2002). However, given that ethical considerations proscribe the use of humans in much experimental research in neuroscience, animal models have of necessity been integral to theory building and testing. As such, any animal model used in building and testing a theory becomes itself integral to that theory. Therefore the validity of the use of a particular animal model becomes an important issue in evaluating the merits of a theory.

Face validity, whilst straightforward, is usually regarded as the weakest form of validity. In animal modelling, face validity comprises the perceived similarity between the construct to be modelled and its appearance in the animal model. The use of the cat as an animal model was, and remains, ubiquitous in research into human sleep, and by extension, into human dreaming (Gottesman 2001). Rationales for using the cat as a model for human sleep include the behavioural similarity of the human and cat in REM sleep, such as rapid eye movement and loss of muscle tone, thus suggesting by analogy that the neuronal activity changes underlying the state of REM sleep in cats also underlie those in humans (Hobson et al. 1998).

However, many of the anatomical structures related to REM sleep processes in cats are arranged differently in humans (and other species) (Siegal 1999) and it is not yet known if REM sleep active cell patterns in the cat occur in humans. Furthermore, a key feature of the AS/AIM dream theory – the PGO waves of

REM sleep found in the cat (Hobson et al. 2000) and postulated to reflect dream generation – are yet to be convincingly demonstrated in humans (Hobson et al. 2000; Lim et al. 2007). Of course, this also reflects the fact that no ethically viable, non-invasive way of detecting possible PGO wave activity in humans has yet been devised (Wehrle et al. 2007).

Although it is quite possible that cats and other animals do dream, animal studies probably will not be able to yield precise information about dream mentation, nor whether it occurs only in REM sleep. For example, Hartmann (1967) describes stimulus deprivation experiments where rhesus monkeys were conditioned to press a bar whenever they saw a visual image on a screen in front of them. The animals were not conditioned to a specific image, but to a variety of different images presented in dark conditions, so that, essentially, they were trained to press a bar whenever they saw something in front of them. During these experiments it was noted that the monkeys actively pressed the bar during REM sleep. These results suggest that visual imagery, of some form, does occur during sleep in these animals. However, whilst these results do provide a basis for a possible methodology for establishing the presence of perceived visual imagery in animals, non-human animals arguably do not share the same levels of conceptual, emotional and linguistic processing with humans, and the functional anatomy and physiology of any dream mentation, even in related animal species, would be both quantitatively and qualitatively different to that of humans. This makes any precise mapping of a theory about the biopsychological mechanisms of sleeping cognitive activity in animals onto a theory about the biopsychological mechanisms of verbal dream reporting in a humans problematic. The construct validity of animal models in neuroscience can refer to the mapping of a theory about the biopsychological mechanisms of a phenomenon in humans onto a theory about the biopsychological mechanisms of that phenomenon in the animal model (Sarter and Bruno 2002). However, the AS/AIM theory can be considered to be an attempt to map incomplete animal models of REM sleep biology onto the phenomenology of human dreaming. Therefore, the currently weak biopsychological construct validity of the cat and other animals as models for human dreaming undermines the PGO-dreaming relationship claims within AS/AIM dream theory.

6.4.2.3 The Assumption that Dreaming Occurs Only in REM Sleep

The third premise in the original AS theory (that human dreaming occurs only during REM sleep) relied upon the initial assumption of early dream researchers (particularly before 1962) that REM sleep and dreaming were isomorphic (Nielsen 2000). Initiated by investigation of NREM dreaming by Foulkes (1962), this assumption came under fire from researchers whose data showed that dreams could also be elicited regularly from awakenings in NREM sleep (Nielsen 2000). However, theoretical orientation can influence the methodologies used in collecting and interpreting sleep mentation, and can thus weaken its external validity. This issue is discussed below.

General Considerations Regarding the Validity of Dream Recording Methodology

The primary underlying threats to the validity of reports of sleep mentation or dream phenomenology stems from three aspects of dream reporting: dream reports are self-reports, made retrospectively and in an awake state. Self-reports may be tainted by respondents' tendencies to say socially desirable things about themselves (Edwards 1990), so that dream self-reports may not provide full accounts of dream material. Because dream accounts are made retrospectively, it may be that the dreams that are recalled may be those with intrinsically more memorable imagery, such as bizarre elements (Cipolli et al. 1993), so that those recorded are not an adequate sample of dream mentation. The ability to recall dreams may be sleep-state dependent (Conduit et al. 2004). In other words, enhanced recall from a particular sleep state may be due to factors inherent in that state such as the level of arousal and therefore of attentional mechanisms (Conduit et al. 2000), or may be due to the capacity available to encode and store the dream in memory so as to enable subsequent recall (Farthing 1992). It is also possible that dream reports are constructed during the awakening process itself, although the subject attributes the dream to sleep (Feinberg and March 1995), or that dream reports mix both dream and waking state phenomenology (Hobson et al. 2000). Dream reports are also limited by the linguistic mode of their report, and the linguistic abilities of participants to report multi-dimensional aspects, such as visual and motoric imagery (Hobson et al. 2000). Can a dream report therefore be tantamount to a dream? No matter how veridical a dream report may be about the phenomenological experience of the dream, it is not a *dream* itself. Therefore, direct conclusions about dreams themselves cannot be drawn from *reports* of dreams. However, at present, retrospective self-reports by awake participants are probably the closest researchers can come to the phenomenological data of sleep mentation and dreaming.

The Validity of Dream Recording Instruments

At the instrumental and procedural level of data collection, differences in researchers' aims and theoretical positions will determine the methodology used to collect dream data in a study, the nature of the data collected, and its utilisation in theory testing. Is the domain to be sampled to be all sleep mentation, or just dreams? If only dreams are of interest, what definition of "dream" will be employed? This definition will be the most important factor in constructing data-collecting instruments such as questionnaires and scales (Farthing 1992; Nielsen 1999).

Sleep mentation attributes have ranged through thinking, reflecting, bodily feeling, vague and fragmentary impressions, sensory hallucinations, visual imagery, emotionality, temporal progression, plot continuity, complexity, to bizarreness, with varying cut-off points distinguishing dreams from thinking. In three studies reviewed by Nielsen (1999), the different degrees of stringency used in dividing dream from non-dream mentation in terms of the detail of description varied inversely with the percentage of dream recall attributed to participants' NREM awakenings; these ranged from 7 to 62%.

For some, any sleep mentation accompanied by perceptual images may be classified as a dream, leading to some asserting that dreaming, of some form, occurs through all stages of sleep (for example, Foulkes (1962), who classified 82% of REM reports and 54% of NREM reports as dreams). Orlinsky (1962) used an eight-point scale to classify sleep-mentation reports, which, for example, ranged from: (0), subject cannot remember dreaming; no report on awakening (43%); to (4), subject remembers a short but coherent dream, the parts of which seem related to each other: for example, a conversation rather than a word, a problem worked through rather than an idea, or purposeful rather than fragmented action (14%); to (7), subject remembers an extremely long and detailed dream sequence of five or more stages; or more than one dream (at least one of which is rated 5) for a single awakening. (Orlinsky (1962); includes percentage of NREM reports from his study of 400 reports from 25 subjects.)

Many researchers classify reports on Orlinsky's scale equivalent to 4 or above as dreams. Orlinsky's data was re-analysed by Farthing (1992), using this definition, with the result that 85% of REM reports were classified as dreams, as were 27% of NREM reports. Clearly, dream definition has a huge bearing on the REM-NREM dreaming debate.

AS/AIM theory is based on a narrower definition of dreaming than described above (in line with the prevalent assumption that dreaming only occurred with REM physiology). In 1977, with the introduction of the AS hypothesis (Hobson 1997) dreaming was defined as:

> ... a mental experience, occurring in sleep which is characterized by (1) hallucinoid imagery, predominantly visual and often vivid; by (2) bizarre elements due to such spatiotemporal distortions as condensation, discontinuity, and acceleration; and by (3) a delusional acceptance of these phenomena as 'real' at the time they occur. (4) Strong emotion may or may not be associated with these distinctive formal properties of the dream, and (5) subsequent recall of these mental events is almost invariably poor unless an immediate arousal from sleep occurs (Hobson 1997, p. 1336, also quoted in Farthing 1992, p. 296).

6.4.3 Dream Mentation in the Updated AS and AIM Models

The AS hypothesis inherently implied that NREM mentation was not dreaming, and thus an explanation of NREM mentation was not provided in the original AS theory (Hobson 1997). In the light of evidence provided by other investigators using scales such as the one by Orlinsky described earlier, it is now conceded in current AS/AIM theory that dream reports may be elicited from other sleep stages, including quiet wakefulness, sleep onset and NREM sleep itself. However, it is still argued that there are significant differences between dream reports obtained from REM and NREM sleep, the most salient being quantitative differences, particularly the longer dream reports obtained after REM awakenings indicating a greater number and intensity of dreamlike features of these dreams, and requiring more words to relate (Hobson et al. 2000). The current definition of dreaming forming the basis of AS/AIM is:

> Mental activity occurring in sleep characterised by vivid sensorimotor imagery that is experienced as waking reality despite such distinctive cognitive features as impossibility or improbability of time, place, person and actions; emotions, especially fear, elation and anger predominate over sadness, shame and guilt and sometimes reach sufficient strength to cause awakening; memory for even very vivid dreams is evanescent and tends to fade quickly upon awakening unless special steps are taken to retain it (Hobson et al. 2000, p. 795).

The AS hypothesis was originally developed within its own paradigmatic definition of dreaming as REM dreaming; by definition, the model could not apply to NREM mentation. The new definition of dreaming quoted above was part of modifications to the theory in response to anomalous data of NREM dreams. The new definition, while not specifying that it was synonymous with the REM dreaming of the early theory, still emphasises those features most common to dreams during REM sleep. By contrast, it is claimed that NREM reports are far more likely to be short, dull and undreamlike (Hobson et al. 2000). Further, it is argued that the dream features specified in the new definition of dreams occur in REM dreams, and that most REM dreams contain these features, whereas they claim that this is rarely the case in NREM mentation. This is stated as "the empirical basis of our contention that all of these features will one day be explainable in terms of the distinctive physiology of REM sleep" (Hobson et al. 2000, p. 799). Thus, in response to findings that some dreaming takes place during NREM sleep, NREM dreamlike mentation was then incorporated into the AS/AIM theory as due to a possible admixture of REM-like phenomena occurring, for instance, in late night stage 2 NREM, or to reactivation occurring within NREM in anticipation of the next bout of REM (Hobson et al. 2000).

In a more recent empirical study by Fosse et al. (2004), REM and NREM reports collected at home were scored for the presence of hallucinations and directed thinking. From this data it was found that generally hallucinations were more frequent in REM sleep and directed thinking was more frequent in NREM for the first 5 h of the night. However, as the night progressed NREM showed an increase in hallucinations and a decrease in directed thinking. This change was so pronounced that late night NREM reports were described as indistinguishable from early REM reports. This then led these researchers to conclude that "as the night progresses, NREM approaches the neurocognitive characteristics of REM" (p. 302). This finding was not directly interpreted in terms of AS/AIM theory by the authors. However, the conclusions of this paper were largely orientated towards explaining NREM dreams as REM-PGO processes occurring during NREM sleep. This therefore can be considered synonymous with Nielsen's (1999, 2000) concept of "covert" REM during NREM sleep as being responsible for NREM dreams (Hobson et al. 2000).

The search for a PGO-correlate during NREM sleep began decades ago, with little relative success in providing consistent indicators of dreaming during NREM sleep in humans (Pivik 1991). In its original conception, the AS hypothesis predicted dreaming was a product of REM sleep processes. Soon after NREM dreaming was accepted, NREM dreaming was then conceptualised as occurring as a product of phasic REM (Pivik 1991) or "covert" REM processes (Nielsen 1999). However, since phasic REM (and NREM) activity in humans is yet to be successfully confirmed as occurring with PGO activity, and "covert" REM is yet to be defined in measurable terms, the current AS/AIM account of NREM dreaming is essentially unfalsifiable, and can be considered ad hoc, and unjustified (Popper 1968). Furthermore, in terms of definitions of dreaming, it can be argued that current AS/AIM theory continues to emphasise differences between NREM and REM mentation within the previous constraints of the AS model in a theory-dependent way. Thus, the definition of dreaming is so stringent that NREM mentation is still mostly, *by definition*, not dreaming, a move which again makes the theory, through employing such ad hoc modifications, potentially unfalsifiable and thus unjustified (Popper 1968). On a more positive note, this modification at least retains the same explanatory mechanism as the original theory, and therefore may be justified on these grounds (Koslowski 1996). This also points to a possible way forward, past the problem that afflicts all theories of dreaming which attempt to link dream mentation to specific stages of sleep: the out-dated, inaccurate and uninformative method of sleep stage scoring used by many dream researchers. The following sections draw upon a review of this issue by Himanen and Hasan (2000), highlighting only a few of the many problems in sleep stage scoring canvassed in that article.

6.4.4 Methodologies Used in Testing the AS and AIM Models

6.4.4.1 Sleep Stage Recording

The standard method of sleep stage scoring, used in most sleep and dream research, is that of Rechtschaffen and Kales, developed in 1968. According to this schema, sleep is divided into stages monitored by electroencephalogram (EEG; the record of the waves of electrical energy detectable at the scalp), electro-oculogram (EOG; the record of eye movements), and electromyogram (EMG; the record of muscle activity, usually under the chin). Sleep is divided into rapid eye movement (REM) and non-REM (NREM) sleep, with NREM further divided into four stages (Rechtschaffen and Kales 1968).

In practice, distinguishing the stages is mainly done by reference to criteria applied to the EEG. Stage 1 (S1) shows a slight slowing of the EEG relative to waking; stage 2 (S2) is characterised by high amplitude K complexes (high-voltage biphasic waves) and low amplitude clusters of spindles (12–14 Hz sinusoidal waves); whilst stages 3 and 4 (S3 and S4) both exhibit the high amplitude, slow

(0.5–2 Hz) delta waves that dominate stage 4. REM sleep is distinguished by its frequent fast eye movements, loss of muscle tone, and a stage 1 low-voltage EEG pattern. Wakefulness is also defined by EEG criteria (Rechtschaffen and Kales 1968).

The rules standardised in a manual edited in 1968 by Rechtschaffen and Kales (1968) (hereafter referred to as the R&K manual) were originally meant as a reference method, to be revised with the accumulation of new knowledge. But this system became and has remained the "gold standard" and the only method of sleep analysis in general use, despite the fact that it was based on the knowledge and technology of that time. This has necessarily restricted development in sleep (and therefore dreaming) research (Himanen and Hasan 2000). Increasing dissatisfaction with the R&K manual has focused primarily on the belief that scoring does not capture any of the microstructure of sleep, for several reasons.

Spatial and Temporal Limitations of Traditional EEG Techniques

In the basic montage recommended by R&K, only the EEG, eye movements and muscle activity of the sleeper are recorded. However, traditional EEG technology and practice have severely limited its spatial and temporal resolution (Gevins et al. 1999). The EEG as specified in the R&K manual has a single derivation, at C4/A1 or C3/A2. These regions were thought at the time of the R&K manual development to have synchronous EEG patterns, although no confirmatory studies took place at the time. It has since been shown that different levels of detection of slow wave activity are recorded from each site (Himanen and Hasan 2000). Such EEGs will also have limited specificity – spatial information cannot be taken into account since electrical potential changes often present a summation of remote events (Kertesz 1994). The temporal limitation inherent in the R&K system is that scoring is done in epochs of equal length (either 20 or 30 s), and each epoch is allocated to one stage. If an epoch contains elements of two stages, it is recorded as that of longest duration, although it may be difficult to ascertain which stage is predominant; while short state changes, when two or three electrophysiologically different states may manifest within a single epoch, are ignored. This low temporal resolution was due to the standard size of one paper EEG page, a limitation clearly technologically indefensible nowadays, with the advent of digital recording techniques (Himanen and Hasan 2000).

Since then, EEG technology has made several advances, by using more electrodes at up to 128 locations, by using signal-enhancing methods, by development of improved source analysis to distinguish components of activity summated from multiple sources in the brain, and by coregistration of data with images obtained by means of other techniques, such as PET, MEG and MRI (Gevins et al. 1999), which clearly would be of assistance in analysing sleep architecture.

In summary, defining sleep architecture in terms of five stages, in units of 20–30 s, is too coarse a description for modern research, and overlooks fundamental information about the microstructure of sleep.

Ambiguity of Rules for Scoring Sleep Stages

There is much evidence to suggest that sleep stages, as currently defined, are not homogeneous, and that division of these at the very least into subcategories is warranted. Observed inconsistencies within stages are often obscured by the R&K manual's rules for scoring stages containing anomalous sequences. These require considerable subjective judgement on the part of scorers, leading to large intra- and inter-observer variability (Himanen and Hasan 2000).

Another problem is that the rules for scoring REM according to the R&K manual are complicated and ambiguous. Alpha activity is common in REM, but no guidelines exist for their scoring – some such epochs may be scored as wakefulness, others as REM sleep. Often the scoring of several epochs in succession is dependent on the change in just one variable. For example, a long stretch of low-voltage EEG may be scored as S1, S2 or REM, depending on the appearance of just one spindle, K complex or REM, yet these could potentially represent some intermediate stages, not yet defined (Himanen and Hasan 2000).

Clearly, the reliability of currently accepted methods of scoring sleep stages may be quite weak, making comparisons across studies and laboratories problematic. If sleep stages are redefined in the future to include sub-stages, or are abolished in favour of another type of classification altogether, then the REM-covert REM/dream isomorphism debate may well disappear as more fruitful and specific correspondences become possible. Patterns of sleep mentation variables may be investigated in relation to features of sleep micro-architecture, through relating sleep mentation to underlying physiological variables. Such investigations would then allow for better empirical testing of phasic or "covert" REM processes purported to underlie the activation-synthesis and AIM models of dreaming (Hobson et al. 2000).

6.4.5 The Use of Convergent Evidence for the AS and AIM Models of Dreaming

It has been argued that the definition of dreaming employed in the updated AS/AIM theory can be considered to be somewhat post hoc, potentially unfalsifiable and thus unjustified. However, recent lesion and neuroimaging evidence has been used to provide convergent evidence for the AS/AIM model, with the authors claiming that this dream definition "lends itself admirably to neurocognitive analysis" (Hobson et al. 2000, p. 795). Thus, nominated features of dreams, such as heightened emotion, may be linked to the activation of brain regions such as the amygdala (an area of the brain implicated in human emotion) reported from neuroimaging studies of REM sleep (Miyauchi et al. 2009). However, there are currently certain methodological limitations inherent in both lesion and neuroimaging studies that may impose limits on their use in this way as supporting evidence.

6.4.5.1 Lesion Studies: Developments and Limitations

Whilst investigations of brain lesion deficits in humans have provided much information about functional neuroanatomy, it has become apparent that caution must be exercised when interpreting data from these studies. As Kertesz (1994) notes, it may be difficult to determine whether any behavioural, cognitive or phenomenal deficits should be ascribed to damage to or destruction of the affected area, to the disconnection of two areas that are part of a network, or to the interruption of any longitudinal pathways which may span the area. Lesions may also produce distant suppressive consequences in functionally related structures – an effect known as *diaschisis*. Alternatively, positive symptoms may become apparent, the outcome of neural activity previously inhibited in the intact brain (Kertesz 1994). Brain plasticity may allow structural and functional reorganisation to provide compensatory pathways, and there may also be deliberate adoption of alternative cognitive strategies for particular tasks. Organisms may use redundant structures, systems or networks for a function, so that the same deficit may therefore arise from lesions at different locations (Bradshaw and Mattingley 1995).

Other methodological problems may also arise because inter-individual variation in neuroanatomy may make it difficult to match for brain lesion location within groups. Even studies involving lesions of the same size or location may not produce the same deficit(s), because different forms of neurological damage (e.g. oedema, haemorrhage, vascular occlusion, tumour growth) may produce different neurological effects (a confounding factor often ignored in neuropsychological research, Kertesz 1994). Such deficits may be unstable, and resolve over time. As recovery occurs, compensatory changes by the whole brain or homologous areas may progressively take place, so time from onset is a crucial variable in lesion-deficit studies (Kertesz 1994). Therefore, theories that rely heavily on data from lesion studies should consider the methodological limitations inherent in these studies, and such data should be viewed cautiously and in the light of other convergent evidence.

6.4.5.2 Neuroimaging: A Developing Technique

Neuroimaging studies have also been utilized as converging evidence for the AS/AIM theories (Hobson et al. 2000). However such research has some inherent methodological difficulties that contrive to make both the testing of hypotheses and comparisons across studies problematic at present. These difficulties arise from three sources: the various levels of validity of the assumptions underlying neuroimaging technology; the methodologies of the technologies; and limitations currently inherent in the technologies.

Validity of Assumptions Underlying Neuroimaging Technology

Positron emission tomography (PET) and functional magnetic resonance imaging (fMRI) are functional imaging techniques that rely upon detecting alteration in regional blood flow or metabolism in the brain. The principle assumption in functional imaging is that active neurons require more oxygen and glucose, and so the brain increases blood flow to these areas to satisfy that need. Increases in blood flow apparent in an area are therefore thought to correspond to increases in synaptic activity in that area. PET monitors the increase in cerebral blood flow (CBF), and in glucose and oxygen consumption, while fMRI detects the degree of blood oxygenation (Magistretti and Pellerin 1999). However, the inter-relationships between mental activity, neuronal activity and blood flow are complex. Blood flow or flow related phenomena may be due to activity of neurons involved in either excitatory or inhibitory processes. More specifically, researchers using large scale neural modelling have provided data that suggest that neural inhibition can either *raise or lower* brain imaging values, depending on levels of local excitatory recurrence or whether the region is being driven by excitation (Tagamets and Horwitz 2001). Therefore, it is possible that the activation seen in functional imaging studies of sleep, while corresponding to *neuronal* activation, do not reflect activation of a specific anatomical region in functional terms (nor need they correlate with the activation aspect of the AS and AIM theories). Additionally, some researchers have argued that greater metabolic activity within a brain region may indicate poor neural efficiency and dysfunction (Bassett and Bullmore 2009; Parks et al. 1989). Thus, greater metabolic activity during sleep might indicate regions of dysfunction or inefficiency of neural systems, rather than indicating areas important for sleep functions.

Validity of Methodologies Used in Neuroimaging

Neuroimaging generates extensive, diverse, complex data, the interpretation of which involves a network of assumptions at every stage. Initial interpretation of this data depends on the system of *informatics* utilised. For example, in MRI, informatics is the network of decisions and assumptions that underlie the design and method by which the imaging instruments capture signals generated by the brain. It includes the specification of the behavioural or cognitive tasks or circumstances to be utilized, as well as the protocols for reconstructing the resulting signals into a three-dimensional representation of the brain, for correcting for and suppressing noise (such as that created by subjects' movement), and for the statistical analysis of the data which enables visualization of the output. Each set of methods will have inherent assumptions and limitations that need to be made explicit, and taken into account in interpreting data; as a consequence, it is difficult to make rigorous comparisons across studies, to conduct meta-analyses and to use diverse findings for theory building, confirmation or refutation (Squier and Domhoff 1998).

In particular, two methodological procedures used in neuroimaging have ramifications for sleep and dream research. First is the selection of a baseline against which regional activations are measured. The control resting condition often used has been that of having the subject lie still in the dark with eyes closed. However, subjects may vary in mental conditions, through boredom, anxiety and so on. Parieto-frontal hyperactivity is a common pattern in the waking state, and is subtracted from the sleep distribution (Maquet and Phillips 1998). If one uses such a waking activation pattern as a baseline, and subtracts it from REM or NREM activation patterns, then what one is left with – the differences in activation patterns – are presumably related to *differences* in the awake and sleep states. But this same subtraction also presumably removes what is *common* to both states, including patterns related to conscious states – whether awake or sleep/dream mentation – which may in fact be the very data one is after.

Secondly, comparisons between states are made by using data averaged across subjects for each baseline and state measured, or by comparison to statistically defined functional maps of the brain based on several subjects. But anatomic variability between individuals can create inaccuracies in intersubject averaging, and outputs tend to reflect the changes that are the most consistent and salient across subjects, perhaps losing other, more subtle, regional changes (Berman et al. 1995). Both of these methodologies, using averaged results, could therefore tend to confound brain activations and sleep stages, especially in the light of the problems in sleep stage scoring discussed earlier. Sleep stage scoring may not necessarily capture fine-grained microstructural distinctions, thus averaging takes place across (micro- and sometimes macro-) sleep states and individuals simultaneously.

Technical Limitations of Neuroimaging

There are currently several other limitations of functional neuroimaging. Firstly, functional imaging does not capture dynamic processing. It depicts sets of activated areas, but gives no information about the flow between them, or other areas; foci of activation may represent only parts of large scale-networks that are converging in one particular area. Secondly, the relationship between rCBF measures and cognitive processing may well be different for more or less frequent processes (Kosslyn 1999). Thirdly, the spatial resolution of PET imaging is relatively poor – 5 mm at best, and its temporal resolution of 1–2 min (Braun et al. 1997) is a relatively long time in terms of cognitive processes, and does not allow the isolation or resolution of multiple phenomenal or physiological components (Berman et al. 1995). In sleep or dream research, these factors currently make precise correlations of PET or MRI activation patterns with other physiological or phenomenological variables unfeasible and rule out any determination of causation.

Neuroimaging Can Only Be Useful if Applied in the Correct Context

Dreaming is now accepted within AS/AIM theory as occurring within NREM sleep (and can be indistinguishable from early REM dreams, Fosse et al. 2004). However, neuroimaging evidence cited in support of this theory draws entirely from studies investigating sleep stage differences in activation between REM and NREM sleep (Hobson et al. 2000). As Solms (1997) has previously highlighted, any future neuroimaging studies aimed at investigating differences between the dreaming and non-dreaming brain must be directed at differences between dreaming and non-dreaming sleep within sleep stages, rather than comparing differences between REM and NREM sleep.

Neuroimaging as a Developing Technology

Functional imaging is itself a developing science, since its most powerful techniques have only become available since the mid-1990s. Its validity is interdependent with the strengths and limitations of its methodologies. Part of the validation of each technique will eventuate through the use of *coregistration* techniques being developed, which combine the superior temporal resolution of EEG and MEG with the spatial resolution of fMRI and PET data (Kosslyn 1999; Maquet and Phillips 1998). This may overcome some of the drawbacks associated with each technique and maximise their advantages (although standardisation of combined protocols would be a formidable undertaking).

Another aspect of the process of validation of neuroimaging as a whole involves investigating the correspondence of activation patterns obtained with the functions of areas believed to be involved in the task being investigated – in other words, by and through the concordance of imaging output with predictions derived from neurobiological theory. But the validity of convergent evidence provided by neuroimaging is also interdependent with the strengths and limitations of neurobiological theory itself. In dream theory, it seems to be the case that imaging studies, which are still essentially exploratory in nature, are being applied post hoc to dream theory. AS/AIM dream theory seeks to explain dream features by reference to brain areas believed to be involved in dreaming. However if the definition of dreaming itself is inaccurate or too narrow, then so the subsequent interpretation of neuroimages will be also. This is reflected in the fact that neuroimaging studies drawn upon as supporting this theory investigate differences between REM and NREM *rather than* dreaming vs. not dreaming.

It is still relatively early in the development of neuroimaging technology, so at this stage activation should only form a basis for hypotheses that a particular structure or region is involved in a particular process. Also, neuroimaging also cannot yet provide strong convergent evidence for dream theories, until it is applied in the appropriate context of comparing states of dreaming rather than sleep stage differences.

Future advances in the spatial resolution of imaging techniques will assist in the valid use of naturally occurring lesions in humans by the accurate identification and measurement of the extent of lesion damage, enabling accurate correlation to behavioural or psychological changes. Because of the current limitations in the use of animal models for human biopsychological theories, accurate data from human lesion studies will become an important source for neurobiological theory testing, and such data, from both human and animal sources, will in turn ultimately assist in the validation of any proposed animal models. Any resulting convergent evidence from these sources will then carry much more weight in neurobiological theory evaluation than it can at present.

6.5 Conclusion

The nature of scientific theories was discussed in this chapter, and a set of criteria for the evaluation of scientific theories was outlined. This set comprised analysis of the structure of the argument of the theory – its logic and the reasonableness of its premises, the validity of its methodologies, the justifiability of interpretation of findings and the management of anomalous data. This analysis was applied to a sequence of neurobiological models of dreaming developed over the past four decades, which we have referred to collectively as the AS/AIM theory of dreaming. We chose this theory because it is considered to be one of the most widely cited and recognized neurobiological dream theories today. However, this theory was found to be wanting at all levels of this analysis. Its status as a theory was considered to be questionable, as no plausible underlying mechanisms can be empirically tested. Rather, evidence supporting the theory consists of a set of possible correlations, with much supporting evidence relying on deficient methodologies.

If, as Laudan (1984) suggests, a theory, or sequence of theory-articulations, may be evaluated as to the progress made relative to the set of aims it was devised to achieve, or as to progress made toward realising or achieving a desirable goal state, then, by the criteria used in this chapter, the AS/AIM neurobiological theory of dreaming has been shown to be inadequate. Although this theory has served to stimulate much research and debate about a fascinating window on human consciousness, some of the key assumptions underlying this theory may have served to *constrain* this research and debate. This critique is offered in the hope that attention is drawn to some of the inconsistencies and methodological problems that are not just inherent in the AS/AIM theory of dreaming, but which also bedevil other dream theories. We hope that this will provide an impetus for the development of new hypotheses and new methodologies. For as methodologies and technologies improve, particularly in the areas of sleep stage recording and neuroimaging, finer-grained analysis of data and detection of correlations between variables will become possible, and new hypotheses will be generated to account for them. From this process of scientific bootstrapping, it is possible that a new paradigm for the neurobiology of dreaming will emerge.

References

Antrobus J (2000a) How does the dreaming brain explain the dreaming mind? Behav Brain Sci 23:904–907

Antrobus J (2000b) Theories of dreaming. In: Kryger M, Roth T, Dement W (eds) Principles and practices of sleep medicine, 3rd edn. W.B. Saunders, Philadelphia, pp 472–481

Aserinski E, Kleitman N (1953) Regularly occurring periods of eye motility, and concomitant phenomena during sleep. Science 118:273–274

Bassett DS, Bullmore ET (2009) Human brain networks in health and disease. Curr Opin Neurol 22(4):340–347

Berman K, Ostrem J, Randolph C, Gold J, Goldberg T, Coppola R, Carson R, Herscovitch P, Weinberger D (1995) Physiological activation of a cortical network during performance of the Wisconsin Card Sorting Test: a positron emission tomography study. Neuropsychologia 33:1027–1046

Bowker R, Morrison A (1976) The startle reflex and PGO spikes. Brain Res 102:185–190

Bradshaw J, Mattingley J (1995) Clinical neuropsychology: behavioral and brain science. Academic, San Diego

Braun A, Balkin T, Wesenten N, Carson R, Varga M, Baldwin P, Selbie S, Belenky G, Herscovitch P (1997) Regional cerebral blood flow throughout the sleep-wake cycle: a H215O PET study. Brain 120:1173–1197

Chalmers A (1982) What is this thing called science? University of Queensland Press, St. Lucia

Chalmers D (1996) Facing up to the problem of consciousness. In: Hameroff S, Kaszniak A, Scott A (eds) Toward a science of consciousness: the first Tucson discussions and debates. MIT Press, Cambridge, pp 5–28

Cipolli C, Bolzani R, Cornoldi C, De Beni R (1993) Bizarreness effect in sleep recall [abstract]. Sleep 16:163–170, Abstract from: PsycINFO Item: 1993-32258-001

Conduit R, Crewther S, Coleman G (2000) Shedding old assumptions and consolidating what we know: toward an attention-based model of dreaming. Behav Brain Sci 23:924–928

Conduit R, Crewther S, Coleman G (2004) Poor recall of eye-movement signals from stage 2 compared to REM sleep: implications for models of dreaming. Conscious Cogn 13:484–500

Dement W (1958) The occurrence of low voltage fast electroencephalogram patterns during behavioural sleep in the cat. Electroencephalogr Clin Neurophysiol 10:291–296

Dement W, Kleitman N (1957) The relation of eye movements during sleep to dream activity: an objective method for the study of dreaming. J Exp Psychol 53:339–346

Domhoff G (2000) Needed: a new theory. Behav Brain Sci 23:928–930

Edwards A (1990) Construct validity and social desirability. Am Psychol 45:287–289

Farthing G (1992) The psychology of consciousness. Prentice-Hall, Englewood Cliffs

Feinberg I, March J (1995) Observations on delta homeostasis, the one-stimulus model of NREM-REM alternation and the neurobiologic implications of experimental dream studies. Behav Brain Res 69:97–108

Fosse R, Stickgold R, Hobson JA (2004) Thinking and hallucinating: reciprocal changes in sleep. Psychophysiology 53:298–305

Foulkes D (1962) Dream reports from different stages of sleep. J Abnorm Soc Psychol 65:14–25

Gevins A, Smith M, McEvoy L, Leong H, Le J (1999) Electroencephalographic imaging of higher brain function. Philos Trans Roy Soc Lond B 354:1125–1134

Gottesman C (2001) The golden age of rapid eye movement sleep discoveries. 1. Lucretius – 1964. Prog Neurobiol 65:211–287

Hartmann E (1967) The biology of dreaming. Thomas, Springfield

Himanen S-L, Hasan J (2000) Limitations of Rechtschaffen and Kales. Sleep Med Rev 4:149–167

Hobson J (1997) Consciousness as a state-dependent phenomenon. In: Cohen J, Schooler J (eds) Scientific approaches to consciousness. Lawrence Erlbaum Associates, Mahwah, pp 379–396

Hobson J, McCarley R (1977) The brain as a dream state generator: an activation-synthesis hypothesis of the dream process. Am J Psychiatry 134:1335–1348

Hobson JA, Stickgold R (1995) The conscious state paradigm: a neurocognitive approach to waking, sleeping and dreaming. In: Gazzaniga MS (ed) The cognitive neurosciences. MIT Press, Cambridge

Hobson J, Stickgold R, Pace-Schott E (1998) The neuropsychology of REM sleep dreaming. Neuroreport 9:R1–R14

Hobson J, Pace-Schott E, Stickgold R (2000) Dreaming and the brain; toward a cognitive neuroscience of conscious states. Behav Brain Sci 23:793–842

Jouvet M, Jouvet D (1963) A study of the neurophysiological mechanisms of dreaming. Electroencephalogr Clin Neurophysiol 24:133–157

Kertesz A (1994) Localization and function: old issues revisited and new developments. In: Kertesz A (ed) Localization and neuroimaging in neuropsychology. Academic, San Diego, pp 1–33

Koslowski B (1996) Theory and evidence: the development of scientific reasoning. MIT Press, Cambridge

Kosslyn S (1999) If neuroimaging is the answer, what is the question? Philos Trans Roy Soc Lond B 354:1283–1294

Kuhn T (1996) The structure of scientific revolutions. University of Chicago Press, Chicago

Laudan L (1984) Science and values. University of California Press, Berkeley

Lim AS, Lozano AM, Moro E et al (2007) Characterization of REM sleep associated ponto-geniculo-occipital waves in the human pons. Sleep 20:823–827

Magistretti P, Pellerin L (1999) Cellular mechanisms of brain energy metabolism and their relevance to functional brain imaging. Philos Trans Roy Soc Lond B 354:1155–1163

Maquet P (2000) Functional neuroimaging of normal human sleep by positron emission technology. J Sleep Res 9:207–231

Maquet P, Phillips C (1998) Functional brain imaging of human sleep. J Sleep Res 7:42–47

Miyauchi S, Misaki M, Kan S, Fukunaga T, Koike T (2009) Human brain activity time-locked to rapid eye movements during REM sleep. Exp Brain Res 192(4):657–667

Nielsen TA (1999) Mentation during sleep: the NREM/REM distinction. In: Lydic R, Baghdoyan H (eds) Handbook of behavioural state control: cellular and molecular mechanisms. CRC Press, Boca Raton, pp 101–128

Nielsen TA (2000) A review of mentation in REM and NREM sleep: "covert" REM sleep as a possible reconciliation of two opposing models. Behav Brain Sci 23:851–866

Orlinsky D (1962) Psychodynamic and cognitive correlates of dream recall – a study of individual differences. Unpublished doctoral dissertation, University of Chicago

Parks RW, Crockett DJ, Tuokko H, Beattie BL, Ashford JW, Coburn KL, Zec RF, Becker RE, McGeer PL, McGeer EG (1989) Neuropsychological "systems efficiency" and positron emission tomography. Neuropsychiatry Clin Neurosci 1:269–282

Pivik RT (1991) Tonic states and phasic events in relation to sleep mentation. In: Ellman SJ, Antrobus JS (eds) The mind in sleep: psychology and physiology, 2nd edn. Wiley, New York

Popper K (1968) The logic of scientific discovery. Hutchinson, London

Rechtschaffen A, Kales A (1968) A manual of standardized terminology, techniques and scoring system for sleep stages of human subjects. Public Health Services, U.S. Government Printing Office, Washington

Rose S (1998) Brains, minds and the world. In: Rose S (ed) From brains to consciousness? Essays on the new sciences of the mind. Princeton University Press, Princeton, pp 1–17

Sanford LD, Silvestri AJ, Ross RJ, Morrison AR (2001) Influence of fear conditioning on elicited ponto-geniculo-occipital waves and rapid eye movement sleep. Arch Ital Biol 139(3):169–183

Sarter M, Bruno J (2002) Animal models in biological psychiatry. In: D'haenen H, den Boer J, Willner P (eds) Biological psychiatry. Wiley, New York

Siegal J (1999) The evolution of REM sleep. In: Lydic R, Baghdoyan H (eds) Handbook of behavioural state control. CRC Press, Boca Raton, pp 87–100

Solms M (1997) The neuropsychology of dreams: a clinico-anatomical study. Lawrence Erlbaum Associates, Mahwah

Solms M (2000) Dreaming and REM sleep are controlled by different brain mechanisms. Behav Brain Sci 23:843–850

Squier L, Domhoff G (1998) The presentation of dreaming and dreams in introductory psychology textbooks: a critical examination with suggestions for textbook authors and psychology instructors. Dreaming 8:149–168

Tagamets MA, Horwitz B (2001) Interpreting PET and fMRI measures of functional neural activity: the effects of synaptic inhibition on cortical activation in human imaging studies. Brain Res Bull 54(3):267–273

Wehrle R, Kaufmann C, Wetter TC et al (2007) Functional microstates within human REM sleep: first evidence from fMRI of a thalamocortical network specific for phasic REM periods. Eur J Neurosci 25:863–871

Chapter 7
Sleep Onset Process as an Altered State of Consciousness

Dean Cvetkovic and Irena Cosic

Abstract Falling asleep is a link between two general states of consciousness, wakefulness and sleep. During the complex process of the wake to sleep transition, various electrophysiological, cognitive and behavioural alterations take place, all linked to states of consciousness. Together, these states have been interpreted as the hypnagogic state. It is no longer believed that sleep onset just acts as a buffer between the awake and sleep processes characterised by the gain and loss of sensory functioning, respectively. On the contrary, the sleep onset phenomenon addresses many issues surrounding some of the most common sleep and mental disorders. What is the moment of sleep onset? How can sleep onset be characterised from the brain, heart and respiration activity during sleep in humans? These are some of the questions that have no clear and definite answers. Neural synchrony and its interaction and coupling with cardio-respiratory synchrony is very much related to its role as a mechanism for integrating brain, heart and respiration, which are all responsible for consciousness, in both humans and animals. The phase synchrony between neural, cardio and respiratory activities and environment provides a signature of subjective experiences, linked to various states of consciousness. Technological advances have led to a gradual shift in the approach adopted by the scientific community towards unravelling and understanding this sleep onset phenomenon as an altered state of consciousness. This chapter presents research from the scientific community, explores some new objective indices that can be adopted to measure sleep onset, and demonstrates how sleep onset can be voluntarily altered and induced using biofeedback technology.

D. Cvetkovic (✉)
School of Electrical and Computer Engineering, RMIT University, Melbourne, VIC, Australia
e-mail: dean.cvetkovic@rmit.edu.au

I. Cosic
College of Science, Engineering and Health, RMIT University, Melbourne, VIC, Australia

D. Cvetkovic and I. Cosic (eds.), *States of Consciousness*, The Frontiers Collection,
DOI 10.1007/978-3-642-18047-7_7,

7.1 Introduction

7.1.1 Hypnagogic Phenomenon

Mavromatis was one of the first to comprehensively explore the hypnagogic phenomenon, which he defined as "hallucinatory events in the state between wakefulness and sleep" (Mavromatis 1987). The term "hypnagogic" comes from the Greek language (*hypnos* means sleep and *agogeus* means conductor or leader) and describes the pre-sleep or sleep onset phenomena. Related terms have been coined: for example, the parahypnagogic state covers "fringe forms of sleep" or trancelike experience bordering on sleep onset but occurring during the brief moments of wakefulness, defined as *daytime parahypnagogia* (*DPH*). A DPH episode typically lasts from a fraction of a second to few seconds, often when one's eyes are open watching TV but not when actively interacting with people. DPH can occur at the moment one falls asleep while watching TV and then quickly re-awakens, being fully conscious of the parahypnogogic state and the images and thoughts experienced in this state, in contrast to the lack of awareness when one falls asleep. Some reports have claimed that subjects who have experienced this state describe it as being different from a typical daydream, which is longer and easily recalled. DPH has been described as vivid, fast, back to reality, being unaware of what is going on, consciousness "hijacking", a flash dream, including pleasant or unpleasant experiences, occurring on border of sleep onset and awake state (Gurtelle and de Oliveira 2004). Overall, DPH differs from any other state of consciousness but may include elements of hypnagogic hallucinatory images or thoughts and creative insights, relaxed wakefulness, mind-wandering, daydreaming, instant self-hypnosis, trance, dissociation, meditation, micro-sleep, waking dreams, etc.

Humans may possibly exhibit states in which parts of the brain are working in offline mode and they are aware of it. Forgetfulness or daydreaming may be examples of this condition. In fact, some studies have revealed that activity in numerous brain regions increases when subjects' minds wander, suggesting that brain areas associated with complex problem-solving are related to daydreaming. This daydreaming state could also be regarded as one of the many states of consciousness (Vetrugno et al. 2009). This switching from an everyday conscious state to a daydreaming state is intriguing. However, some studies have highlighted the complexity of sleeping and waking states and challenge the current concepts of consciousness and the function of sleep. Severe state-switching, where elements of one state of being (awake, NREM and REM sleep) sometimes spill over into another state, is known as *status dissociatus* (SD) and is a pathological condition that occurs in patients normally suffering from REM sleep behaviour disorder (when dreamlike experiences are acted out in real-life situations, as in sleepwalking and more extreme violent acts). There are many other disorders associated with SD, such as narcolepsy, hypnagogic or hypnopompic hallucinations, sleep paralysis, sleep inertia and others.

While sleep paralysis results in hallucinatory experiences, during Recurrent Isolated Sleep Paralysis (RISP) episodes one is convinced that these experiences are real, generated from external sources, such as paranormal or supernatural experiences (Terrillon and Marques-Bonham 2001). These experiences can last for up to an hour as sequential episodes of hypnagogic and hypnopompic hallucinations (visual, auditory, tactile and olfactory). RISP often occurs when one is drifting from wake to sleep or upon waking from sleep and most commonly while lying on one's back. The behavioural changes that occur at the beginning of these RISP episodes are typically an increasing feeling of one's own heaviness as the muscle atonia (chin ECG) appears. This progressive feeling of heaviness is often perceived as an invisible force applied on top of subject lying down or as an entity that is sitting on his or her chest. In these situations, breathing becomes difficult to maintain, fear is experienced, a loud buzzing or ringing sound is sensed, and one suddenly realizes that it is impossible to move or to speak or to cry out to free oneself. The hypnagogic hallucinations within RISP episodes may also be regarded as revealing one of the many states of consciousness where state-switching occurs.

Human daydreaming, *status dissociatus* and the hypnagogic hallucinations that accompany sleep paralysis conditions may in fact contradict Buzsaki's claim that "a main function of sleep is to isolate the brain from the body and the environment which is the reason why we call it a default state" (Buzsaki 2006, pp. 186–187).

7.2 Objective Measures of Sleep Onset

Falling asleep is a complex process (Ogilvie 2001). A common sleep disorder, insomnia, is often associated with disturbance of this complex hypnagogic process. The advanced Yoga and Japanese Zen masters have demonstrated a hypnagogic progression of electroencephalographic (EEG) states during their meditation exercises (Mavromatis 1987). However, the EEG activity is quite different in the waking, sleeping and meditating states. During meditation, there is alpha EEG coherence across the cortex hemispheres and alpha-theta activity is predominant. During sleep onset, the EEG alpha-to-theta transition is exhibited. The correlations between hypnagogic phenomena and sleep onset EEG rhythms were firstly reported in 1938 (Davies et al. 1938). The hypnagogic state was described as beginning just before the subject's objective EEG alpha rhythm vanishes or when subjects are "losing their alpha EEG" (Dement and Kleitman 1957; Rechtschaffen and Kales 1968) and subjective dreamlike visions, fantasies and feelings appear. In the late 1960s, sleep onset was characterized as alpha, fragmented alpha, and appearance of theta. Physiologically, sleep onset occurs with: a decrease in the electromyographic (EMG) muscle tone activity; decrease in heart rate and respiration; shift to hypo-metabolic parasympathetic activity; increase in rectal temperature; decrease in skin conductivity; shift in EEG activity from low amplitude high frequency to high amplitude low frequency. Drowsiness, which may necessarily be the same as sleep onset, is characterized by slow eye movements (SEM). Subjects often go in and out of this state, gradually remaining in it for increasingly longer periods, up to 10 s, as drowsiness deepens. The incidence of SEM occurs a certain time before the onset of EEG drowsiness, which increases as the number of repeated drowsy period increases. The state of drowsiness has been described as "the onset of occipital alpha-blocking" where the muscle tone diminishes very soon after the alpha pattern is lost (Liberson and Liberson 1966). Subjects would often recognize the initial drowsiness with a delay of 20 s or more. Other EEG studies have revealed that drowsiness is characterized from progression of occipital alpha activity to alpha-blocking accompanied by a more anterior (frontal brain region) low amplitude theta activity, which gradually increase in amplitude leading to high voltage vertex waves. There are two alpha oscillations that characterize the sleep onset process, such as alpha prior to sleep onset and alpha activity at sleep onset and during sleep (Ogilvie et al. 1991). For example, there is a slowing in responsiveness to external stimuli as alpha activity decreases in the sleep onset period. However, when alpha re-appears at sleep onset, this responsiveness ceases. The results indicate that lower delta and higher alpha predominance is the cause of the distorted sleep onset process that is responsible for insomnia.

Alpha activity is probably one of the main indicators of sleep onset, but there are other EEG bands that can characterize this process. There have been indications of: changes in delta and sigma frequencies (Lubin et al. 1969); an increase in delta, theta and alpha bands at the beginning of stage 1 and continued 10 min into sleep (Hori 1985); theta increase as the alpha and beta decrease (Ogilvie 1991); beta decrease as delta increases (Merica and Fortune 2004); decrease in narrow alpha band (9–11 Hz) at occipital region and increase at 3 and 4 Hz at the frontal and central cortex regions (Badia et al. 1994). Also, there are fluctuating changes, mainly in delta.

Sleep onset is characterized by three phases: the initial process of alpha-related change (beginning of sleep onset and the initial transition into stage 1); the intermediate process of theta and vertex waves (regarded as middle sleep onset); and the terminal process signifying the end of wakefulness, known as sigma or the sleep spindle process. The intermediate process of sleep onset has not been investigated in great detail. In fact, most automatic detection developed so far by the biomedical engineering and sleep community has been on sleep spindle (a marker of the end of sleep onset and well understood), complex transients (i.e. K complexes or CAP transients), slow wave sleep (SWS), and alpha, beta and gamma bands. This probably has to do with the fact that vertex and theta activity are regarded as wake transients to pre-sleep onset and are not considered to characterize sleep onset or even have any influence over its progression. The author (Cvetkovic) has characterized sleep onset from measures indicating that alpha decreases as delta increases with a theta increase to a lesser degree. Considering that delta or SWS is more predominant and evident (as a sharp marker) in this sleep onset progression, it has been utilized in the author's studies of automated detection of sleep onset and biofeedback operation. It would be of great importance to explore these intermediate processes of sleep onset. There is little information concerning the changes that occur during the wake to sleep transition or sleep onset. The definition, the electrophysiological marker, and the entire process of sleep onset are not clear enough and at times rather complicated to specify. Sleep onset does not occur at one time, it cannot be predicted (or at least this has not been investigated), and there are oscillations in vigilance that occur before stable sleep. Besides the inter-subject variability, the other reason why a discrete point of sleep onset cannot be specified is because the transition from wakefulness to sleep is gradual and not a precise behavioural or physiological point of sleep onset (Rechtschaffen 1994). The true wake-to-sleep transition could be said to begin from stages 1 to 2 and the arrival of spindles (Ogilve 2001). The actual sleep onset may occur within waking stages 1 and 2 and it remains in a loop or cyclic mode for a certain time, until it transits into light and then deeper sleep. Likewise, sleep depth does not move directly to deeper sleep and lighter sleep. This is due to oscillatory dynamics present in sleep regulation (Huupponen et al. 2003).

7.3 Sleep Onset and Depth Biomarkers

Automated computer-based sleep detection and analysis technology aims at providing a reproducible and accurate description of the sleep process and microsleep (Penzel and Conradt 2000). Spindle frequencies range from 10.5 to 16 Hz during sleep intervals of 0.5–2.0 s, occurring with the highest density during light sleep and lowest density during deep sleep (Himanen et al. 2002). Spindles are generated in the thalamus and comprise activity of the thalamic reticular nuclei and thalamo-cortical and cortico-thalamic neurons (Steriade et al. 1993a, b). To this day, no studies have succeeded in accurately assessing spindle frequency. One study simulated the spindle signal with the dominant 13 Hz using different methods but it remains challenging to apply these methods using real EEG signals (Huupponen et al. 2006). Another study developed four sleep spindle detection methods, consisting of sigma index for spindle detection based on the FFT spectrum, a fuzzy detector, fixed spindle amplitude analysis based on FIR filtering, and a combination detector utilizing sigma index and spindle amplitude (Huupponen et al. 2007). However, flexible automated spindle detection methods are still needed to study spindle characteristics in sleep EEG.

Primary insomnia is often considered a disorder of hyperarousal, which means that the patient has an arousal that is incompatible with falling asleep or maintaining sleep (Riemann et al. 2010). The concept of hyperarousal is quite complex and it still holds a number of questions, such as what is meant by arousal, how it becomes elevated, and whether hyperarousals and sleep are mutually inclusive. The current standard definitions of the sleep arousal indexes are based on a sudden change in theta, alpha band and/or frequencies that are greater than 16 Hz but excluding sleep spindles. To score these arousals, a subject has to be asleep for at least 10 s before the arousal is detected and for at least 3 s as the EEG frequency changes. The sleep arousal index is used to quantify sleep fragmentation, derived as the number of arousals per hour of sleep. Other biomarkers that may still need improvement are: cyclic alternating pattern (CAP) rate (% of NREM sleep occurrence by CAP sequences); and paradoxical breathing (a phase-out of thoracic as opposed to abdominal respiratory traces). Sleep quality can be measured with the "weighted-transition sleep fragmentation index" (Swarnkar et al. 2009). This index has the disadvantage that it does not focus on the dynamic measure of multiple phases during the sleep onset.

The sleep EEG transients associated with non-REM sleep micro and macro-structure are called the cyclic alternating pattern (CAP). These transients are briefly present during sleep and consist of fluctuations from 20 to 90 s. The CAP is EEG activity that may signify sleep instability, sleep disturbance or both, not covered by standardised sleep fragmentation indexes using Rechtschaffen and Kales (Rechtschaffen and Kales 1968) or the ASDA (Terzano et al. 2001; Bonnet et al. 1992). The CAP analysis is not meant to replace standard sleep stage or arousal scoring, but rather to extend quantitative sleep via automated sleep analysis. The CAP is identified by periodic EEG transient events and patterns which may include: delta bursts, vertex sharp transients, K-complex sequences with or without spindles, polyphasic bursts, K-alpha, intermittent alpha and EEG arousals. In addition to the detection of sleep onset process, electroculography (EOG) activity can also be utilized for detection of slow eye movements (SEM). Further feature extractions and signal decompositions of various PSG stages in the frequency and time domains can be conducted using adaptive time-frequency approximation algorithms such as Malat's matching pursuit and pulse transient time (PTT) (to recognize arousals) and sleep-related respiratory events. These signal-processing algorithms are often used to recognise and detect microarousals and the micro and macro structure of sleep EEG over finer time scales.

There have been a number of attempts to develop a biomarker that can measure overall sleep depth and not just sleep onset. One such technique uses an EEG power frequency variable that can be computationally derived from the spectral analysis of the EEG signal, such as spectral edge frequencies (SEF), for example SEF 95, SEF 90 and SEF 50 (median frequency). These are the frequency (in Hz) markers derived from the "centre of gravity" of the EEG band spectrum, known as the spectral edge frequencies and referred as frequencies below 95, 90 and 50% of the total EEG spectrum power. An effective method takes the mean frequency of the EEG spectrum and has been used to investigate the application of adjustable windowing before spectrum estimation, optional median smoothing of analysis results and use of the amplitude spectrum instead of the power spectrum (Penzel et al. 1991). Other methods used to quantify sleep depth are: Hjorth parameters, based on the amplitude, time-scale and complexity of the EEG in the time domain (Hjorth 1970); EEG synchronization through its mean frequency inside a certain frequency band (12.5 or 36.5 Hz) (Saastamoinen et al. 2006; Huupponen et al. 2004); mean frequency based on the spectral components of EEG delta, theta and alpha bands, by finding the centre of gravity of the spectrum (Huupponen et al. 2003). There have been considerable advances in the development of these biomarkers to monitor and estimate optimal sleep depth.

7.3.1 Sleep Onset: Study 1

The authors' pilot study was conducted to objectively measure the wake-to-sleep transition or drowsiness period from the sleep EEG activity (Cvetkovic and Cosic 2008). It was presented as new method for estimation of sleep onset transition. The study consisted of only one human subject who was recruited for an overnight polysomnographic (PSG) recording at St. Lukes Hospital (Sydney, NSW, Australia). The full PSG was recorded using Bio-Logic System and Adult Sleepcan Vision Analysis (Bio-Logic Corp., USA). The recorded data was visually scored by the sleep technician according to Rechtschaffen and Kales (R&K) from 30 s epochs. The estimation of sleep onset transition was computed in EEG signals by applying various coherence, power frequency and band power methods (Fig. 7.1).

Earlier studies reported a drop in occipito-frontal and centro-parietal delta coherence after alpha disappearance. Generally, sleep EEG was often seen to be synchronized in the anterior and waking in the posterior region. The EEG inter-hemispheric coherence was computed to investigate the functional connectivity between the left and right brain hemispheres (C3–C4 EEG electrode sites). Coherence estimates the linear cross-correlation between two signals as a function of frequency. EEG coherence also estimates the degree of synchrony between the electrical activities of the two brain regions, concentrating on a certain frequency or EEG band. A well-known non-parametric spectral estimation algorithm known as Capon's approach or the minimum variance distortion-less response (MVDR) (Benesty et al. 2005) was applied to EEG signals for the first time (Fig. 7.2). The MVDR spectrum is often considered as an output of a bank of filters with each filter centred on one of the analysis frequencies. The MVDR bandpass filters are both data and frequency dependent in comparison to parametric periodogram approach, which is independent of both data and frequency.

The mean frequency estimate was based on the developed weighted mean frequency of the EEG spectrum, defined as the "brain-rate" (Pop-Jordanov and Pop-Jordanova 2004; Pop-Jordanova and Pop-Jordanov 2005; refer to Chapter 8 of this book). The brain-rate (arousal indicator) is used to reveal the patterns of sensitivity/rigidity of EEG spectrum and is defined as a mean frequency of the brain oscillations weighted over all bands of the EEG power spectrum. The formula for calculation of the brain-rate is based on a similar algorithm used for determining the centre of gravity (mass) of the EEG spectrum. The spectral power estimate of each EEG band was performed using the short-time Fourier transform (STFT).

No statistical analysis was performed on the processed data for any significant evidence because only one subject was tested. However, the results indicated a sharp suppression in coherence at 4 Hz from awake to stage 1 and recovered right after that period. The "brain-rate" decreased from 7 Hz (awake) and during stages 1 and 2 it decreased further to 5.5 Hz. The EEG relative and absolute band power results revealed a clear suppression in alpha band from stage 1 onset and gradually decreasing during stages 2–4. In terms of relative delta band, there was an increase in delta relative power from wake to stages 1–4 transition. There was a noticeable

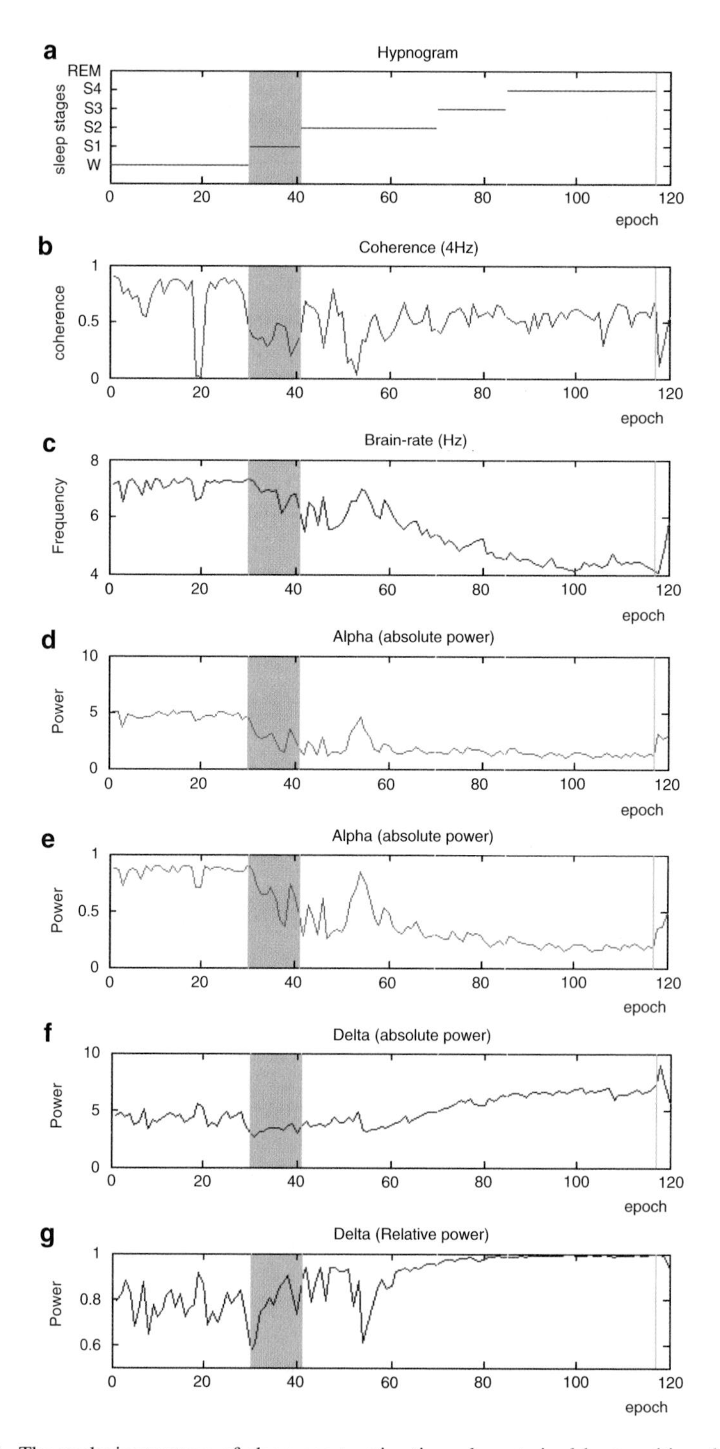

Fig. 7.1 The analysis structure of sleep onset estimation, characterised by transitions between wake (W), stage 1 (S1), stage 2 (S2), stage 3 (S3), stage 4 (S4) and REM, is expressed by the

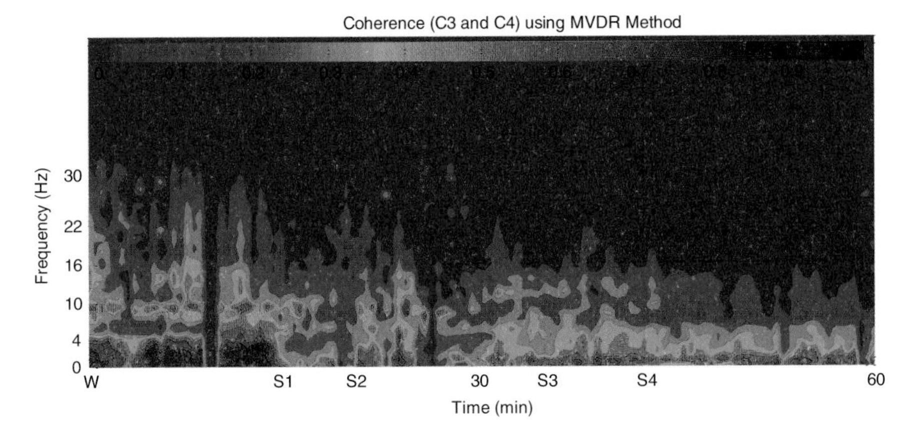

Fig. 7.2 The inter-hemispheric MVDR coherence (C3–C4) was computed over a 60-min period (30 s segments) consisting of W, S1, S2, S3 and S4. The *colour bar* represents the coherence intensity (0–1)

increase in absolute delta band from the middle of stage 2. All of the three estimates (power, frequency and coherence) have shown potential as the solid indicators in sleep onset and sleep depth detection.

7.4 Brain, Heart and Respiration Interaction at Sleep Onset

There are many unknowns about how the sleep EEG interacts with heart rate variability (HRV) and in particular with sleep onset (Jurysta et al. 2003). One of the first studies to measure the cardiac activity as a function of EEG-defined state alterations over the sleep onset period (Burgess et al. 1999) assessed the sleep onset of cardiac activity in humans through the measure derived from the spectral analysis of cardiac beat-to-beat intervals and the respiratory sinus arrhythmia (RSA). Both of these measures reflect the autonomic control of cardiac activity and RSA indicating parasympathetic nervous system (PNS) activity and the sympathetic nervous system (SNS) activity.

The EEG delta and cardiac autonomic activity is altered in humans during sleep onset (Shinar et al. 2003). One group found that the standard definition of sleep onset coincides with an alpha power decrease (confirmed with most studies) below 28% of total EEG energy (Shinar et al. 2006). This sharp decrease in alpha power at sleep onset is *synchronous* with a reciprocal strong increase in delta power. In terms of ECG, all HRV parameters decreased sharply with sleep onset: R-R intervals

Fig. 7.1 (Continued) hypnogram (**a**). The processed and analysed signals include delta 4 Hz coherence with its intensity range from 0 to 1 (**b**); the brain-rate power frequency represented in Hz (**c**); the absolute (**d**) and relative (**e**) alpha power; and the absolute (**f**) and relative (**g**) delta power. The *shaded area* shown in all sub-plots indicate S1 or the sleep onset period

(RRI) significantly increased, revealing a decrease in heart rate (HR) during sleep onset; very low frequency (VLF) power decreased (by a factor of 2.5) and this decrease was evident 2 min before the sleep onset; the ratio of LF/HF power decreased significantly during the sleep onset. This significant decrease in autonomic balance, reflected by the LF/HF ratio, indicates the shift from predominant sympathetic activity when awake to parasympathetic activity after sleep onset. In healthy subjects, when the heart rate exhibits low frequency (LF) oscillations, it has the predominant sympathetic activity that is often present during REM sleep. By contrast, the high frequency (HF) oscillations are biomarker of vagal activity, present during NREM sleep. The respiration (mean frequency) increased slightly, the EMG amplitude and variability was reduced (by a factor of 2 and 3) after the sleep onset. It was evident that a number of muscle twitches of a certain magnitude occurred before the sleep onset. However, there is a slight contradiction in the way that HRV changes during sleep onset. While Shinar's findings show that as delta EEG increases, most HRV bands decrease, Jurysta's study demonstrates that delta EEG band and HRV high frequency (HF) changes are the same (i.e. delta and HF increase or decrease together). However, there is a time delay of around 20 s between changes in cardiac vagal activity and delta EEG activity.

The dynamic interaction between sleep EEG and HRV may be influenced by respiratory and breathing. But the nature of its interdependency would still be unknown. In sleep onset, there are a number of awake to sleep transitions, with a number of spontaneous arousals. The breath-to-breath measurements consist of automated detection of breath inspiration and expiration. The minimum inspiratory R-R intervals and maximum expiratory R-R for each breath are often calculated. RSA value for each breath is calculated based on the difference between the maximum and minimum R-R intervals. R-R respiratory measurements are also typically conducted using thermistor sensor recordings from the inspiration-to-inspiration interval. The thermistor sensor works either with a mouthpiece or with a breathing mask and a nose-clamp. In one study, an existing cardio-respiratory algorithm was used to reconstruct respiration signals from inter-heart beat intervals, based on high frequency (HF) HRV and the RSA effect (Hamann et al. 2009). These extracted respiratory signals were then utilised to investigate the relationship between respiratory and heartbeat phases or the cardio-respiratory synchronization during the main sleep stages. For this investigation, an automated cardio-respiratory synchronisation method known as a synchrogram was developed. The cardio-respiratory synchronization is calculated when heartbeats are in phase with the breathing cycle. The respiration influences the sympatho-vagal autonomous nervous system (ANS) differently during the inspiration and expiration. Sympathetic activity and heart rate is increased during inspiration and decreased with activation of cardiac vagal activity during expiration.

Based on clinical evidence, there is a delay in the sleep onset between the EEG and respiration (Ogilvie and Wilkinson 1984). Disordered breathing, related to sleep apnoea, is first to appear, followed by the EEG sleep spindle transients in stage 2. It was reported that an increase in upper airway resistance causes the ventilation to drop as sleep begins, which in turn actuates the EEG to shifts from predominant alpha to theta and back to alpha activity (Trinder et al. 1992). The results revealed that the ventilation substantially and rapidly decreases for each breath during alpha-to-theta transition and ventilation increases during theta-to-alpha transition. The origin of these changes is thought to be an alteration in the regulatory control of the respiratory system at sleep onset (Bulow 1963) with the emphasis on the role of state in determining respiratory instability during stage 1 sleep (alpha-theta band fluctuation). The reduction in ventilation as sleep begins generally has to do with the decrease in respiratory pump output and upper airway muscle activity (Naifeh and Kamiya 1981). This theory indicates that sleep onset mechanisms are related to brain and respiration interaction.

The transition from wakefulness to NREM sleep is characterised by the inactivation of the wakefulness "telencephalic" control mechanism and the release of the automatic control mechanism. The irregular breathing at sleep onset consists of regular waxing and waning of breathing amplitude (called periodic breathing). Similar waxing and waning is related to the sleep EEG spindle effect. Periodic breathing is defined as oscillations in breathing amplitude, which regularly decreases and increases to low or high amplitudes. This phenomena occurs in 40–80% of normal subjects and varies with age groups. The pattern of periodic breathing resembles either Cheyne-Stokes respiration, with a progressive decrease and increase in breathing amplitude before the apnoea or hypopnoea, or Biot's breathing. The apnoeas observed during drowsiness are of central type, with occasional obstructive apneas. The sleep onset process is synchronous with a progressive decrease in breathing amplitude. If there is an abrupt decrease in breathing during this sleep onset period, it activates arousal and increases EEG alpha activity. The periodic breathing lasts between 10 and 20 min and up to 60 min and oscillates throughout the sleep onset, between arousals and stage 1 or 2, and disappears when stable stage 2 is reached. This transition (stages 1 and 2 of NREM sleep in humans) is characterised by breathing instability and the appearance of respiratory and circulatory periodic phenomena. Regular breathing sets in with deep NREM sleep (stages 3 and 4 in humans), when breathing is driven by the automatic control mechanism.

7.5 Voluntary Altering Hypnagogic State of Consciousness

The human sensory information processing system shuts down gradually as the sleep onset progresses. Subjects' responses to stimuli (consisting of auditory tones presented irregularly, starting with the highest volume and decreasing in volume as the sleep onset lengthens) gradually decrease and reaction times increase. The subjective state is altered where subjects' control over their own thoughts and images, and awareness of their external environment, decreases. There have also been reports that early sleep onset was possible when the eyes were still open. Even the responsiveness to external stimuli is limited, which adds to the difficulty in testing for reaction time at sleep onset. The decline in response to auditory stimulation was reported to lag behind the decline in response to visual stimulation. Studies in the

1980s confirmed that responses to auditory stimulation are fairly successful if the subjects' alpha is not reduced. However, responses to photic stimulation were not entirely successful even with a small reduction of alpha rhythm. These results indicate a number of possibilities during the sleep onset, one being that the visual information processing part of the brain is quicker to block any external stimuli while falling asleep than the auditory information processing part, which may further suggest that auditory stimulation might be better utilized in evaluating the sleep onset process due to its responsiveness accuracy. The complexity of determining and defining the specific moment of sleep onset is further highlighted by individual differences in sleep onset behaviour patterns. Therefore, it might be better to use not just one measure but a series of physiological and behavioural measures to better evaluate this sleep onset dynamics. Behavioural tasks such as correlating reaction times using auditory and photic stimuli with electrophysiological measures might also enhance investigations into the sleep onset phenomenon.

External auditory stimulation has an excitatory or arousal effect. It is unknown why acoustic stimuli are processed differently during sleep and wakefulness. There is no neuronal mechanism to demonstrate this phenomenon. The results have shown that acoustic stimulation during sleep was accompanied by an increase in EEG lower delta band (below 2.0 Hz) as compared to no evident changes in waking EEG (Czisch et al. 2002). Depending on whether auditory tones are frequent (repetitive) or infrequent (deviant), the auditory evoked potentials are different during sleep and awake states. These results further support the theory that sensory information processing varies during sleep stages (Nielsen-Bohlman et al. 1991). Research on the auditory stimulus repetition rates on the steady-state auditory evoked potentials in the gamma range (30–50 Hz) has been well established in the studies. However, there is a void in research on EEG synchronization to auditory stimuli with repetition rates below 10 Hz and especially in delta and theta bands, which is the preferred tempo in listening to and making of music. Synchronization between periodic auditory stimuli and EEG responses was analysed with stimulus-locked inter-trial coherence during awake states only (Will and Berg 2007). The inter-trial coherence or the "phase-locking factor" was computed at 3 s epochs (each epoch start aligned with the stimulus onset) at each frequency and latency window. The results revealed formerly unknown synchronization responses of EEG activity to repetitive auditory stimuli. The auditory stimulation with 1–8 Hz repetitions resulted in increased phase synchronization in all EEG frequency bands. However, the stimulation in the lower range (1–5 Hz) exhibited inter-trial coherence in its corresponding single-frequency bands (1–5 Hz), which characterize delta frequency entrainment. These results suggest that delta entrainment is possible but will need to be tested in a sleep environment. If successful, slow wave sleep (SWS), which plays an integral part in sleep depth, may be entrained (suppressed or enhanced), depending on its behaviour. The SWS is often at its strongest when the subject is sleep deprived. SWS increases from stage 1 to 4 but the rate of increase decreases in subsequent phases during an overnight sleep period. Optimising this SWS activity may in fact control the sleep offset (sleep to wake transition). Another study has shown that 6 h of waking-auditory stimulation were followed by an increase in SWS, a shortening of the latency between SWS periods and longer sleep onset latency (Cantero et al. 2002). REM sleep parameters remained unaffected by the waking-auditory stimulation. These results indicate that sleep architecture depends on auditory demands during the prior wakefulness.

Sleep spindles are an important component of any sleep micro-architecture that blocks the transfer of sensory information in the cerebral cortex at the level of the thalamus. In other words, most of the external sensory information is blocked by this state, enabling the transition from awake to sleep to occur and proceed to other states and phases. However, to what degree is this blocking possible and what happens if this spindle activation is optimized or altered? Would the sleep architecture change considerably and would the sleep onset be affected as a result? During spindle and slow wave sleep (SWS) oscillations, thalamo-cortical neurons are hyperpolarized and have an increased membrane conductance, which make them unresponsive to signals from the external environment (Llinas and Steriade 2006). There is evidence that complex sound stimulation changes the spindle-related brain state (Miller and Schreiner 2000). It was demonstrated that spindle power activity in anaesthetized cats can be suppressed by multi-frequency sound at an EEG gamma range tone. As a result, by suppressing spindles, it also suppresses beta activity and delta to some extent (Britvina and Eggermont 2008).

Human EEG activity, induced by and synchronous with intermittent photic stimulation at frequencies other than that of the alpha EEG rhythm, were first observed in the 1930s (Adrian and Matthews 1934). The findings indicated that the alpha rhythms could be driven beyond their natural rate by sensory stimulation. This relationship between alpha rhythm and flicker potentials and synchronisation had many challenges which are even encountered to this day in attempts to explain the occurrence of flicker potentials at frequencies above and below that of alpha rhythm (Toman 1941; Walter and Walter 1949). The photic driving response has been defined as an occurrence of EEG waves of the same frequency or of EEG waves that are harmonically related (phase-locked) to the frequency of a rhythmic photic stimulus. The photic driving response is also referred to as the steady-state evoked potential (SSEP). This expression implies that the brain is presumed to have achieved a steady state of excitability. The photic driving can be elicited when the stimulus is applied at the same frequency as the subject's spontaneous EEG rhythm (i.e. alpha or beta band). According to classical theories, the photic driving response is a resonance between spontaneous rhythms of the brain and visually evoked responses. A number of studies indicate the possibility of entraining specific frequencies of brain waves by presenting subjects with frequency-specific flickering lights. A prior history of naturally occurring hypnagogic imagery was found to be related to the incidence of visual imagery experienced in rhythmic photic stimulation (Freedman and Marks 1965). In EEG functional tests, an EEG photic driving response is routinely used to uncover potential epileptiform abnormalities, as a sensitive neurophysiological measure used to assess drug effects to treat Parkinson disease and sleep deprivation and schizophrenia. One approach used in clinical practice presumes that a patient's EEG baseline dominant frequency within some band (e.g. alpha and beta) may represent the presenting pathological state.

Thus therapy consists of attempting to drive the EEG away from the dominant frequency to new frequency values. Clinical practices using the photic driving response are often measured using hemispheric and asymmetric methods, which have often shown alterations in the functional states of the brain (Erol 1999). It is speculated that "patients with schizophrenia show severe disruption of EEG measures of neural synchronization to periodic auditory and visual stimulation." (O'Donnell et al. 2002). Beta and gamma EEG power/amplitude and phase synchronization are altered. This deficit in EEG synchronization could contribute to the behavioural disturbances of perceptual and temporal integration observed in schizophrenia. This technique may be applied in testing of impact of brain states on information processing and to treat functional states in models of brain disease or disorder. The occipital neural networks in the mammalian brain typically respond to photic stimulation with 12 Hz oscillations (Basar 2008), whereas the temporal auditory region responds to auditory stimulation with 10 Hz oscillations. Both of the brain regions contain neurons that fire upon expected input, neurons that respond to actual stimulus, and neurons that fire whenever there is a discrepancy between stimuli. Besides monitoring sleep onset automatically and investigating its oscillatory coupling through synchronization, it would be interesting to incorporate a reaction time module. In cases of primary insomnia, there is evidence that patients are in a constant state of hyperarousal where metabolic rates and levels of the stress hormone cortisol are high. Auditory and photic stimulation may be used to reduce or alter these hyperarousals.

In patients with primary insomnia, a constant state of hyperarousal is incompatible with the initiation or maintenance of sleep. Music via biofeedback induces an arousal effect that can lead to a relaxation effect, which has been proven valuable for the behavioural treatment of insomnia. Biofeedback technology has been critically reviewed in a practice parameter published by the American Academy of Sleep Medicine (AASM). Auditory and photic stimulation may be used to reduce or alter states of arousal. For auditory stimuli such as music, the speed or tempo of the rhythm appears to be one of the main determinants of the cardiovascular and respiratory responses, which show a clinically strong correlation between tempo and respiratory rhythm. The entrainment or forcing of the respiratory and sympathetic oscillators by auditory driving input results in increased respiratory frequency and sympathetic outflow. However, the exact nature of the central respiratory pattern generated remains open to debate. Some have classified it as a kind of oscillator that can be entrained by external or internal inputs.

7.6 Biofeedback and Entrainment Technology

7.6.1 Entrainment: Study 2

Frequency entrainment works as long as the frequency of the second or third harmonic of the light flashes does not deviate by more than 1–2 Hz from the free-running (i.e. spontaneous) alpha rhythm (Buzsaki 2006). The author performed a preliminary examination of EEG entrainment modelling and simulation (Cvetkovic et al. 2009). A model of the thalamocortical system was employed to generate simulated EEG signals and was tested in various configurations using cross-correlation and coherence methods in the search for entrainment under very simple conditions. It was shown that the stimulated signal shape could have a significant impact on the entrainment characteristics of the system. It was also found that an incident entraining signal of much higher frequency is unsuitable for entrainment.

The EEG coherence for inter-hemispheric (left vs. right brain) and intra-hemispheric (front vs. back of the cortex) has been investigated by the authors under various extremely low frequency (ELF) photic or visual stimulations (Cvetkovic and Cosic 2009a). The results showed the functional significance of EEG alpha rhythms at parietal and occipital cortex, which respond to oscillatory 13 and 16.66 Hz photic stimulation, respectively. The use of EEG coherence and asymmetric spectral estimates in particular EEG frequency bands was found suitable to identify and characterise brain regions and to evaluate the strength of responses due to rhythmic stimulation. The coherence function quantifies the association between pairs of EEG signals as a function of frequency. Intra-hemispheric and inter-hemispheric coherences were studied in photosensitive subjects to investigate the mechanism and functional connectivity during photic stimulus (Wada et al. 1996). The aim of the authors' earlier study was to investigate whether the visual stimulation (VS) at ELF could possibly induce changes in the corresponding EEG frequency bands by examining the functional connectedness between brain regions (Cvetkovic and Cosic 2009a). This connectedness and power between brain regions was evaluated by applying the improved objective measures (non-parametric spectral estimation coherence algorithm) and protocol. The authors' earlier results revealed a substantial increase in intra-hemispheric coherence in alpha rhythms at right centro- and parieto-occipital and parietal and inter-hemispheric coherence at parietal and occipital regions. Due to the protocol design, the results could not reveal whether the decreased complexity (relating to increased coherence) of beta waves resulted from the alpha entrainment phenomenon induced by photic stimulation. This study may also increase awareness of the utility of EEG visual driving response studies in EEG clinical practice to uncover potential neurophysiologic abnormalities.

7.6.2 Biofeedback

Biofeedback research was popular in the late 1960s and 1970s, but its marketing and commercialization was much faster, before the proper research could be conducted, so the research declined by the end of 1970s. Unfortunately, it did not deliver what it promised (Buzsaki 2006). However, several years ago, one author (Cvetkovic) decided to dust it off and apply some of the latest biomedical engineering technology to find a scientific treatment for insomnia and voluntarily induce sleep in subjects. This has been author's ongoing research since the start of his Ph.D. studies in 2002. Some of the earlier experiments date back to the 1940s, where a group of researchers attempted to induce a hypnagogic state by controlling physiological variables using induced biofeedback (Kubie and Margolin 1942). At that time, it was revealed that the hypnagogic progression was correlated with decrease in EEG alpha, towards the alpha-theta and eventually to pure theta band. The same progression was also known to be correlated with decrease of ventilation and a switch from abdominal to thoracic breathing. The method basically consisted of an apparatus that fed back sensory signals (tones) to subject, via electronic biosignal data acquisition system and headphones, based on respiratory rhythms, thus inducing a hypnagogic state. This method was used for psychotherapeutic purposes, where the subjective experience from these tests had one subject reporting vivid images of himself as a child. The biofeedback studies are conducted on subjects to train them to attain and maintain a hypnagogic state by monitoring their physiological arousals. Another biofeedback study involved a 15 min alpha–theta feedback, followed by a 30 min theta feedback, where subjects were asked to report their subjective experiences (Green and Green 1978). Once the subjects reached prolonged alpha and theta rhythms, audio signals of 400 and 900 Hz respectively were fed back to them via headphones. This method was used for subjects to practice at prolonging their hypnagogic state on their own. Mavromatis found that most researchers in biofeedback reported their subjects "oscillating in and out of hypnagogia" based on physiological signals. Subjective data showed that hypnagogia was realized but perhaps in a lighter stage. Hypnagogic induction may result in induction not only of hypnagogic states but also certain adjacent states and mental processes, which may still remain unknown. The ultimate goal of one author (Cvetkovic), and apparently of many scientists previously, is to be able to steer subjects' physiological rhythms towards the desired state of consciousness of hypnagogia.

In the late 1980s, a study was conducted to determine whether it is feasible to obtain operant conditioned responses via auditory biofeedback during REM sleep (Sockeel et al. 1987). This study revealed that phasic and tonic components of REM sleep can be altered in a specific manner by continuous or interrupted continuous white noise auditory stimulation. Phasic REM sleep stages are intermittent (rapid eye movements and muscle twitches) and tonic REM sleep stages are continuous (desynchronized EEG and voluntary muscle inhibition). NREM has four stages, whereas REM has phasic and tonic stages. The experimental setup consisted of horizontal EOG for biofeedback control, a REM detector and a sound generator. The auditory stimulations were delivered via headphones under two biofeedback conditions. The first condition consisted of a white noise beep of 40 ms duration once REM episode occurred at intensity 70 and 95 dB. The second condition was a continuous white noise at 70 dB. The REM detection consisted of REM counting and online REM burst detection. The results showed that under the first biofeedback condition, the duration and number of REM sleep episodes increased.

In order to explain how biofeedback works and how it can alter consciousness, we need to know the physiological mechanisms. But they are still unknown. The sleep induction mechanisms can be based on von Economo's prediction (more than 80 years ago) that the region of the hypothalamus near the optic chiasm contains sleep-promoting neurons. The hypothalamus mediates the role of the brainstem in the activation and relaxation of cerebral networks. The hypothalamus also mediates the interaction of biofeedback with the brainstem nuclei. Biofeedback is used to train the arousal regulation and the modulation of the activation and relaxation of cerebral network dynamics. All sensory information, except that of smell, must pass through the thalamus in order to gain access into other brain regions. Auditory and photic stimulation excites electrical potentials within the thalamus, entraining and amplifying a signal which is then distributed throughout the cerebral cortex via thalamocortical loops. In contrast, some studies have reported that it is the subcortical neuromodulators that keep the thalamocortical system awake rather than sensory stimuli from the body and the environment. This thalamocortical system consists of organised sequences of sleep stages and is influenced to alter its sleep architecture. Therefore, both external and internal stimuli need to be considered in altering the thalamocortical system.

7.6.3 Biofeedback: Study 3

In the previous studies, the authors were able to investigate the influence of photic stimulation on the post-processed EEG signals. The authors' next project was related to the development of real-time (online) EEG and ECG signal processing with similar simultaneous photic and audio and non-simultaneous ELF magnetic stimulation (Cvetkovic 2005; Cvetkovic and Cosic 2009b). The aim of this research was to implement multiple online EEG and HRV feature extraction methods and interface them with hardware stimulators for online sensory stimulation. The main applications of this system are primarily for biofeedback and neurofeedback treatments. Typically, biofeedback is understood as detecting EEG and ECG signal features and transforming them into audio tones and flickering light so that the subject can recognise and monitor the occurrence of alpha waves in real time. Alpha and other EEG bands can be fed back to subject, so after a while a subject can drive his or her brain activity towards the alpha state and others. It is known that low frequency flashes can enhance the occipital alpha oscillation.

The biofeedback recording methodology was to record the EEG and ECG signals using the mobile biosignal acquisition device g.MOBIlab (g.tec Medical Engineering GmbH, Austria). The EEG and ECG biosignals were read and processed in a Matlab Simulink (Mathworks, USA) environment online using the g.tec Simulink highspeed online processing blockset. MOBIlab's analog channels were set to read two EEG and one ECG channel. The EEG signals were recorded using C3-A2 and C4-A1 referential montage with ECG recording using Lead-I montage. The EEG and ECG data was further processed and analysed using Matlab

Simulink software. The main online processing consisted of implementation of multiple feature extraction techniques, modelled using customized Simulink embedded functions. Some of the computed EEG feature extractions included: Lempel and Ziv complexity; coherence at each frequency band; mean and median frequencies, brain-rate frequency and 90 and 95% spectral edge frequencies (SEFs); absolute and relative spectral powers at delta, theta, alpha, sigma, beta and gamma bands; and alpha, theta and delta (slow wave) indices (AWI, TWI and DWI). Most EEG extracted features could be selected from various parametric and non-parametric spectral functions. Lempel and Ziv (LZ) complexity was proposed as a useful measure that could characterise the degree of order or disorder and development of EEG patterns (Lempel and Ziv 1976; Zhang and Roy 2001). In the first step, the EEG signals were transformed into binary sequences, then normalized and the rate of new pattern investigated with LZ computational values ranging from near 0 (order) to 1 (disorder or randomness). The online brain-rate (arousal indicator) was determined together with the mean power frequency (MPF) and median frequency (MF), which are two of the most commonly used single indices of the power spectral alterations. The SEF is usually expressed as *SEF x*, which stands at the frequency below which x percent of the total power of a given signal is located (typically x is in the range 75–95%). With power spectral analysis of any signal, single variables could be derived from the native signal, for example SEF 90 or SEF 95, the frequency below which 90 or 95%, respectively, of the total signal power is located (Levy et al. 1980).

The online ECG features were initially extracted from R-R intervals and the following methods were applied to compute them: HRV including mean NN, SDNN, NN50, pNN50, RMMSD; Poincare HRV (SD1, SD2 and SD1/SD2); normalised LF, HF and LF/HF index; and detrended fluctuation analysis (DFA) (alpha). For HRV analysis, the initial step was to detect the R peaks of the ECG signal with a window length of 5 s. To obtain a resolution of 1 s, the window was overlapped by 4 s. Detection times of R peak and intervals between successive R of QRS complex were computed (Pan and Tompkins 1985). The length (SD2) and the width (SD1) of the Poincaré plot images represent short and long-term variability of any nonlinear dynamic system (Brennan et al. 2001). DFA was utilized to investigate the long-range correlations of the R-R interval time series data (Peng et al. 1993).

A decision module was also developed to determine the dynamic behaviour of features such as the increase, decrease and threshold, and to detect the beta, alpha, theta and upper and lower delta bands that characterize the wake and stage 1 sleep as well as lighter and deeper sleep. An algorithm was developed to detect each band and wake and stage 1 (partially) according to the AASM Manual (2007) where the wake stage is scored when more than 50% of the 30 s epoch has an alpha rhythm and stage 1 sleep is characterized by its amplitude decrease and appearance of mixed frequency activity (theta band in 4–7 Hz range) for more than 50% of the 30 s epoch. The hardware part of the biofeedback system was designed to continuously acquire dynamic streamed data from Simulink output(s) to drive the light emitting diodes (LEDs), stereo audio and/or electric solenoid coil(s) at ELF, for selective photic, audio and magnetic field stimulation, respectively. The system consists of three stages: Simulink interface; ADuC microcontroller and custom driver hardware unit. The hardware stimulators consisted of a four-per-eye LED matrix, an audio amplifier to drive the headphones and a solenoid coil.

The work on mapping the extracted features with the stimulator parameters is currently in progress. Once this mapping is complete, the biofeedback system will then be tested on human subjects for its effectiveness and online processing performance. In the meantime, EEG and ECG can be monitored online (as opposed to offline in the past) to determine the effects of photic, audio and magnetic stimulation and at the same induce relaxation and sleep.

7.7 Physiological Synchronisation, Oscillator Interaction and Coupling

Synchronisation is a universal phenomenon that arise observed in many fields of science and applied in electronic and mechanical engineering areas. It is now gaining more recognition for its role in biological and physiological systems.

It is quite clear that the electrical activity in the brain differs considerably from wake to sleep states. When awake, the EEG activity consists of low-amplitude oscillations in a wider frequency band. Once in the sleep state and approaching deeper levels, the amplitude increases in a narrower frequency band. The mean EEG frequency shifts from higher to lower frequencies and from lower to higher amplitudes. We believe that this reflects the increased synchronization of brain activity as sleep depth increases. As wakefulness moves towards sleep, the EEG activity moves from desynchronized to synchronized activity.

For example, positive amplitude-frequency cross-modulation is observed in alpha amplitude and theta phase interaction during light sleep and sigma amplitude and alpha frequency interaction during deep sleep. As the alpha EEG decreases during light sleep, its alpha oscillator shifts into theta band. Conversely, during deep sleep, alpha increases and alpha oscillator shifts to the opposite side, to the sigma band. Also, the negative cross-modulation within the same bands and locations indicates the importance of increasing frequency corresponding to decreasing amplitude. This behaviour can be observed in a typical wake-sleep transition, characterized by a simultaneous decrease of low-amplitude high-frequency alpha band activity and an increase of high-amplitude low-frequency theta and delta bands.

As in a musical symphony, one instrument cannot play all the music. The symphony requires other instruments, which, when played, need to be harmonically in tune. If sleep architecture can be described as a symphony, its vital electrophysiological oscillators (i.e. neuronal, cardiac and respiratory) need to be harmonically coupled to act as one harmonic oscillator. This is why these signals need to be synchronised and explains how they determine the oscillatory characteristics of sleep switching and the transition from waking to sleeping states. The synchronisation of these three signals in the transition may enable us to quantify their interaction. The dynamics of neuronal-cardio-respiratory oscillations during sleep transitions is not yet well enough understood for such quantification. Once the interaction is quantified, we may be able to design a novel biomarker for future research.

The understanding of such basic terms as synchronisation, locking and entrainment differs, depending on the background and individual viewpoint of the researcher. Before defining synchronisation, it is important to define the physiological oscillator(s) used. It needs to be determined whether each of the physiological oscillators are isolated and continue to oscillate in their own rhythms, where such oscillators are said to be autonomous and are in a class of nonlinear models known in physics and nonlinear dynamics as self-sustained or self-oscillatory systems. The intensity of oscillation is determined by its amplitude and a quantity called phase. Next, synchronisation needs to be examined for entrainment of the self-sustained oscillator(s) by external (auditory or photic) or internal (the three physiological oscillators) drivers. A weak external force cannot influence the amplitude but can shift the phase of the oscillator (i.e. synchronisation is observed not as equality of phases but as the onset of a fixed phase difference, called phase locking, with the external force and oscillator frequencies in an $m:n$ ratio). If a very small force (small detuning) can entrain the oscillator and yet pull the frequency of the system towards its own frequency, this is called frequency locking.

A future project may include a similar cross-coupling method over time and in particular over frequency, in order to investigate the cross-frequency coupling of EEG and HRV bands (Cohen 2008). A cross-frequency coupling method could be applied in synchronisation of oscillatory EEG signals across different frequency bands. It could identify dynamic changes in synchronisation over time and frequency. More precisely, Cohen has a method for statistically assessing the degree to which fluctuations in amplitude or power of the higher frequency band are synchronised with the activity in the lower frequency band. Like many other methods, it has its limitations, but to date there are no cross-frequency coupling methods that are flexible for both time and frequency. The cross-frequency coupling between oscillations is used to determine the coupling interaction between frequency bands, such as the coupling interaction between certain EEG bands of higher frequencies and HRV of very low frequencies. Cross-frequency phase interactions ($n:m$ coupling, where n and m are the two frequencies) have two types, for phase locking and amplitude locking. Bispectral and wavelet methods are among the existing methods for analysing cross-frequency coupling.

The applied theory of synchronisation has so far been restricted to periodic oscillators. However, the synchronisation of irregular and non-stationary oscillators such as human heart and respiratory systems has become a popular field of study in the last 10 years since it was first observed in the late 1990s (Schafer et al. 1998). Before that it was understood that cardiac and respiratory rhythms in humans are unsynchronised. The respiratory modulation of the heart rate is a well known phenomenon, according to which the heart rate increases during inspiration and decreases during expiration. In relation to modulation of heart rate by respiration, there is not exactly an entrainment between these two rhythms. It is understood that there is a non-phase-locked weak coupling between cardiac and respiratory activity (Schafer et al. 1998). There is an interaction between them and they are subject to the influence of other physiological systems contributing to cardiovascular control and to external perturbations. If an additional physiological oscillatory system (i.e. neural activity) is added to this weak coupling with external sensory stimulation, it may be possible to change the stability or even the existence of phase-locked solutions, which may alter any phase transitions between these synchronised and non-synchronised states.

The concept of globally coupled oscillators may be applied for the investigation of respiratory-cardio-neural interaction (Pikovsky et al. 2001). Each oscillator can be described as a self-sustained oscillator that can interact with all others and be influenced by a common external force. These oscillators may also share their frequencies, and if they are close to each other or in different frequency bands, then another external force can synchronise all the oscillators by a mean field coupling between all the oscillators. As a result, the oscillators could oscillate coherently with the same frequency but with a possible phase shift. It was reported how selectively distributed oscillatory networks (for delta, theta, alpha and gamma EEG responses) can be activated by sensory (and cognitive) events. These events or stimuli can evoke or induce other oscillations with various amplitudes, phases and synchronisations (Basar et al. 2001). Given that alpha and theta oscillations originate from neocortical alpha and hippocampal theta pacemakers respectively, it is unknown whether each of these pacemakers functions dependently or independently of other alpha and theta pacemakers. Considering the sheer complexity of all the unknowns related to how these alpha and theta pacemakers interact with wake-to-sleep mechanisms, further experiments with stimulus-driven synchrony, from internal cardio-respiration and external photic-auditory stimulators, may be needed to develop a sleep onset pacemaker. To date, very few rhythm combinations in the brain have been studied or understood. Desynchronisation is a well studied but little understood phenomenon. The term refers to the quantitative reduction of alpha activity upon presentation of a stimulus. However, the stimulation often induces or enhances gamma frequency oscillations, which may be the cause of the reduced alpha power. Coupling of oscillators having similar frequency but arising from different architectures, such as the hippocampus and neocortex, provides special challenges. Coupling oscillators of two or more frequencies can generate complex envelopes of population activity and synchrony. Two or more oscillators at different frequencies can relate to each other not only through power modulation of the faster signal but also by phase-coupling of the different oscillators. Such cross-frequency phase synchrony has been observed between alpha and gamma oscillations during mental arithmetic tasks (Palva and Linkenkaer-Hansen 2005).

To decide which of two oscillators is the driver and which is the driven, i.e., to determine the causality between two oscillations, the Granger causality test has been found to be useful, as it uses a probabilistic concept of causality, based on prediction (Granger 1969). The Granger causality test is also effective in determining causality among the simultaneous interaction between multiple signals. The Granger causality measures are defined in terms of coefficients of the multivariate autoregressive (MVAR) model, in both the time and frequency domains. There are two symmetric Granger measures, MVAR coherence and partial coherence, and there are asymmetric or directed Granger causality measures.

Phase synchronisation and the relationship between the phases of neural, cardio and respiratory signals have been investigated in the past, but doing so in real time can be very challenging and problematic in terms of short interval synchronisation, noise and amplitude variations (Van Quyen and Bragin 2007). The phase synchronisation techniques are important and useful in monitoring normal and disordered long-range synchronization, including schizophrenia, epilepsy, dementia and Parkinson's disease.

Phase synchronisation occurs when two oscillators are weakly coupled. One of the most important characteristics of any coupled oscillatory model system is the transition in the synchronisation behaviour. This synchronisation can be altered if noise is applied to the oscillatory system. It is well understood that heart rate and breathing act like two weakly coupled oscillators, with the breathing to heartbeat coupling direction (Bartsch et al. 2007). In the case of cardio-respiration synchronisation, an uncorrelated noise can be applied from the brain which would increase the synchronisation. Conversely, correlated noise from the brain would decrease cardio-respiratory synchronization. During REM sleep, the cardio-respiratory synchronization is typically three times smaller when compared to wakefulness. During NREM sleep, the same synchronization is 2.4 times greater compared to that when awake.

The neuronal activity, acting as noise, can influence the cardiac and respiratory oscillators under the long-term correlations during REM but not NREM sleep. Could the short-term correlations be involved in neural activity on both of these oscillators?

To analyse this question, an algorithm was developed to automatically detect such synchronisation from respiratory and cardiac activity simultaneously. The automated algorithm analysed recordings of the relative position of respiratory inspiration within the corresponding cardiac cycle or cardio-respiratory synchrograms, during wakefulness for phase synchronization. The algorithm took recordings of the times of heartbeats and mapped them to the cumulative phase of the respiratory signal using a Hilbert transform. The ratio of 3 heart beats to 1 breathing cycle was adopted (generating three parallel synchrograms). It was concluded that the correlation stability between REM and NREM in heartbeats and breathing fluctuations was due to the interaction of the central and autonomous nervous systems (CNS and ANS). This is interpreted as heartbeat and breathing generators acting like two weakly coupled oscillators, enhanced by the uncorrelated noise applied from brain. More importantly, this coupling is unidirectional, with respiration influencing the heart beats. This is mainly due to the 1:3 ratio of breaths to heartbeats.

To test the hypothesis that the characteristic alterations exist in both cardio-respiratory and cerebral oscillator parameters and that the couplings vary with depth of anaesthesia in rats, a phase dynamics approach was developed to extract the instantaneous frequencies and phases of heartbeat, respiration and slow delta waves (0.5–3.5 Hz) (Musizza et al. 2007). The Morlet mother wavelet transform was used to extract the amplitudes of delta and theta EEG bands and a Hilbert transform was applied to obtain the instantaneous delta frequency. Also, nonlinear dynamics and information theory were employed to find the causal relationship between the cardiac, respiratory and slow delta oscillations. It was demonstrated that respiration drives the cardiac oscillator during deep anaesthesia and there is a unidirectional interaction of respiration with delta oscillations, whereas during the shallow anaesthesia, the direction either reverses or the cardio-respiratory interaction is insignificant. There is also evidence of interaction between respiratory and delta oscillations, with respiration driving cortical activity during deep anaesthesia. This was modelled with respiratory, cardiac and cortical (delta band) oscillators and mutual interactions. The Musizza group concluded that nonlinear dynamics and information theory can be used to identify different stages of anaesthesia and the effects of different anaesthetics. It is also known that anaesthesia is another state where neuronal oscillations synchronise and desynchronise. The authors believe that similar neuronal oscillatory synchronisation and desynchronisation can be observed throughout the sleep onset process.

One of the biggest challenges in matching nonlinear model predictions to real physiological data is the assumption that the signal is stationary and the need to correct for noise by distinguishing the signal from artefacts and noise-like processes originating in the EEG activity. A method is needed that can reduce the noise and then compute a statistic that can measure the reduced artefact processes. One such method is the DFA algorithm (Peng et al. 1992). The authors implemented an automated DFA method which was applied to the time-series of R-R intervals for quantifying correlations in non-stationary signals (i.e. physiological signals). This method estimates locally detrended scaling exponents and can discriminate between random and long-range correlations. A study was undertaken to compare the HRV spectral analysis and DFA in order to discriminate sleep stages and sleep apnoea (Penzel et al. 2003). The HRV LF component decreases from wake to light sleep and further to deep sleep. However, it increases during REM sleep to LF values similar to those exhibited during wakefulness. The HR component increases from waking to light sleep and further to deep sleep and decreases during REM. The study stated that "short-range correlations due to the effects of breathing on heart rate . . . could be related to effects of slower brain functions seen in sleep regulation on heart rate". This explains and highlights the strong interaction between neuronal, cardio and respiratory systems during sleep.

In order to fully "understand the dynamics of a complex system, one should thus identify its components and quantify their interactions" (Gans et al. 2009, p. 098701–1). This quantification is related linear cross-correlation and nonlinear phase synchronization interactions between many physiological oscillators or many quasi-oscillatory processes in the brain alone. However, there are limitations in applying these quantifying measures. Some methods are only applicable to narrow or wide frequency bands, or short-term or long-term interactions between oscillators. Therefore, for each of these conditions, a particular method was needed.

The instantaneous amplitude and frequency for each component was extracted using a Hilbert transform. Another method involved an empirical mode decomposition (EMD), proposed by Huang and Hilbert, to preprocess and reduce the noise in respiratory experimental signals (Wu and Hu 2006). This study found that the frequency of alpha activity was positively modulated with the amplitude of delta activity. Conversely, the frequency of theta activity was negatively modulated with the amplitude of delta activity. The interactions between these positive and negative modulations was lower during REM as compared to NREM (light and deep sleep).

To study long-term correlations of EEG signals of amplitudes and frequencies in all bands, the centred moving average (CMA) technique was preferred over the autocorrelation function. Autocorrelation is typically not used due to the non-stationary nature of EEG signals and statistical reliability. CMA is equivalent to DFA. The results revealed that the frequency-frequency modulation was synchronous in the alteration of alpha and beta bands. However, this synchronisation was shown only during wakefulness and not during sleep. The authors therefore suggest that stochastic nonlinear dynamics and information theory be applied to observe the interaction, to measure and identify different phases (not stages) of sleep onset process.

7.8 Discussion and Conclusion

So is there an altered state of consciousness related solely to the sleep onset process and the hypnagogic state?

We believe that wakefulness and sleep are not independent states of consciousness and that they might be mutually inclusive. The three principal states of consciousness are defined as wakefulness, REM sleep and slow wave sleep (SWS). We believe that the wake-to-sleep transition should be regarded as an altered state of consciousness. Consider that SWS characterizes a significant part of sleep depth and that as the SWS increases, the synchronization of the EEG increases (as it moves from desynchronized to synchronized activity). However, this is not the case for sleep onset. The synchronization is dynamic and varies between the neural dynamics and the heart rate variability during the sleep onset. In turn, this makes quantifying sleep onset as an altered state of consciousness even more challenging due to this sleep onset complexity and instability.

The waking state and REM are characterized by desynchronized EEG activity, suggesting that the gamma band may be the signature of this waking and REM sleep EEG synchrony. Interestingly, these gamma-band oscillations are present in the process of meditation too. However, the sleep onset and meditation processes are quite different from each other. The synchronization of neuronal oscillations has been identified in mental processes to be a basis of human and animal attention, learning, conscious perception and working memory. Ever since Berger's discovery of EEG, the functional significance of synchronization activity remains unknown. What is also unknown is the location in the brain where the sleep onset and consciousness processes run. Perhaps there is no fixed location. They could be dynamic. If so, can sleep onset be characterized or modelled? It is reported that sleep onset does not necessarily sit in any single region of the brain, that "structures in the brainstem regulate our states of consciousness" (Zeman 2001, p. 1268) and that the suprachiasmatic nucleus of hypothalamus is a "timekeeper" of consciousness. While there have been advances in research to localise sleep induction in the anterior hypothalamus and basal forebrain, the exact moment of sleep onset has not been established or characterised, mainly due to variability within and between subjects. One author (Cvetkovic) suggests that the entire brain may be characterised as a global oscillator, which may contain multiple thalamocortical (including other brain architecture) oscillators responsible for sleep stage activity (including sleep onset), as well as cardiac and respiratory oscillators. These oscillators may also be referred to as pacemakers. One of the features of biological oscillators is their ability to synchronise to, or be entrained by, external inputs. While the exact nature of the central respiratory pattern generating system remains open to debate, it may broadly be classified as a kind of oscillator system.

The first author (Cvetkovic) is continuing his research on the quantification and treatment of insomnia, which is often related to sleep onset instability. The computational biomarker algorithm will need to be developed to incorporate the automated (instantaneous) measurement of degree of synchronization and the strength of interaction and directionality of coupling between the neural, cardio and respiratory oscillators during the sleep onset process. The accuracy and efficiency of this biomarker algorithm will greatly depend on the ability to discriminate between normal synchronization, interaction and coupling and the instabilities that may characterize insomnia and other neurological disorders. The instabilities associated with insomnia will need to be instantaneously corrected using sensory (auditory and photic) driving responses or a biofeedback steady-state evoked potential approach (being developed by Cvetkovic), in order to create the seamless sleep onset associated with healthy sleep (yet to be fully explored).

References

Adrian ED, Matthews BHC (1934) The Berger rhythm: potential changes from the occipital lobes in man. Brain 57:355–384. doi:10.1093/brain/57.4.355

Badia P, Wright KP Jr, Wauquier A (1994) Fluctuations in single-hertz EEG activity during the transition to sleep. In: Ogilvie RD, Harsh JR (eds) Sleep onset: normal and abnormal processes. American Psychological Association, Washington

Bartsch R, Kantelhardt JW, Penzel T, Havlin S (2007) Experimental evidence for phase synchronization transitions in the human cardio-respiratory system. Phys Rev Lett 98:054102

Basar E (2008) Oscillations in "brain-body-mind": a holistic view including the autonomous system. Brain Res 1235:2–11

Basar E, Basar-Eroglu C, Karakas S, Schurmann M (2001) Gamma, alpha, delta, and theta oscillations govern cognitive processes. Int J Psychophysiol 39:241–248

Benesty J, Chen J, Huang Y (2005) A generalized MVDR spectrum. IEEE Signal Process Lett 12(12):827–830

Bonnet M, Carley D, Carskadon M, Easton P, Guilleminault C, Harper R, Hayes B, Hirshkowitz M, Periklis K, Keenan S, Pressman M, Roehrs T, Smith J, Walsh J, Weber S, Westbrook P (1992) EEG arousals: scoring rules and examples, a preliminary report from the sleep disorders atlas task force of the american sleep disorders association. Sleep 15:173–184

Brennan M, Palaniswami M, Kamen P (2001) Do existing measures of Poincaré plot geometry reflect nonlinear features of heart rate variability? IEEE Trans Biomed Eng 48:1342–1347

Britvina T, Eggermont JJ (2008) Multi-frequency auditory stimulation disrupts spindle activity in anesthetized animals. Neuroscience 151:888–900

Bulow K (1963) Respiration and wakefulness in man. Acta Physiol Scand 59:1–110

Burgess HJ, Kleiman J, Trinder J (1999) Cardiac activity during sleep onset. Psychophysiology 36:298–306

Buzsaki G (2006) Rhythms of the brain. Oxford University Press, New York

Cantero JL, Atienza M, Salas RM (2002) Effects of waking-auditory stimulation on human sleep architecture. Behav Brain Res 128:53–59

Cohen MX (2008) Assessing transient cross-frequency coupling in EEG data. J Neurosci Meth 168:494–499

Cvetkovic D (2005) Electromagnetic and audio-visual stimulation of the human brain at extremely low frequencies. PhD Thesis (Biomedical Engineering), RMIT University, School of Electrical and Computer Engineering, Melbourne, Australia

Cvetkovic D, Cosic I (2008) Sleep onset estimator: evaluation of parameters. In: 30th annual international conference IEEE engineering in medicine and biology society (EMBS), Vancouver, Canada, pp 3860–3863

Cvetkovic D, Cosic I (2009a) EEG inter/intra-hemispheric coherence and asymmetric responses to visual stimulations. Med Biol Eng Comput 47(10):1023–1034. doi:10.1007/s11517-009-0499-z

Cvetkovic D, Cosic I (2009b) Alterations of human electroencephalographic activity caused by multiple extremely low frequency magnetic field exposures. Med Biol Eng Comput 47(10): 1063–1073. doi:10.1007/s11517-009-0525-1

Cvetkovic D, Powers R, Cosic I (2009) Preliminary evaluation of EEG entrainment using thalamocortical modeling. Expert Syst 26(4):320–338. doi:10.1111/j.1468-0394.2009.00493.x

Czisch M, Wetter TC, Kaufmann C, Pollmacher T, Holsboer F, Auer DP (2002) Altered processing of acoustic stimuli using sleep: reduced auditory activation and visual deactivation detected by a combined fMRI/EEG study. Neuroimage 16:251–258. doi:10.1006/nimg.2002.1071

Davies H, Davis PA, Loomis AL, Harvey EN, Hobart G (1938) Changes in human brain potentials during the onset of sleep. Science 86:448–450

Dement WC, Kleitman N (1957) The relation of eye movements during sleep to dream activity: an objective method for the study of dreaming. J Exp Psychol 53:339–346

Erol B (1999) Brain function and oscillations. II. Integrative brain function. Neurophysiology and cognitive processes. Springer, Berlin, 129–142

Freedman SJ, Marks PA (1965) Visual imagery produced by rhythmic photic stimulation: personality correlates and phenomenology. Brit J Psychol 56(1):95–112

Gans F, Schumann AY, Kantelhardt JW, Penzel T, Fietze I (2009) Cross-modulated amplitudes and frequencies characterize interacting components in complex systems. Phys Rev Lett 102: 098701

Granger CWJ (1969) Investigating causal relations by econometric models and cross-spectral methods. Econometrica 37:424–438

Green E, Green A (1978) Beyond biofeedback. Bell, London, UK

Gurtelle EB, de Oliveira JL (2004) Daytime parahypnagogia: a state of consciousness that occurs when we almost fall asleep. Med Hypothesis 62:166–168

Hamann C, Bartsch RP, Schumann AY, Penzel T, Havlin S, Kantelhardt JW (2009) Automated synchrogram analysis applied to heartbeat and reconstructed respiration. Chaos 19:015106

Himanen S-L, Virkkala J, Huhtala H, Hasan J (2002) Spindle frequencies in sleep EEG show U-shape within first four NREM sleep episodes. J Sleep Res 11:35–42

Hjorth B (1970) EEG analysis based on time domain properties. Electroencephalogr Clin Neurophysiol 29:306–310

Hori T (1985) Spatiotemporal changes of EEG activity during waking-sleeping transition period. Int J Neurosci 27:101–114

Huupponen E, Himanen S-L, Hasan J, Varru A (2003) Sleep depth oscillations: an aspect to consider. J Med Syst 27(4):337–345

Huupponen E, Himanen S-L, Hasan J, Varri A (2004) Automatic quantification of light sleep shows differences between apnea patients and healthy subjects. Int J Psychophysiol 51: 223–230

Huupponen E, De Clercq W, Gomez-Herrero G, Saastamoinen A, Egiazarian K, Varri A, Vanrumste B, Vergult A, Van Huffel S, Wim Van Paesschen J, Hasan J, Himanen S-L (2006) Determination of dominant simulated spindle frequency with different methods. J Neurosci Meth 156:275–283

Huupponen E, Gomez-Herrero G, Saastamoinen A, Varri A, Hasan J, Himanen S-L (2007) Development and comparison of four sleep spindle detection methods. Artif Intell Med 40: 157–170

Iber C, Ancoli-Israel S, Chesson AL, Quan SF (2007) The AASM manual for the scoring of sleep and associated events: rules, terminology and technical specifications. American Academy of Sleep Medicine, Westchester, IL

Jurysta F, van de Borne P, Migeotte P-F, Dumont M, Lanquart J-P, Degaute J-P, Linkowski P (2003) A study of the dynamic interactions between sleep EEG and heart rate variability in healthy young men. Clin Neurophysiol 114:2146–2155

Kubie L, Margolin S (1942) A physiological method for the induction of states of partial sleep and securing free associations and early memories in such states. Trans Am Neurol Assoc 68: 136–139

Lempel A, Ziv J (1976) On the complexity of finite sequences. IEEE Trans Inform Theor 22(1): 75–81

Levy WJ, Shapiro HM, Maruchak G, Meathe E (1980) Automated EEG processing for intraoperative monitoring. Anesthesiology 53:223–236

Liberson WT, Liberson CW (1966) EEG records, reaction times, eye movements, respiration, and mental content during drowsiness. In: Wortis J (ed) Recent advances in biological psychiatry, vol 8. Plenum, New York

Llinas RR, Steriade M (2006) Bursting of thalamic neurons and states of vigilance. J Neurosci 95:3297–3308

Lubin A, Johnson LC, Austin MT (1969) Discrimination among states of conciousness using EEG spectra. Psychophysiology 6:122–132

Mavromatis A (1987) Hypnagogia. Routledge and Kegan Paul, London, UK

Merica H, Fortune RD (2004) Physiological review, state transitions between wake and sleep, and within the ultradian cycle, with focus on the link to neuronal activity. Sleep Med Rev 8: 473–485

Miller LM, Schreiner CE (2000) Stimulus-based state control in the thalamocortical system. J Neurosci 20:7011–7016

Musizza B, Stefanovski A, McClintock PVE, Palus M, Petrovcic J, Ribaric S, Bajrovic FF (2007) Interactions between cardiac, respiratory and EEG-δ oscillations in rats during anaesthesia. J Physiol 580:315–326

Naifeh KH, Kamiya J (1981) The nature of respiratory changes associated with sleep onset. Sleep 4:49–59

Nielsen-Bohlman L, Knight RT, Woods DL, Woodward K (1991) Differential auditory processing continuous during sleep. Electroencephalogr Clin Neurophysiol 79:281–290

O'Donnell BF, Wilt MA, Brenner CA, Busey TA, Kwon JS (2002) EEG synchronization deficits in schizophrenia spectrum disorders. Int Congr Ser 1232:697–703

Ogilvie RD (2001) Physiological review, the process of falling asleep. Sleep Med Rev 5(3): 247–270

Ogilvie RD, Simons IA, MacDonald T, Rustember J (1991) Behavioural event-related potential and EEG/FFT changes at sleep onset. Psychophysiology 28:54–64

Ogilvie RD, Wilkinson RT (1984) The detection of sleep onset: behavioural and physiological convergence. Psychophysiology 21:510–520

Ogilvie RD, Simons IA, MacDonald T, Rustember J (1991) Behavioural event-related potential and EEG/FFT changes at sleep onset. Psychophysiology 28:54–64

Palva S, Linkenkaer-Hansen K (2005) Early neural correlates of conscious somatosensory perception. J Neurosci 25:5248–5258

Pan J, Tompkins WJ (1985) Real time QRS detector algorithm. IEEE Trans Biomed Eng 32(3): 230–236

Peng CK, Buldyrev S, Goldberger AL, Havlin S, Sciortino F, Simons M, Stanley HE (1992) Long-range correlations in nucleotide sequences. Nature 356:168–170

Peng CK, Mietus J, Hausdorff JM, Havlin S, Stanley HE, Goldberger AL (1993) Long-range anticorrelations and non-gaussian behavior of the heartbeat. Phys Rev Lett 70:1343–1346

Penzel T, Conradt R (2000) Computer-based sleep recording and analysis. Sleep Med Rev 4(2): 131–148

Penzel T, Stephan R, Kubicki S, Herrmann WM (1991) Integrated sleep analysis, with emphasis on automatic methods. In: Degen R, Rodin EA (eds) Epilepsy, sleep and sleep deprivation, 2nd edn. Elsevier, Amsterdam, pp 177–200 (Epilepsy Res. Suppl. 2)

Penzel T, Kantelhardt JW, Grote L, Peter J-H, Bunde A (2003) Comparison of detrended fluctuation analysis and spectral analysis for heart rate variability in sleep and sleep apnea. IEEE Trans Biomed Eng 50(10):1143–1151

Pikovsky A, Rosenblum M, Kurths J (2001) Synchronisation a universal concept in nonlinear sciences, vol 12, Cambridge Nonlinear Science Series. Cambridge University Press, UK, pp 126–129

Pop-Jordanov J, Pop-Jordanova N (2004) Quantum interpretation of mental arousal spectra. Contrib Sec Biol Med Sci MANU XXV(1–2):75–94

Pop-Jordanova N, Pop-Jordanov J (2005) Spectrum-weighted EEG frequency ("brain-rate") as a quantitative indicator of mental arousal. Contrib Sec Biol Med Sci MASA 2(2):35–42

Rechtschaffen A (1994) Sleep onset: conceptual issues. In: Ogilvie RD, Harsh JR (eds) Sleep onset: normal and abnormal processes. American Psychological Association, Washington, pp 3–18

Rechtschaffen A, Kales A (1968) A manual of standardized terminology techniques and scoring system for sleep stages of human subjects. US Government Printing Office, Washington, DC, Publication No. 204

Riemann D, Spiegelhalder K, Feige B, Voderholzer U, Berger M, Perlis M, Nissen C (2010) The hyperarousal model of insomnia: a review of the concept and its evidence. Sleep Med Rev 14(1):19–31

Saastamoinen A, Huupponen E, Varri A, Hasan J, Himanen S-L (2006) Computer program for automated sleep depth estimation. Comput Meth Programs Biomed 82:58–66

Schafer C, Rosenblum MG, Kurths J, Abel H-Henning (1998) Heartbeat synchronized with ventilation. Nature 392:239–240

Shinar Z, Baharav A, Akselrod S (2003) Changes in autonomic nervous system activity and in electro-cortical activity during sleep onset. Comput Cardiol 30:303–306

Shinar Z, Akselrod S, Dagan Y, Baharav A (2006) Autonomic changes during wake-sleep transition: a heart rate variability based approach. Auton Neurosci Basic Clin 130:17–27

Sockeel P, Mouze-Amady M, Leconte P (1987) Modification of EEG asymmetry induced by auditory biofeedback loop during REM sleep in man. Int J Psychophysiol 5:253–260

Steriade M, Nunez A, Amzica F (1993a) A novel slow ($<$1Hz) oscillation of neocortical neurons in vivo: depolarizing and hyperpolarizing components. J Neurosci 13:3252–3265

Steriade M, McCormick D, Sejnowski T (1993b) Thalamocortical oscillations in the sleeping and aroused brain. Science 13(262):679–685

Swarnkar V, Abeyratne UR, Huskins C, Duce B (2009) A state transition-based method for quantifying EEG sleep fragmentation. Med Biol Eng Comput 47:1053–1061

Terrillon J-C, Marques-Bonham S (2001) Does recurrent isolated sleep paralysis involve more than cognitive neurosciences? J Sci Explor 15(1):97–123

Terzano MG, Parrino L, Sherieri A, Chervin R, Chokroverty S, Guilleminault C, Hirshkowitz M, Mahowald M, Moldofsky H, Rosa A, Thomas R, Walters A (2001) Atlas, rules, and recording techniques for the scoring of cyclic alternating pattern (CAP) in human sleep. Sleep Med 2:537–553

Toman J (1941) Flicker potentials and the alpha rhythm in man. J Neurophysiol 4:51–61

Trinder J, Whitworth F, Kay A, Wilkin P (1992) Respiratory instability during sleep onset. J Appl Physiol 73:2462–2469

Van Quyen ML, Bragin A (2007) Analysis of dynamic brain oscillations: methodological advances. Trends Neurosci 30(7):365–373

Vetrugno R, Alessandria M, D'Angelo R, Plazzi G, Provini F, Cortelli P, Montagna P (2009) Status dissociates evolving from REM sleep behavior disorder in multiple system atrophy. Sleep Med 10(2):247–252

Wada Y, Nanbu Y, Kadoshima R, Jiang Z-Y, Koshino Y, Hashimoto T (1996) Interhemispheric EEG coherence during photic stimulation: sex differences in normal young adults. Int J Psychophysiol 22:45–51. doi:10.1016/0167-8760(96)00011-6

Walter VJ, Walter WG (1949) The central effects of rhythmic sensory stimulation. Electroencephalogr Clin Neurophysiol 1:57–86. doi:10.1007/s11517-008-0321-3

Will U, Berg E (2007) Brain wave synchronization and entrainment to periodic acoustic stimuli. Neurosci Lett 424:55–60

Wu M-C, Hu C-K (2006) Empirical mode decomposition and synchrogram approach to cardiorespiratory synchronisation. Phys Rev E 73:051917

Zeman A (2001) Consciousness. Brain 124:1263–1289

Zhang X-S, Roy RJ (2001) EEG complexity as a measure of depth of anesthesia for patients. IEEE Trans Biomed Eng 48(12):1424–1433

Chapter 8
Brain Rate as an Indicator of the Level of Consciousness

Nada Pop-Jordanova

Abstract As a general activation of the mind, mental arousal characterizes the level of consciousness, irrespective of its content. Recently, a parameter called the brain rate (expressing the mean frequency of brain rhythms) has been introduced. It appears that the brain rate, which is equal to an EEG spectrum weighted frequency, can serve as a diagnostic indicator of general mental activation (level of consciousness) and serve in addition to heart rate, blood pressure, and temperature as a standard indicator of general bodily activation. Empirically, it has been shown that brain rate measurements can be used to discriminate between the groups of under-arousal and over-arousal disorders, and to assess attention deficit hyperactivity disorder, inner arousal, and the quality of sleep, as well as to indicate the IQ changes caused by some environmental toxins. Brain rate is also suitable for revealing the patterns of sensitivity/rigidity in the EEG spectrum, including frequency bands related to the permeability of corresponding neuronal circuits, based on which individually adapted biofeedback protocols can be specified.

8.1 Introduction

The cerebral cortex is, without doubt, the most complex structure formed during evolution. Moreover, it represents the ultimate physical substrate from whose activity human consciousness emerges.

A number of definitions and proposals concerning the nature of consciousness have been discussed, identifying it as: subjective experience or awareness or wakefulness or the executive control system of the mind. In medicine (e.g. anesthesiology) consciousness is simply regarded as wakefulness and is assessed using the Glasgow coma scale, by observing a patient's alertness and responsiveness.

N. Pop-Jordanova (✉)
Department of Psychophysiology, Paediatric Clinic, Faculty of Medicine, University of Skopje, Vodnjanska 17, Republic of Macedonia

D. Cvetkovic and I. Cosic (eds.), *States of Consciousness*, The Frontiers Collection,
DOI 10.1007/978-3-642-18047-7_8, © Springer-Verlag Berlin Heidelberg 2011

Summarizing, the standard definition of consciousness as awareness of self and the surroundings can be inferred.

The states of consciousness concern subjective experience, which may imply three sorts or characteristics of states (Chalmers 2000): (a) being conscious, (b) the background state of consciousness, and (c) the contents of consciousness.

The first characteristic of a state is just that it is conscious as opposed to not being conscious. The related neural correlates of consciousness (NCC) are specific systems in the brain whose activity is the direct cause of the state of conscious experience, so that consciousness appears to be a property of the activity of these neural areas.

The background state of consciousness is an overall characteristic and is also related directly to the NCC. The person is awake, asleep, dreaming, under hypnosis, etc. The background state of consciousness may be understood as a finer-grained version of the first characteristic (being conscious) and is also dependent on the state of the neural system.

The third characteristic of consciousness is its content. The content of a state might include the experience of a particular visual image, sound pattern, thought, etc. The NCC of the content of consciousness is presumed to be related to the activation of specific areas in the cortex.

On the other hand, mental arousal is primarily defined as a "general activation of the mind" resulting from a person's interaction with the environment (Kahnemann 1973). Thacher and John (1977) defined mental arousal as a "general operation of consciousness", while Damasio (2003) defined it as a "simple response of the brain related to increases in activity", lying beneath auto-regulating homeostasis, balancing all bodily functions.

In all these definitions, mental arousal appears as a general, basic and integral characteristic of a mental state, characterizing the background state, i.e. the level of consciousness, irrespective of its content (Pop-Jordanov and Pop-Jordanova 2009).

The main objective of this chapter is to consider mental arousal as representing the level of consciousness, with a focus on its quantification and applications.

8.2 Empirical Evidence

It is a well-established empirical fact that the level of mental arousal is correlated with EEG frequency. Specifically, beta waves are correlated with an eyes-open alert state, alpha waves are produced in an eyes-closed and quiet but waking state, while slow waves like theta and delta are produced in a drowsy state or in sleep. This can be expressed in a basic classification of EEG activity (Table 8.1) and a more elaborate subdivision of bands (Table 8.2).

It is important to mention that the actual electric activity of the brain is polychromatic, represented by a time-changing EEG spectrum. Thus in infants and young children the dominant EEG frequencies are slow (delta and theta) waves. The increase in EEG frequency during a lifetime reflects the process of brain

Table 8.1 Textbook classification of EEG activity and the level of arousal (Pritchard and Alloway 1999)

Classification	Frequency (Hz)	Level of Arousal
Beta waves	14–30	Alert, eyes open
Alpha waves	8–14	Quiet waking, eyes closed
Theta waves	4–8	Drowsy, sleep stages 1 and 2
Delta waves	0.5–4	Deep sleep, stages 3 and 4

Table 8.2 Extended classification of EEG activity and the mental states (Bendorfer 2001)

Brain waves	Frequency (Hz)	States
Gamma	35+	Association with peak performance
High Beta	22–25	High correlation with anxiety
Mid Beta	15–20	Active, external attention
SMR	12–15	Relaxed state, boby stillness
Alpha	8–12	Relaxed, passive attention
Theta	4–7	Very relaxed, inwardly focused
Delta	0.5–3	Sleep, deep mediation

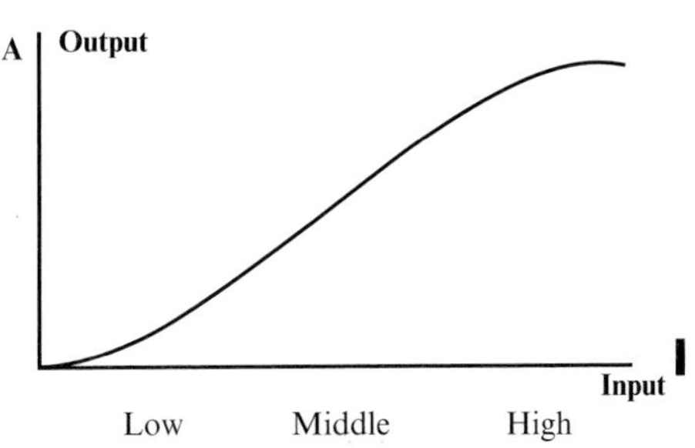

Fig. 8.1 Schematic representation of the dependence of arousal (A) on input agents

maturation. If we calculate the brain rate, we obtain that it rises with the age too. In particular, the alpha peak frequency varies as a function of age, neurological diseases, memory performance, brain volume and task demands. At the age of 12 years, a well-developed alpha peak at about 10 Hz can be distinguished.

On the other hand, standard approaches to neuronal activity consider each of the brain systems (sensory, executive, etc.) as a neuronal network, where the transfer operation performed by a neuron or the whole network is represented by a sigmoid dependence of the overall output activity (associated with arousal) on the input that drives the system (Kropotov 2009). This means that the output varies from poor activation by a low input, through a nearly linear change for a moderate input, to a ceiling (plateau) at high inputs (Fig. 8.1).

Combining the empirically established correlations from Tables 8.1 and 8.2 with Fig. 8.1, the qualitative dependence of mental arousal on EEG frequency is summarized in Fig. 8.2.

The corresponding brain activity can be classified in three groups: obligate unconscious, facultative unconscious, and obligate conscious (Table 8.3). Here,

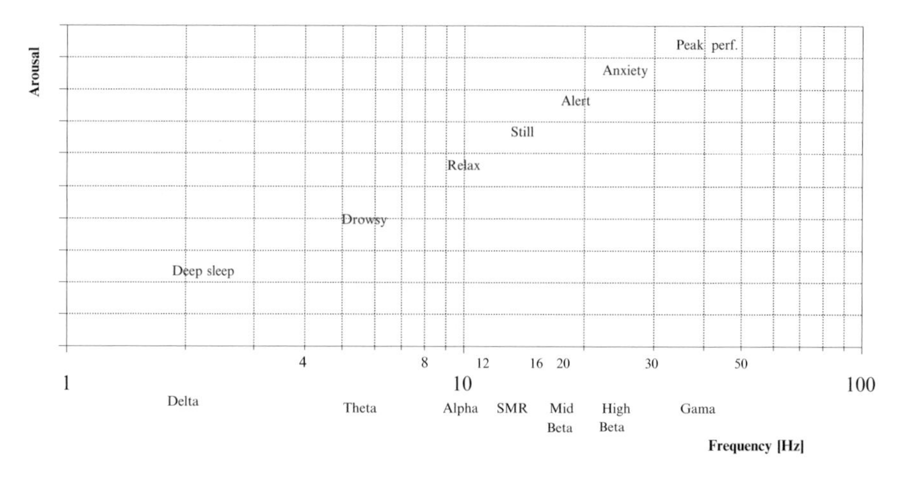

Fig. 8.2 Standard classification of dominant EEG activity and the correlated mental states, i.e. arousal (adapted from Tables 8.1, 8.2 and Fig. 8.1). Note the log scale on the abscissa and the absence of scale on the ordinate

Table 8.3 Unconscious and conscious brain activities (McFadden 2006)

Obligate unconscious	Facultative unconscious	Obligate conscious
Endocrine control	Breathing	Reading
Temperature homeostasis	Maintaining balance during locomotion	Writing creative activity
	Learned activities, such as driving or riding a bicycle	Learning conversing
	Eating	Arithmetic
	Recalling information	Memorizing information

the facultative activities can be performed in parallel but the last group appears to require serial performance.

Based on detailed integral analyses of all the empirical evidence listed in this chapter, a substantial still unanswered question can be formulated: what is the possible theoretical explanation of this arousal-frequency correlation? Its consideration is beyond the scope of this chapter [more details and one possible solution based on quantum transition probabilities can be found in Pop-Jordanov and Pop-Jordanova (2009, 2010)].

8.2.1 *The Algorithm for Brain Rate Calculation*

Bearing in mind the origin and the meaning of the empirical data presented in Fig. 8.2, the spectrum-weighted mean frequency of the EEG rhythms may serve as a quantitative indicator of the general brain activation or consciousness level (independently of the starting approach, whether it be classical or quantum).

In an earlier work, Pop-Jordanova and Pop-Jordanov (2005) have called this parameter the *brain rate* (by analogy with, e.g., heart rate, defined as the mean frequency of heart rhythms). As such, it can contribute to a gross, initial assessment, without replacing the subtle, differential investigations of the various disorders corresponding to the same general level of activation.

Since it is defined as the mean frequency of brain oscillations weighted over the all bands of the EEG potential (or power) spectrum, the brain rate (f_b) maybe calculated as:

$$f_b = \sum_i f_i P_i = \sum_i f_i \frac{V_i}{V}, \quad \text{with } V = \sum_i V_i, \tag{8.1}$$

where the index i denotes the frequency band (for delta $i = 1$, for theta $i = 2$, etc.) and V_i is the corresponding mean amplitude of the electric potential (or power). Following the standard five-band classification, one has $f_i = 2$, 6, 10, 14 and 18, respectively.

Of course, a more precise calculation of f_b can be performed using the corresponding integral form, where the frequency bands are merged in the limit:

$$f_b = \frac{1}{V}\int fV(f)df, \quad V = \int V(f)df. \tag{8.2}$$

In order to illustrate the applications, in the following sections we use (8.1) based on the frequency bands (whose amplitudes are explicitly presented numerically by the corresponding EEG instruments).

As can be seen, the formula for calculating the brain rate f_b is based on the same algorithm as that used for determining characteristic parameters of a spectral system. Namely, the parameter characterizing the state and the changes (shifts) of any system with spectral properties is the spectrum's center of gravity, i.e. the mean value that comprises weighted contributions of all spectral components (bands). Similarly, for example, the physical phase (solid $\leftrightarrow$ liquid $\leftrightarrow$ gas) of an aggregate depends on the integral parameter temperature T, the stability of a boat on the center of gravity Xc, and the criticality of a reactor cell on the neutron spectrum averaged reaction rate R.

The analogy with the temperature is the most striking, since both concepts are based on similar physical models. Namely, the temperature $\left(T = 2/3\sum_i E_i P_i\right)$ represents the mean energy weighted over the particle distribution spectrum (with P_i being the probability of having energy E_i), while the brain rate $f_b = \sum_i f_i Pi$ represents the mean EEG frequency weighted over the brain potential (or power) distribution spectrum (with P_i being the probability of having frequency f_i). Alternatively, in the case of neutron spectra, there are slow (thermal), medium (resonant) and fast (fission) regions, which can be compared with the slow (delta/theta), medium (alpha) and fast (SMR/beta) regions (bands) of the EEG spectrum.

Despite the different contents, these analogies may suggest the use of similar mathematical representations. Consequently, the brain rate f_b can indicate the states of under-arousal (UA) or over-arousal (OA), in the same way as the other mentioned indicators (Xc, T, R) differentiate the levels of activation of the corresponding physical systems.

The brain rate as an indicator of mental arousal can be calculated using EEG parameters at any site on the scalp (Cz, Fz, etc.). For preliminary diagnosis, we propose Cz as a central and most informative point. In medical practice, the brain rate can be used for checking mental arousal, in the same way that temperature, pulse, blood pressure, etc. are used as indicators of bodily functions.

In the following sections, the concept of brain rate as a consciousness level indicator is applied for estimating subgroup differentiation, attention deficit hyperactivity disorder, inner arousal and sleep quality, as well as the biofeedback efficiency and individual band sensitivity.

8.3 Subgroup Differentiation and ADHD

Specific correlations between the brain rate and the manifested level of clinical arousal can be inferred. Slow waves or UA can be noted in depression or autism, while OA and fast waves are registered in anxiety, alcoholism or caffeine intake. So-called mixed disorders (UA/OA) can be seen in some forms of headache, obsessive compulsive disorder, or attention deficit hyperactivity disorder (ADHD), where subgroups of the disorder are defined from the prevalence of fast or slow waves.

If we start the evaluation of a patient by calculating the brain rate, we can make well founded decisions about continuing the treatment protocol (psychotherapy, medication or neurofeedback). In mental UA states (depression, autism, mental retardation) the main aim of the treatment must be accelerating the mental activity. Conversely, in OA states (anxiety, drug abuse) the aim is to slow down the mental activity. In the mixed (OA/UA) states, such as ADHD, the protocol must be adapted in correlation to the mental arousal state, evaluated by the brain rate formula.

ADHD is a common mental problem in schoolchildren, with an incidence estimated between 1 and 12% in different world regions. Main characteristics of the disorder are: inattention, impulsivity, distractibility and hyperactivity. The clinical classification is according to the Diagnostic and Statistic Manual for Mental Diseases (DSM-IV). Modern techniques like positron emission tomography (PET) or functional magnetic resonance (fMRI) show structural differences on the frontal lobe of the neocortex or in subcortical structures like the basal ganglia and thalamus. Neurochemical changes in the central nervous system (CNS) are also involved, concerning the level of transmitters like dopamine, serotonin and catecholamine. Likewise, using the quantitative electroencephalogram (q-EEG) we can point out significant differences between ADHD and normal children. In this context, q-EEG is a very favorable diagnostic tool for ADHD.

The q-EEG studies in the ADHD group demonstrated increased theta activity predominantly in frontal regions, and decreased beta activity in comparison to normal children (Mann et al. 1992). In this context, increased theta/beta ratio is reported as a typical finding in ADHD children (Lubar 1991; Monastra et al. 2001; Muller 2006). Consequently, the typical ADHD finding can be under-arousal. The cluster analysis identified several subgroups of ADHD based on different EEG topographies. As a consequence, four distinct subgroups of ADHD with regard to electrophysiology are defined (Kropotov 2009).

The four subgroups correspond to excesses of activity in corresponding cortical regions: (1) delta and theta centrally-frontally; (2) theta frontal midline; (3) beta frontally; and (4) alpha posterior, central and frontal. Figure 8.3 shows the q-EEG spectra of an adult patient (treated by the author) belonging to the first ADHD subtype. As can be seen, the low frequency values of peak power are obtained frontally and centrally (6.10 Hz at Fz and 5.86 Hz at Cz).

If we introduce brain rate as a general indicator of mental arousal in the ADHD example, we can see that the first two subtypes are correlated with lower brain rate, the third subtype with higher brain rate, and the fourth subtype with an excess of

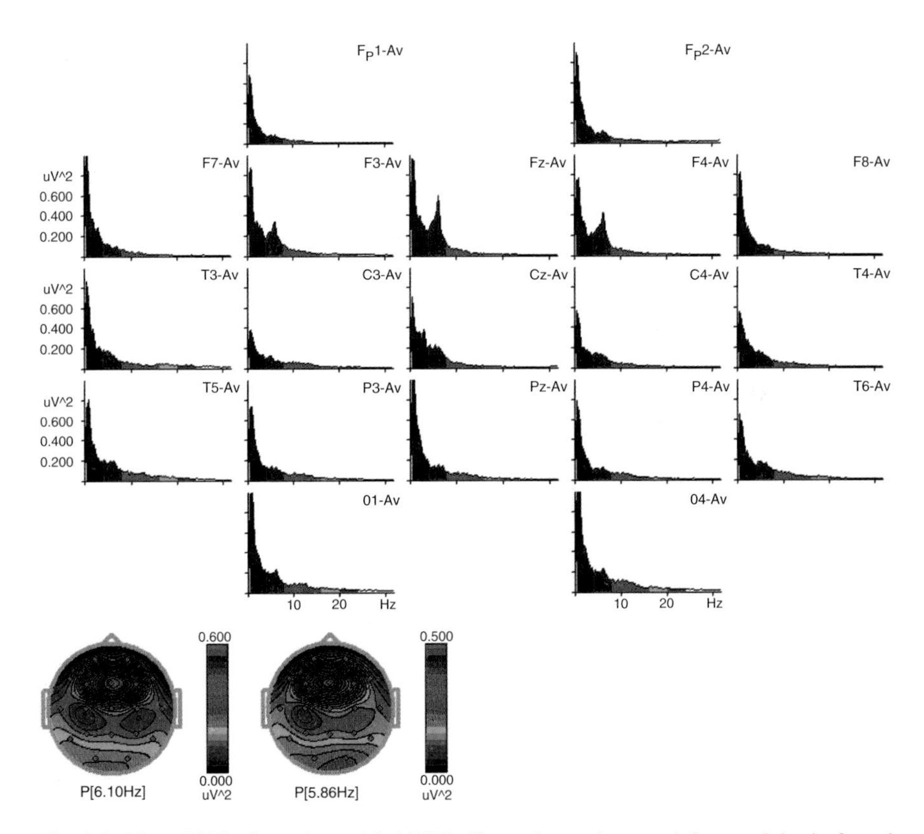

Fig. 8.3 The q-EEG of a patient with ADHD (first subtype, increased theta activity in frontal regions)

Table 8.4 Brain rate using reference databases, calculated with the spectral data for ten adults with ADHD (from White 2003)

	NeuroRep	SKIL	EureKa3!
Theta–beta ratio	1.49	1.70	1.46
Brain rate	9.73	8.51	8.94

alpha activity and a "normal" arousal state. In the case of ADHD, the neurofeedback protocol for UA and OA is clear, while for normal arousal it is not.

In addition, if we calculate brain rate using the reference databases (White 2003) we obtain that the brain rate in Cz for adult ADHD is between 8.51 and 9.73 (Table 8.4). Note that Cz is a central point in 10/20 EEG recording system. It was chosen as the most representative location for the activation of the whole brain cortex.

Since they are in the low alpha range, the results for f_b suggest that in this case ADHD Cluster 3 (Kropotov 2009) in fact prevails, indicating the presence of "inward attention" (Cooper et al. 2003). This is not visible from the theta/beta ratio.

8.4 Inner Arousal

As was mentioned, the brain rate includes all frequency bands and it is an indicator for mental arousal, i.e. the level of consciousness. Our findings in EEG evaluation show that normally in the eyes-open condition (EO) the brain rate is higher than in the eyes-closed condition (EC), which can be attributed to the influence of environmental stimuli. We analyzed EEG data obtained in Cz for different disorders in both EO and EC conditions. The results are shown in Table 8.5.

It can be seen that in some clinical cases (anxiety, PTSD, panic attack, stuttering and OCD) brain rates appeared to be higher in EC than in EO, which we interpret as an indicator of increased inner arousal. Conversely, at *pavor nocturnus*, tics and ADHD brain rate are higher in EO conditions, which means that inner arousal is practically absent. That is, inner arousal implies that higher band frequencies are more present than in the normal EC relaxed state. This can be a sign of inner tension, anxiety, rumination, etc. If we analyze the obtained data, it is logical that in anxiety, PTSD, panic attack or OCD inner arousal is high.

For clarity, we explain the main clinical characteristics of the evaluated disorders.

- Anxiety is an unpleasant emotional state ranging from mild unease to intense fear. An anxious person usually feels a sense of impeding doom although there is no obvious threat.
- Post-traumatic stress disorder (PTSD) is a specific form of anxiety that comes after a stressful or frightening event. The symptoms include recurring memories or dreams of the event, a sense of personal isolation, disturbed sleep, and bad concentration.

Table 8.5 Brain rate as indicator of inner arousal

Disorder	Number of patients	Mean age (years)	EO	EC
Anxiety (adults)	2	45	8.56	10.54
PTSD	10	12	6.27	7.54
OCD	6	14.5	6.82	7.52
Panic attacks	4	13.5	7.58	8.21
Anxiety (children)	8	14.6	7.57	8.19
Stuttering	8	11	8.27	8.50
Autism	5	3.5	5.68	5.86
Ticks	3	12.5	8.48	8.47
ADHD	50	8.1	7.86	7.60
Nightmares	2	7.5	8.48	8.13

- Panic attack comprises a brief period (a few minutes) of acute anxiety, often dominated by an intense fear of dying or losing one's reason. Panic attacks occur unpredictably at first, but tend to become associated with certain places (e.g. when overcrowded).
- Obsessive-compulsive disorder is characterized by persistent ideas (obsessions) that make sufferers carry out repetitive, ritualized acts (compulsions). Anxiety could be an accompanying manifestation.
- Tics are repeated, uncontrolled, purposeless contractions of a group of muscles, most commonly in the face, shoulders or arms. Tics are motor release of the high anxiety state.
- Stuttering is a speech disorder in which there is repeated hesitation and delay in uttering words, unusual prolongation of sounds, and repetition of word elements. It is often stress-related and comorbid with anxiety.
- Nightmares are unpleasant vivid dreams, usually accompanied by a sense of suffocation. Nightmares occur during REM (rapid eye movement) sleep in the middle and later parts of the night, and are often clearly remembered if the dreamer awakes completely.

8.5 Brain Rate, Heart Rate and Sleep

Another specific application of brain-rate is presented in a study of Kaniusas et al. (2007), where brain rate (f_b) has been compared with spectrum-weighted frequencies from the heart rate variability (f_{hv}). The authors conclude that the brain rate is approximately related to the sleep profile, having predictive value in identifying shallow sleep vs. deep sleep, and helping to assess the quality of sleep. The inverse relation between f_b and f_{hv} changes is also noted. The results obtained in this study are shown in Table 8.6.

Table 8.6 Spectrum weighted frequencies from EEG (brain rate) and from heart rate variability (HRV) for different sleep stages (Kaniusas et al. 2007)

Sleep stages	$f_b = f_w$ from EEG (Hz)	$f_{hv} = f_w$ from HRV (Hz)
Awake	6.85	0.076
REM	5.34	0.076
S1	–	–
S2	4.18	0.116
S3	2.72	0.128
S4	2.45	0.132

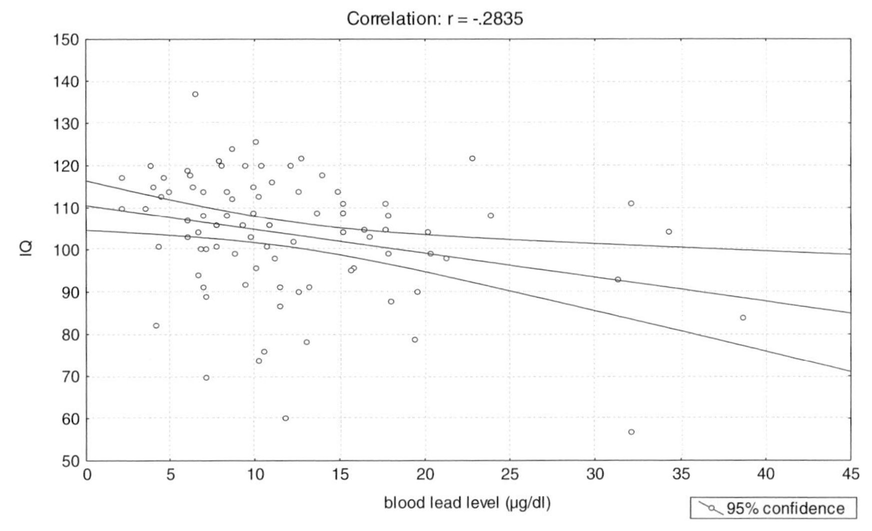

Fig. 8.4 Correlation between blood leads level and intelligence scores

8.6 Blood Lead Level and Brain Rate

A selected group of 31 children (mean age 12.8 years) with increased blood lead levels, who live near Veles ironworks in the Republic of Macedonia, were examined for cognitive and behavioral problems. Cognitive psychological tests (Raven Progressive Matrices and Bender Gestalt Test) showed a diminution of intelligence scores and graphomotor abilities with the increase of blood lead level (Fig. 8.4). This finding was confirmed by the values obtained for theta/beta ratio (Fig. 8.5) and brain rate (Fig. 8.6).

As can be seen, a positive correlation ($r = 0.47$) between blood lead level and theta/beta ratio, along with a strong negative correlation ($r = 0.80$) between theta/beta and brain rate were found. As a result, brain rate appears to be a sensitive indicator of spectral shift related to mental activation changes induced by lead pollution.

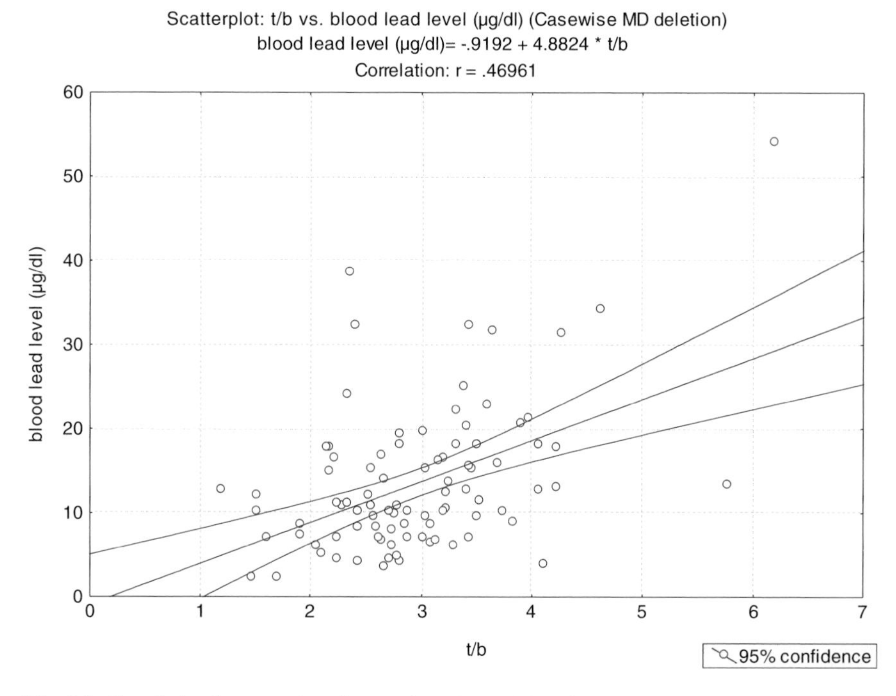

Fig. 8.5 Correlation between theta/beta ratio and blood lead level

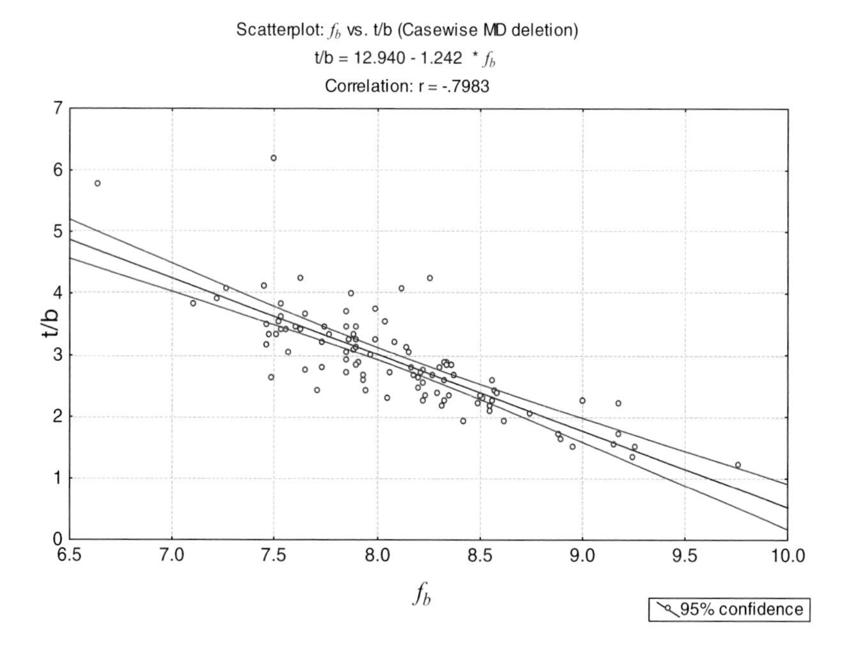

Fig. 8.6 Correlation between brain rate and theta/beta ratio

8.7 Brain Rate as a Biofeedback Parameter

Biofeedback has arisen as a modern computer-related operant conditioning technique used for assessment and therapy of many psychophysiological disorders, especially the stress-related ones. Its objective is to increase voluntary control over the physiological processes that are otherwise outside awareness, using information about them in the form of an external signal. In this sense, biofeedback can be considered as a technique relevant to consciousness.

Various biofeedback modalities are increasingly used worldwide as non-pharmacological and cost-effective research and therapeutic tools. A significant increase in research has documented the efficiency of biofeedback for children and adolescents that manifest behavioral, emotional, and cognitive problems (Culbert et al. 1996; Scott 1998; Schwartz 1987; Pop-Jordanova 1999).

Biofeedback modalities can be divided into peripheral (based on electromyography, electrodermal response, heart rate, temperature, blood volume pulse) or central (based on electroencephalography, i.e. neurofeedback).

Electrodermal response (EDR) is a complex reaction with a number of control centers in the CNS. Three systems, related to arousal, emotion, and locomotion, are responsible for the control of electrodermal activity (Bouscin 1992). The reticular formation controls EDR related to states of arousal, the limbic structures (hypothalamus, cingulate gyrus, and hippocampus) are involved in EDR activity related to emotional responses and thermoregulation, while the motor cortex and parts of the basal ganglia are involved in locomotion. In particular, skin potential and skin conductance are used as parameters in EDR biofeedback and are related to both sympathetic and parasympathetic arousal (Andreassi 2000; Mangina and Beuzeron-Mangina 1996).

Treatment by EDR biofeedback is generally based on training patients in strategies for lowering arousal and maintaining a healthful sympathetic/parasympathetic tone. Consequently, EDR biofeedback modality is a first choice for introvert persons, where high inner arousal is a typical finding and biofeedback training is supposed to lower sympathetic activity. Changes in electrodermal activity can be reliably detected within one second of stimulus presentation, often following a single event (Kropotov 2009). It is important to know that electrodermal conductance precedes any other signals related to neuroimaging, e.g. positron emission tomography (PET), blood oxygen level-dependent functional magnetic resonance (BOLD), single photon emission computerized tomography (SPECT), etc. In other words, the changes of electrodermal activity can be registered before the changes obtained by the other neuroimaging techniques.

Neurofeedback (NF), i.e. EEG biofeedback, refers to a specific operant-conditioning paradigm where an individual learns how to influence the electrical activity (frequency, amplitude, or synchronization) of his or her brain. It involves teaching skills by rewarding the experience of inducing EEG changes reflected in a perceivable signal (light or sound). Neurofeedback has been shown to be particularly useful in reference to pathologies characterized by dysfunctional regulation of cortical arousal, such as epilepsy and ADHD (Lubar 1991, 1997; Birbaumer et al. 1999; Monastra et al. 2001; Pop-Jordanova et al. 2005). We also used EEG biofeedback in anorectic girls (Pop-Jordanova 2000, 2003), PTSD (Pop-Jordanova and Zorcec 2004), headaches (Pop-Jordanova 2008), as well as for optimal school and music performance (Pop-Jordanova et al. 2008; Markovska-Simoska at al. 2008).

All neurofeedback interventions can be roughly reduced to the need of achieving flexibility in increasing or decreasing the general mental activation, i.e. mental arousal. In practice, whenever a certain band is trained, the other bands are affected too it may even appear that, e.g. "the changes that occurred as a result of stimulating in the alpha frequency were not in alpha but were in beta" (Pop-Jordanova and Pop-Jordanov 2005). Therefore, the brain rate variable could be employed as a complementary biofeedback parameter, characterizing the whole EEG spectrum (as distinct from, e.g. the theta–beta ratio). The rationale is that, according to the mentioned empirical results, the EEG frequency shifts are related to mental activation or deactivation, as the main objective of the treatment. In the conditions with under-arousal, the neurofeedback protocol aims to raise mental activity or arousal. Conversely, if the brain rate indicates over-arousal, the training must be directed toward lowering the mental state.

Using brain rate as a neurofeedback parameter for ADHD children ($N = 50$ mean age 11.11 years) (Pop-Jordanova et al. 2005, 2008) obtained before treatment a brain rate value of 7.80 ± 0.47, and after treatment 8.22 ± 0.63. So a shift of the spectrum from under-arousal to normal mental arousal occurred, and this corresponded to improved attention and cognition as well as better school performance. This change of brain rate (i.e. arousal level) appeared to be more realistic in respect to the changes of psychological state of children than the drastic reduction of theta/beta ratio, which appeared to be as much as halved (Fig. 8.7).

A detailed analysis of q-EEG after the neurofeedback training with brain rate as a parameter determined which bands had been most changed. For instance, in some cases shifting the brain rate to higher values resulted in increasing high alpha or beta frequencies; in others, the same change appeared after diminishing the power of theta or delta bands. As a result, the q-EEG comparison before and after the brain rate training can be informative for assessing the individual spectrum band sensitivity.

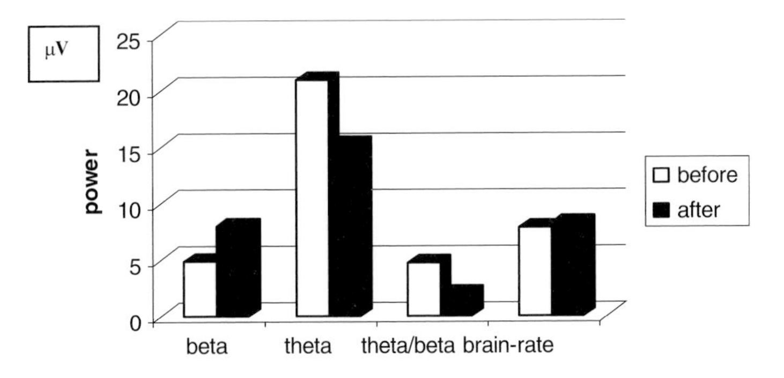

Fig. 8.7 Changes of biofeedback parameters (before and after NF training)

8.8 Conclusions

Brain rate (the mean frequency of brain rhythms, i.e. EEG-spectrum weighted frequency) can be considered as an integral brain state attribute correlated with the brain's electric, mental and metabolic activity. In particular, it can serve as a preliminary diagnostic indicator of general mental activation (i.e. mental arousal representing consciousness level), in addition to heart rate, blood pressure or temperature as one of the standard indicators of general bodily activation.

As a measure of arousal, brain rate can be applied to discriminate between the groups of UA and OA disorders, as well as the subgroups within mixed disorders, in particular attention deficit hyperactivity disorder. Also, by comparing EC and EO brain rate values, the diagnosis of inner arousal can be simply achieved.

Brain rate values can be correlated with the sleep profile, helping to assess the quality of sleep. Also, brain rate can serve as a preliminary indicator of possible mental activation changes caused by some environmental toxins.

In biofeedback treatment, brain rate can be efficiently used as a multiband biofeedback parameter, complementary to few-band parameters and skin conduction. In this context, it is especially suitable for revealing the patterns of sensitivity or rigidity of the EEG spectrum, including frequency bands related to the permeability of corresponding neuronal circuits, based on which individually adapted biofeedback protocols can be specified.

Further studies of the advantages and limitations of the brain rate concept for diagnostics and treatment are needed.

References

Andreassi JL (2000) Psychophysiology. Human behaviour and physiological response. LEA, London

Bendorfer (2001) Alpha-theta neurofeedback: Its promises & challenges. BFE 5th Annual Meeting, Prien

Birbaumer N, Roberts LE, Lutzenbrger W, Rockstroh B, Elbert T (1999) Area-specific self-regulation of slow cortical potential on the sagittal midline and its effects on behaviour. Electroencephalogr Clin Neurophysiol 84:352–361

Boucsein W (1992) Electrodermal activity. Plenum, New York

Chalmers D (2000) What is a neural correlate of consciousness? In: Metzinger T (ed) Neural correlates of consciousness. MIT, Cambridge, MA

Cooper NR, Croft RJ, Dominey SJ, Burges AP, Grizelier JH (2003) Paradox lost? Exploring the role of alpha oscillations during externally vs. internally directed attention and the implications for idling and inhibition hypotheses. Int J Psychophysiol 47(1):65–74

Culbert TC, Kajander RL, Reaney JB (1996) Biofeedback with children and adolescent: clinical observations and patient perspectives. J Dev Pediatr 17(5):342–350

Damasio A (2003) Looking for Spinoza: joy, sorrow, and the feeling brain. Harcourt, Orlando, FL, p 30

Kahnemann D (1973) Attention and effort. Prentice-Hall, New Jersey

Kaniusas E, Varoneckas G, Alonderis A, Podlipsky A (2007) Heart rate variability and EEG during sleep using spectrum-weighted frequencies – a case study, COST B27. EU/ESF, Brussels

Kropotov J (2009) Quantitative EEG, ERP's and neurotherapy. Elsevier, Amsterdam

Lubar JF (1991) Discourse on development on EEG diagnostics and biofeedback treatment for attention deficit/hyperactivity disorders. Biofeedback Self-Regul 16:201–225

Lubar JF (1997) Neurocortical dynamics: Implications for understanding the role of neurofeedback and related techniques for the enhancement of attention. Appl Psychophysiol Biofeedback 22(2):111–126

Mangina CA, Beuzeron-Mangina JH (1996) Direct electrical stimulation of specific human brain structures and bilateral electrodermal activity. Int J Psychophysiol 22:1–8

Mann C, Lubar J, Zimmerman A, Miller C, Muenchen R (1992) Quantitative analysis of EEG in boys with attention-deficit/hyperactivity disorder: A controlled study with clinical implications. Paediatric Neurology 8:30–36

Markovska-Simoska S, Pop-Jordanova N, Georgiev D (2008) Simultaneous EEG and EMG biofeedback for peak performance in musician. Prilozi 1:239–253

McFadden J (2006) The CEMI field theory. In: Tuszynski JA (ed) The emerging Physics of Consciousness, Springer-Verlag, Berlin Heidelberg, pp 387–406.

Monastra VJ, Lubar JF, Linden M (2001) The development of quantitative electroencephalographic scanning process for attention deficit-hyperactivity disorder: reliability and validity studies. Neuropsychology 15(1):136–144

Müller A (2006) Neurobiological diagnostics and therapy in ADHD, in: news in pediatrics. University of Skopje, Skopje, pp 135–147

Pop-Jordanov J, Pop-Jordanova N (2009) Neurophysical substrates of arousal and attention. Cogn Process 10(Suppl.1):S71–S79

Pop-Jordanova N (1999) Electrodermal response based biofeedback in pediatric patients. Paediatr Croat 43:117–120

Pop-Jordanova N (2000) Biofeedback mitigation for eating disorders in preadolescents. Int Pediatr 1:76–82

Pop-Jordanova N (2003) Eating disorders in the preadolescent period: psychological characteristics and biofeedback mitigation, Chapter III. In: Swain P (ed) Focus on eating disorder research. Nova Biomedical books, New York

Pop-Jordanova N (2006) Biofeedback modalities for children and adolescents. In: Columbus F (ed) New research on biofeedback. Nova Biomedical Book, New York

Pop-Jordanova N (2008) EEG spectra in pediatric research and practice. Prilozi 1:221–239

Pop-Jordanova N (2009) Biofeedback application for somatoform disorders and attention deficit hyperactivity disorder (ADHD) in children. Int J Med Sci 1(2):17–22

Pop-Jordanova N, Cakalaroska I (2008) Biofeedback modalities for better achievement in high better achievement in high school students school students. MJM 2:25–30, also Revista Espanola de Neuropsicologia, 1:97–98

Pop-Jordanova N, Pop-Jordanov J (2005) Spectrum-weighted EEG frequency ("brain rate") as a quantitative indicator of mental arousal. Prilozi 2:35–42

Pop-Jordanova N, Zorcec T (2004) Child trauma, attachment and biofeedback mitigation. Prilozi 1–2:103–114

Pop-Jordanova N, Markovska-Simoska S, Zorcec T (2005) Neurofeedback treatment of children with attention deficit hyperactivity disorder. Prilozi 1:71–80

Pop-Jordanov J, Pop-Jordanova N (2010) Quantum transition probabilities and the level of consciousness. Journal of Psychophysiology 24(2):136–140

Pritchard, Alloway (1999) Medical Neuroscience, Frence Greek Publishing, LLC, Madison, Connecticut, p.397

Schwartz MS (1987) Biofeedback: a practitioner's guide. Guilford Press, New York

Scott F (1998) EEG biofeedback for children and adolescent: a pediatrician's perspective. Biofeedback 26(3):18–20

Thacher RW, John ER (1977) Functional neuroscience: foundations of cognitive processing. Erlbaum, Hillsdale, NJ

White NJ (2003) Comparison of EEG reference databases in basic signal analysis and in evaluation of adult ADHD. J Neurother 7(3/4):161

Chapter 9
On Physiological Bases of States of Expanded Consciousness

Emil Jovanov

Abstract Altered states of consciousness provide valuable insights essential for a better understanding of the phenomenon of consciousness. Improved understanding of physiological correlates of states with expanded consciousness may allow possible use of biofeedback as a tool for achieving of those states at will. As a consequence, this approach may facilitate improvement of efficiency, creativity, and spiritual growth. This chapter presents a survey of relevant physiological correlates and present examples of two specific techniques: slow yogic breathing and chanting. We hypothesize that stabilization of physiological rhythms, such as breathing, heart rate variability, or blood pressure variability, creates favorable conditions in which states of expanded consciousness may arise.

9.1 Introduction

Consciousness remains the ultimate secret of human existence and contemporary science. Scientists even disagree about the possibility of correlating any of the physiological signals with subjective experience. Chalmers calls this the "hard problem":

> CONSCIOUSNESS, the subjective experience of an inner self, poses one of the greatest challenges to neuroscience. Even a detailed knowledge of the brain's workings and the neural correlates of consciousness may fail to explain how or why human beings have self-aware minds (Chalmers 2002).

Altered states of consciousness are essential for a better understanding of the phenomenon of consciousness (Tart 1972). Taking the system to extreme modes of operation or suspending elements of the functionality (e.g. sensor inhibition or overload, drug induced changes, etc., Vaitl et al. 2005) may improve our understanding of the states of expanded consciousness and provide valuable insights into

E. Jovanov (✉)
Department of Electrical and Computer Engineering, University of Alabama, Huntsville, AL, USA
e-mail: emil.jovanov@uah.edu

D. Cvetkovic and I. Cosic (eds.), *States of Consciousness*, The Frontiers Collection,
DOI 10.1007/978-3-642-18047-7_9,

possible mechanisms for improved efficiency, creativity, and spiritual growth. Csíkszentmihályi (1990) emphasizes that people are most happy when they are in a state of flow – a state of concentration or complete absorption with the activity at hand and the situation. The flow is a temporary state of heightened concentration that enables peak performance, while the "zone" usually refers to a similar state in athletes.

The idea of flow is identical to the feeling of being in the zone or in the groove. The flow state is an optimal state of intrinsic motivation, where the person is fully immersed in what he or she is doing, characterized by a feeling of great absorption, engagement, fulfillment, and skill – and during which temporal concerns (time, food, ego-self, etc.) are typically ignored. Children at play often exhibit a similar state of full immersion. Csíkszentmihályi described flow as:

> being completely involved in an activity for its own sake. The ego falls away. Time flies. Every action, movement, and thought follows inevitably from the previous one, like playing jazz. Your whole being is involved, and you're using your skills to the utmost (Csíkszentmihályi 1990).

It has been shown that the mindfulness, meditation, yoga, and martial arts seem to improve a person's capacity for focused attention, characteristic for the flow state (Vaitl et al. 2005). All of these activities provide the opportunity for prolonged improvement of attention and concentration.

There are two principal approaches to the investigation of consciousness. Connectionists describe the conscious processing of each stimulus and subjective experience as specific responses in the brain, activation of individual neurons, their hierarchical processing, and associations with other brain regions (Chalmers 2002; Crick 1994). For example, recognizing somebody's face includes hierarchical processing of signals coming from eyes, the emergence of face features through hierarchical processing, and collection of meaning through associative response in other parts of the cortical network. The second approach is based on fields and assumes a synergistic effect of cerebral electromagnetic fields in addition to the neural connections. It is hypothesized that the field serves as a global integrating medium of neural activity (Cosic et al. 2006; Rakovic 1991; Penrose 1994).

We hypothesize that the states of expanded consciousness (sometimes called "higher consciousness") allow access to individual and collective unconsciousness as a mechanism of insights and creativity. We assume collective consciousness as a physical mechanism described by Carl Gustav Jung as:

> a second psychic system of a collective, universal, and impersonal nature which is identical in all individuals. This collective unconscious does not develop individually but is inherited. It consists of pre-existent forms, the archetypes, which can only become conscious secondarily and which give definite form to certain psychic contents (Jung 1991).

The connectionist approach would attribute collective unconscious and archetypes to genetically inherited organization of our nervous system and the seamlessly infinite knowledge database of our genetic information. However, this approach can hardly explain complex psychological phenomena, which are still hard to document, such as synchronicity (Jung 1991). The field based approach may provide possible explanations for those phenomena through interaction with the "field of consciousness" (Penrose 1994; Rakovic 1991, 1995).

Our body functions through a hierarchy of interrelated rhythms. We believe that every thought or action creates "ripples" throughout the hierarchy of our rhythms. Science may or may not be able to explain the ultimate mechanisms of consciousness (the hard problem) and the physiological correlates of conscious processes. However, by monitoring physiological correlates as ripples in our stream of consciousness and providing results with minimum latency, we might facilitate personalized insights into physiological correlates of various altered states of consciousness. For example, alerting the user about detected anxiety (increased heart rate and decreased heart rate variability) may allow user to gain new insight into the particular content in their stream of thoughts. If the person was thinking about swimming, for example, an indication of emotional stress and anxiety might indicate traumatic childhood experience hidden deeply in the personal unconsciousness. The user might further explore the experience in one or more sessions. Ultimately, the user can resolve the issue by safely exploring the unconscious mind and hidden experiences. As another example, it has been shown that major discoveries often appear during a relaxed, but alert, state. We call it the *eureka* effect, in reference to the Archimedes, who proclaimed "Eureka!" when he stepped into a bath and suddenly realized that the volume of water displaced must be equal to the volume of the submerged parts of his body. By providing feedback about individual physiological correlates of conscious processes, we might a person to re-create the physiological basis of a particular state of consciousness at will and explore it in the quest for insights that would allow expanding of their consciousness.

This process may work not only in the short term but also have lasting effects. It is known that the brain has the ability to rewire itself in the presence of the appropriate sensory input, even in later life after maturation of the nervous system. Recent evidence includes long term changes in brain electrical activity in meditators (Tei et al. 2009) and following integrative body-mind training (Tang et al. 2009).

Recent studies indicate the possible use of biofeedback for voluntary control of physiological processes, allowing researchers to observe perceived changes in the state of consciousness as correlates of physiological states (Benson 1984; Cade and Coxhead 1979; Schwartz and Andrasik 2003; Vaitl et al. 2005). Various methods have been proven effective in control over physiological processes, such as temperature, heart rate, blood pressure, vasomotor responses and muscular tension. Recent scientific studies indicate that elevation of theta over alpha activity could enhance music performance (including ratings of interpretative imagination) (Egner and Gruzelier 2003). Also, improvements in attention and semantic working memory in medical students have been reported. Moreover, if EEG/MEG represents correlates of the state of consciousness, inducing changes of brain activity might lead to voluntary changes in consciousness.

Comparative analyses of traditional spiritual techniques for stimulation of altered states of consciousness indicate similar changes in basic physiological signals. We hypothesize that stabilization of physiological rhythms, such as breathing, heart rate variability, blood pressure variability, etc., creates favorable conditions in which states of expanded consciousness may arise. In this chapter we present changes in autonomous nervous system during meditation, slow yogic breathing exercises, and chanting as commonly used mechanisms to facilitate changes in the state of consciousness.

9.2 Physiological Correlates of Consciousness

The study of consciousness has entered an intense experimental phase (Tononi and Koch 2008). The most frequently monitored parameter is brain electrical activity (electroencephalography, EEG) for analysis of activation of neurons and groups of neurons and functional magnetic resonance (fMRI) for monitoring of activation (metabolism) of brain regions. Experiments involving monitoring of individual neurons and small regions require insertion of electrodes in the brain, and therefore can be used only for animal experiments. However, the assessment of higher cognitive functions, such as object recognition, is limited and requires a very specific experimental setup. Therefore, subjective experience during altered states of consciousness is a valuable tool for understanding the nature of consciousness (Rakovic et al. 1999; Tart 1972).

One of the safest methods of generating altered state of consciousness is meditation (Banquet 1973; Hirai 1960; Saraswati 2004). The traditional yogic path to experiences of expanded consciousness includes a combination of breathing exercises (pranayama, Saraswati 2004; Swara Yoga 1983) and meditation techniques:

> Through those practices (pranayama and meditative practices), the prana can be controlled. In this manner one is freed from sorrow, filled with divine ecstasy and becomes enraptured with the supreme experience (Saraswati 2004, p. 432).

Rhythmic drumming has been used for centuries alone or in combination with dance and/or song as a method of achieving an altered state of consciousness and is a frequently cited technique for the shamanic journey (Eliade 1964; Harner 1990; Maxfield 1990). Although drumming is widely used, the exact physiological mechanism behind the changes it achieves is still unknown. We present a brief review of the main physiological changes induced by rhythmic music and some fundamental technical issues in the analysis of physiological signals during entrainment.

Music is traditionally used as a medium for brain stimulation. However, the frequency range of many physiological rhythms (e.g. 3, 4, and 10 Hz cycles) is not in the audible range. Therefore, two basic approaches are applied:

- *Rhythmic entrainment*, such as drumming or repetition of high frequency tones at lower frequencies.
- *Binaural beats*, which arise from the difference in pitch between two audible sounds of higher frequencies (Oster 1973). Although individual sounds can have higher frequency (pitch), their difference is in the range of target physiological

frequencies and there is evidence that this generates brain entrainment at the difference of pitch frequencies (Lane et al. 1998; Stevens et al. 2003).

In addition to auditory driving, stimulation techniques frequently use multiple stimulation modalities. The most frequently used are optical and vibration stimulation (Thompson 2006; Vibroacoustic, http://vibroacoustic.org/). Optical stimulation can be implemented by providing light modulated by the sound. Changes in sound intensity may produce changes in light intensity or light frequency (color). As a result, the vision processing cortical areas in the brain start receiving direct electrical stimulation from optical nerves. Vibration stimulation can be achieved by exposing the body to sonic vibration. For example, low-frequency bass speakers might be embedded in a special chair or bed to deliver vibration directly to tactile sensors in our body. As a result, the music can be perceived not only through hearing, but also through the visual and tactile senses. This effect might be amplified by body movements. Multi-sensory stimulation very often produces synergetic effect where the overall effect is more than the sum of individual effects.

Shamanic drumming mostly consists of a steady, monotonous beat of 3–4.5 beats/s. It is used to facilitate entry into an altered state of consciousness and travel to other realms and realities, and to interact with the spirit world for the benefit of their community. It is interesting to note that those frequencies correlate with the delta frequency band of brain electrical activity. In electrophysiology this state is an indicator of the first phase of deep sleep and coma. Maxfield found more theta activity while subjects were listening to rhythmic monotonous and patterned drum beats than when they were listening to unstructured beat sequences (Maxfield 1990).

During the shaman's journey, he or she may travel to the "upper world" or the "lower world." Images that are traditionally associated with an entry to the upper world include climbing a mountain, tree, cliff, rainbow or ladder, ascending into the sky on smoke; flying on an animal, carpet or broom, and meeting a teacher or guide. The upper world journey may be particularly ecstatic. In the lower world journey, the shaman may experience images of entering into the earth through a cave, hollow tree stump, water hole, tunnel, or tube. Powerful animals or animal allies (guardian spirits) or other figures representing the spirit realm may be encountered. The lower world is traditionally a place of psychological or emotional tests and challenges (Eliade 1964; Harner 1990). Similar images and metaphors are also common features in mainstream Western psychological guided imagery and hypnotic induction.

To accomplish the shamanic journey, the shaman enters into a specific type of altered state of consciousness that requires that he or she remain alert and aware. In this state, the shaman can move at will between ordinary and non-ordinary reality. Michael Harner designates this pro-active state as a Shamanic State of Consciousness (SSC). There are various techniques for entering into the SSC, including sensory deprivation, fasting, fatigue, hyperventilation, dancing, singing, chanting, exposure to extremes of temperature, the use of hallucinogenic substances, the set and setting dictated by the beliefs and ritualized ceremonies of the culture, and, most relevant to this study, the rapid and sustained use of percussive sound (Ludwig 1968; Tart 1972; Vaitl et al. 2005).

With the present state of technology, investigation of these types of practices can be undertaken in a much more precise and dynamic way than ever before. Modern technology can help provide a precise experimental setup including sound generation and real-time feedback with significant improvements in monitoring systems.

9.2.1 What We Measure

Activation of certain functions might be monitored directly or indirectly. Direct monitoring allows monitoring of physical correlates of physiological functions, such as electrical activity of neurons. The most frequently used direct parameters represent brain electrical activity with an electroencephalogram (EEG) or magnetoencephalogram (MEG), heart electrical activity with an electrocardiogram (ECG), muscle electrical activity with an electromyogram (EMG), and electrical activity of eyes with an electrooculogram (EOG).

Indirect monitoring is frequently used to assess the increased metabolism of the affected region. For example, activation of certain brain regions might be monitored as increased amplitude of brain electrical activity or the increase of blood flow to the region of interest. However, the parts of the brain that control emotional response are buried deeply inside the brain and hard to monitor. Therefore, indirect monitoring may be employed. The most frequently used indirect parameters include body temperature, galvanic skin resistance (GSR), pulse pressure (photoplethysmogram, PPG), blood pressure (BP), functional near infrared brain imaging (fNIR), positron emission tomography (PET), functional magnetic resonance imaging (fMRI), and others.

Electrical signals are amplified and converted to digital format for archiving and processing. Each signal is characterized by intensity (amplitude of the signal in volts), characteristic patterns in time, and spectral content. For example, Fig. 9.1 presents electrical activity of the heart as an ECG, with characteristic patterns that represent individual heart beats. Each signal can be decomposed to a set of components at different frequencies. The frequency of each component specifies how many times a single wave occurs over a period of 1 s and is cited as the number of cycles per second or hertz (Hz). The most frequently used transformation from time to frequency domain is called a Fourier transform. It is used to transform or decompose a time signal into frequency components. Each component is represented by amplitude and phase, and collectively they represent the *spectrum* of the signal.

9.2.2 Brain Activity

Investigation of the brain's electrical activity (EEG) began with the publications of Hans Berger in 1924. Berger described the basic rhythms of the brain's electrical

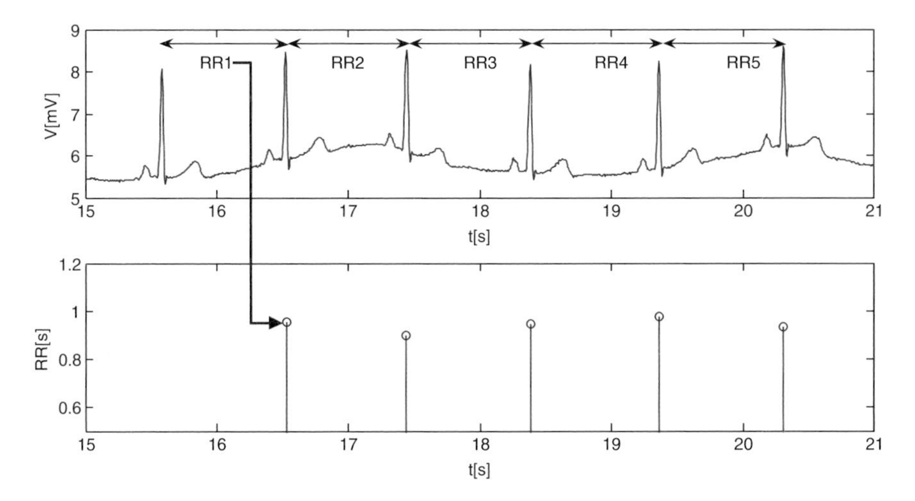

Fig. 9.1 ECG waveform (*upper plot*) and interbeat intervals (R–R intervals in *lower plot*)

activity. It is interesting to note his claim that "we have to assume that the central nervous system is always, and not only during wakefulness, in a state of considerable activity." This concept is similar to the recent concept of the default mode network (DMN) or the brain's dark energy (Raichle 2010). Brain states can be considered as separate global functional states over time that show extended periods of quasi-stability separated by rapid and major changes of state (Lehmann et al. 2009). Microstate analysis of EEG data indicates that the human brain's electrical state is quasi-stable for a fraction of a second, then rapidly reorganizes into another state. Different classes of thoughts in a daydreaming condition were found to belong to different classes of microstates of about 120 ms duration that may represent psychophysiological building blocks or "atoms of thought" (Lehmann et al. 2009).

Changes of electrical potential on our scalp are generated by the synchronized firing of large populations of neurons as a response to sensory stimulation or as a result of mental processing (Basar 1980, 1988; Mormann and Koch 2007; Nunez and Srinivasan 2006). Unfortunately, the skull attenuates the signals. As a result we measure signals with amplitudes of tens of microvolts. In addition, the signal is contaminated with power line interference and modulated by changes of conductivity of the scalp caused by changes in the blood vessels that form a somewhat independent process. In spite of all the issues described above, the EEG has been routinely used as a diagnostic tool for more than half a century.

The brain's electrical activity in various frequency bands is often understood as a correlate of certain states or a correlate of specific processing. Both the intensity and phase of waves at different brain locations indicate possible correlates of conscious processing. The conscious perception of a stimulus triggers widespread cortical interactions that indicate the binding of remote cell groups to form cell assemblies (Nunez and Srinivasan 2006).

The most frequent spectral components in EEG/MEG are (Guyton and Hall 2006):

- *Gamma* rhythm (30–100 Hz) is widely accepted to represent the binding of different populations of neurons to perform a certain function.
- *Beta* rhythm (12–30 Hz) is associated with active attention and focus on the exterior world. Beta is also present during states of tension, anxiety, fear and alarm.
- *Alpha* rhythm (8–12 Hz) is the basic rhythm amplified by closing the eyes and by relaxation. Consciousness is alert but unfocused, or focused on the interior world.
- *Theta* rhythm (4–8 Hz) are usually associated with drowsy, near-unconscious states, such as the threshold period just before waking or sleeping. They have also been connected to states of reverie and hypnagogic or dream-like imagery. Often these images are startling or surprising. For many people, it is difficult to maintain consciousness during periods with increased theta activity without some sort of training, such as meditation. Awake theta is associated with reports of relaxed, meditative, and creative states. Long-term meditation is characterized by increased theta EEG activity over the frontal region. The intensity of the blissful experience correlates with increases in theta power in anterior–frontal and frontal–midline regions.
- *Delta* rhythm is associated with deep sleep or unconsciousness.
- *Slow cortical potentials* (SCP) represents activation of a group of neurons approximately every 10 s. Note that traditional EEG amplifiers discard all rhythms slower than 0.5 or 1.5 Hz.

9.2.3 Limbic System Activity

Variation of periods between consecutive heartbeats provides unique insights into the state of autonomous nervous system. This parameter is very attractive since it is very easy to monitor, even in ambulatory settings. For example, Fig. 9.1 shows the ECG of heart activity recorded at the chest. In the upper plot, six heartbeats are clearly visible, denoted by sharp R-peaks (Guyton and Hall 2006). Interbeat intervals are represented as time differences between consecutive heartbeats. The lower plot represents five interbeat intervals from the upper plot. For convenience, we represent the value of interbeat intervals at the moment of the current heartbeat, although we could represent it in the middle of the interbeat interval. The value of the interbeat interval can be represented as instantaneous heart rate. For example, the first R–R interval is $RR1 = 0.956$ s, which is equivalent to a heart rate of $60/0.956 = 62.76$ beats/min. The mean value of five intervals in Fig. 9.1 is 0.94 s, which is equivalent to a heart rate of $60/0.94 = 63.6$ beats/min. Therefore, interbeat variability, R–R variability, and heart rate variability (HRV) can be used interchangeably to describe the same phenomenon.

The diagnostic value of HRV has been known for more than 40 years. HRV provides insights into the operation of the autonomic nervous system (ANS) with huge potential for early diagnosis and analysis of trends. In 1965, Hon and Lee demonstrated that changes in interbeat interval precede changes of heart rate in the case of fetal distress. Several studies in the last 20 years have demonstrated a correlation between post-infarction mortality and reduced HRV (1996) and mortality of emergency room patients (Cook et al. 2006).

New computer and sensor technology has made possible spectral analysis of the sequence of R–R intervals to reveal characteristic rhythms in heart rate variations. The main component of HRV is respiratory sinus arrhythmia (RSA), generated by a combination of respiration-induced biochemical changes, changes in intrathoracic pressure, and central vagal stimulation (Song and Lehrer 2003; Zhang et al. 1997). This change can be seen in Fig. 9.1, since the basic period of R–R interval changes is around 4 s, which is caused by breathing at 15 breaths/min. The main spectral components of HRV are:

- *High frequency band* (0.15–0.4 Hz) represents changes in the ANS, mostly caused by changes in breathing. The effect is called respiratory sinus arrhythmia (RSA).
- *Low frequency band* (0.04–0.15 Hz) represents changes caused by regulation of blood pressure and the sympathetic branch of the ANS.
- *Ultralow frequency band* represents changes mostly caused by the regulation of body temperature.

The frequency of the RSA component falls in the high frequency (HF) range for more than 9 breaths/min. Slower breathing generates peak power on the HRV in the low frequency (LF) range. The highest oscillation amplitudes are measured in the range 0.055–0.11 Hz and generate sinusoidal oscillations of R-R intervals by resonance among various oscillatory processes in the cardiovascular system (Vaschillo et al. 2002). Song and Lehrer found that a respiration rate of 4 breaths/min produced the highest amplitudes on the HRV, while even lower rates (3 breaths/min) generated smaller amplitude (Song and Lehrer 2003). Trained individuals (e.g., using yogic breathing techniques) can breathe at very slow rates, down to 1 breath/min or less (Jovanov 2005; Lehrer et al. 1999). Very slow breathing with frequency of less than 0.04 Hz (more than 25 s per breath) generates a dominant respiration component in the very low frequency (VLF) band (0.003–0.04 Hz) (Jovanov 2005). An example of spontaneous resonance caused by slow breathing is shown in Fig. 9.2.

Practitioners often report a state of calm, clear, and expended consciousness after practice. Therefore, it is frequently used as a preparation for longer meditation and to facilitate quieting of mind and detachment of everyday worries and non-essential mind processes.

Studies during bilateral and 15 min of unilateral nostril breathing indicate that in addition to cardiovascular changes (e.g., increased RSA), beta-2 EEG activity in the frontal leads shows lowered peak power during unilateral nostril breathing, indicating a relaxation-specific effect and a homolateral relationship between the nostril airflow and EEG theta activity. This is attributed to a lateralized modulation of the subcortical generators of EEG theta-band during unilateral nostril breathing. As a result, altered states of consciousness can be evoked by changes in the respiratory pattern and they share common basic mechanisms of respiratory, circulatory, and electrocortical interaction (Vaitl et al. 2005).

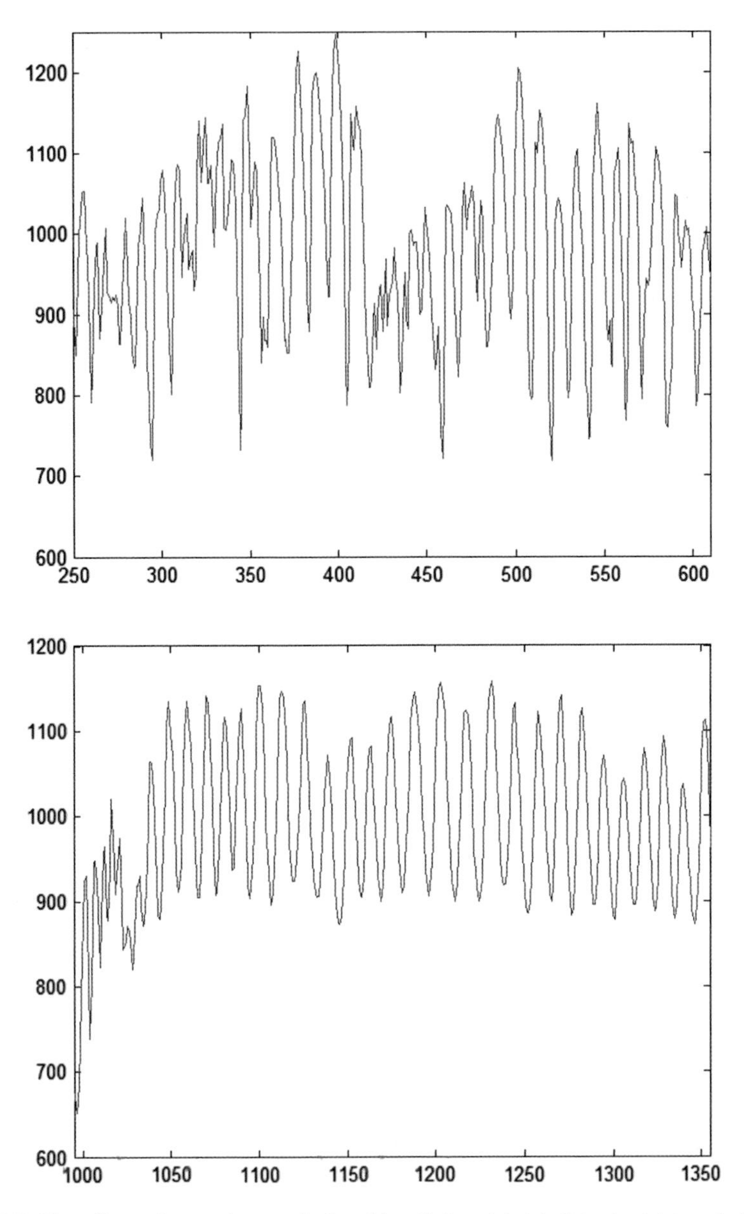

Fig. 9.2 The effect of very slow yogic breathing (1 breath/min): Inter-beat intervals before (*upper*) and after (*lower*) very slow breathing exercise (Jovanov 2005). Inter-beat intervals in milliseconds are represented as a function of time in seconds

Voluntary control of cardiac variability through biofeedback may have important effects on autonomic health and has been used in treatment of asthma patients (Vaschillo et al. 2006).

Reliable estimates of slow rhythms require very long sequences of R-R intervals, with typical durations of 5–10 min. However, this implies that every spectral estimate represents mostly the state in the middle of the processing window, which in turn implies a processing latency of about 5 min. This might be a problem for real-time analysis of cardiac variability, e.g., for emergency interventions where early detection of hemorrhaging is crucial for the survival of subjects (Jovanov et al. 2006). Minimal latency is also essential for biofeedback applications to allow users to evaluate the effectiveness of different techniques.

Changes in ANS activity are very well understood and documented (Heart Rate Variability 1996). Research has confirmed that such practices as yoga and meditation produce changes in the electrical activity of the brain, leading to a baseline increase in alpha and/or theta rhythms, and enhanced baseline theta and the maintenance of theta waves during meditation are found to be characteristic of long-term meditators. These meditators are able to keep their self-awareness intact and remain alert in this "twilight" state of consciousness.

In research on imaging, meditation, and relaxation techniques with inexperienced practitioners, it is a common observation that maximal or optimal physiological response occurs within the first 10–15 min of stimulation, and after 25 min a diminishing return sets in. Audio material for meditation and relaxation are often limited to 30 min in length for this reason (Cade and Coxhead 1979).

Low frequencies may facilitate communication with the personal unconsciousness and collective unconsciousness, as defined by Jung (1991). In Jungian terms, the role of rhythmic entrainment at lower frequencies (in the spiritual and religious cultures of the world) may be to facilitate spiritual experiences in which individual unconscious material is connected to the greater sphere of collective consciousness. In Western psychological terms, these experiences may facilitate the integration of incongruous or difficult psychological material into a person's overall sense of self, producing a genuine state of lessened psychological stress and anxiety that can lead to feelings of deep connection with other individuals. The strengthened connections may remain as a lasting effect of these spiritual experiences, and as we suggested earlier, can have long-lasting and measurable psychophysical restorative properties.

9.3 Experimentation Techniques

Experiments involving physiological correlates of conscious experiences are relatively rare. Most of the published papers evaluate the overall physiological status of a certain state of consciousness, e.g., during meditation. Monitoring techniques mostly include EEG, fMRI, and heart rate. The main problem with human experiments is making exact correlations and mappings between physiological correlates and psychological states. Users can rarely associate a specific state with a particular moment when a specific correlate has been recorded. Every action performed to mark an event creates a disturbance in the flow of consciousness and disturbs the current state by initiating a physical action. The approach used by Lehman et al. (1995) was to interrupt the session immediately after a certain physiological correlate was recorded and annotate the event with the subject's description of the content of consciousness and emotional state. However, this approach interrupted the initial event and the flow.

This section describes physiological states of altered consciousness caused by meditation, rhythmic music, and chanting. The literature cited in this section presents the state of the art in our understanding of physiological correlates of altered states of consciousness.

9.3.1 Physiological Changes Caused by Meditation

Most of the research associated with changes in brain electrical activity in altered state of consciousness is related to meditation. Meditation exercises mostly aim at increased awareness of ongoing experiences through sustained attention and detachment as a conscious effort not to analyze or judge those experiences. Long-term meditation practice is believed to affect the ability to increase awareness and achieve greater detachment during non-meditative states, as evidenced by analysis of the sources of electrical activity in the brain (Tei et al. 2009). However, different types of mediation may produce different states and different physiological correlates.

The most important features of EEG changes related to meditation are:

- Establishing alpha activity even with open eyes (Hirai 1960)
- Increased amplitude of alpha activity (Banquet 1973; Hirai 1960; Wallace and Benson 1972)
- Slower frequency of alpha rhythm (Banquet 1973; Hirai 1960; Wallace and Benson 1972)
- Rhythmic theta waves (Banquet 1973; Hirai 1960; Wallace and Benson 1972)
- Increased synchronization (hypersynchronization) (Banquet 1973)
- Dissociation of perception from the external sense organs (Hirai 1960; Ray 1988)
- Transcendent signal (Ray 1988; Ray et al. 1999)
- Occasional fast wave activity during meditation (Banquet 1973; Cahn et al. 2010; Ott 2001; Ray 1988)

The first four changes are reported during one of the first studies of EEG changes related to Zen meditation (Hirai 1960). Kasamatsu and Hirai ranked the changes in this order and find out that the changes depend directly on mental state and experience in meditation. During *zazen* (Zen meditation), alpha was slowing to 7–8 Hz, and rhythmical theta waves at 6–7 Hz appeared in the last phase attained only by skilled monks with long meditation experience.

Increased awareness is often evidenced as increased gamma activity. Ray calls this "focused arousal" at a frequency of 38 Hz during the *Dharana* stage of *Rajayoga* (Ray 1988). *Dharana* means holding the mind at a certain point.

Changed perception during meditation is frequently reported. Subjects usually define this as a relaxed awareness with stable reception. We defined this state as dissociation of perception from the external sense organs. Quantitative investigation is possible by analyzing evoked responses, such as brain electrical activity evoked by sound stimulation (e.g., clicks). It is normal to see a diminished response to frequent stimuli (habituation). This approach is commonly used in the operating room to estimate the depth of anesthesia. However, Hirai found alpha block dehabituation during meditation (Hirai 1960).

Fast wave activity was occasionally reported. Banquet identified synchronous beta waves from all brain regions of almost constant frequency and amplitude (Banquet 1973). This activity was found in four advanced meditators during their subjectively reported deepest meditation.

To the best of our knowledge, EEG changes related to the healing process are rarely investigated. Zhang reported the EEG alpha activity during the Qi Gong state that occurred predominantly in the anterior regions (Zhang et al. 1988). Qi Gong represents a system of physical and mental training for "energy cultivation" and the manipulation of intrinsic energy that can be used for healing. The peak frequency of EEG alpha rhythm was slower than the resting state, and the change of EEG during Qi Gong between anterior and posterior readings had negative correlation. It can be seen that reported changes are very similar to the previously described changes during the meditation (Jovanov 2005; Rakovic et al. 1999).

Prominent changes in ANS function have been frequently reported, mostly as an increase of cardiac variability (Peng et al. 1999; Ray et al. 1999). This change is mostly attributed to very slow breathing.

9.3.2 Physiological Changes Caused by Rhythmic Music

Music is frequently used as an effective means of achieving altered states of consciousness (Drury 1989; Eliade 1964; Harner 1990). As melody or rhythm, music has the power to move participants on subconscious level. After experiencing the shamanic journey Nevill Drury writes:

> One thing never ceases to amaze me – that within an hour or so of drumming, ordinary city folk are able to tap extraordinary mythic realities that they have never dreamed of (Drury 1989).

The literature reports that listening to or playing sustained, repetitive drumming, is associated with an ASC or a qualitative shift in mental functioning (Neher 1961, 1962). The phenomenology of the interactions between drumming and states of consciousness has been well documented, but much remains unexplained. Possible explanations include: (a) Rhythmic drumming might act as a focus for concentration, used in combination with sensory deprivation, fasting, fatigue, and mental imagery to achieve an ASC; (b) the rhythm or monotony of the drumming facilitates an ASC; (c) evoked responses in brain electrical activity create a resonance with the external stimulation and induce an ASC.

Music has been also used to create an emotional state that would facilitate an ASC (Kreutz et al. 2008; Rakovic et al. 1999; Ray 1988).

9.3.3 Physiological Changes Caused by Chanting

Rhythmic chanting can be seen as a devotional practice that synchronizes the hierarchy of a subject's bodily rhythms and strengthens the coupling between them. Therefore, during this practice we can expect to see an increased interaction of breathing and HRV, or larger amplitude of the RSA and larger variation of interbeat intervals.

We investigated changes of heart rate and HRV caused by prolonged chanting. Participants reported a quiet and peaceful state after prolonged collective chanting. We measured heart rate as interbeat intervals of two subjects before, during, and after the session. An example of changes in heart activity is shown in Fig. 9.3. The

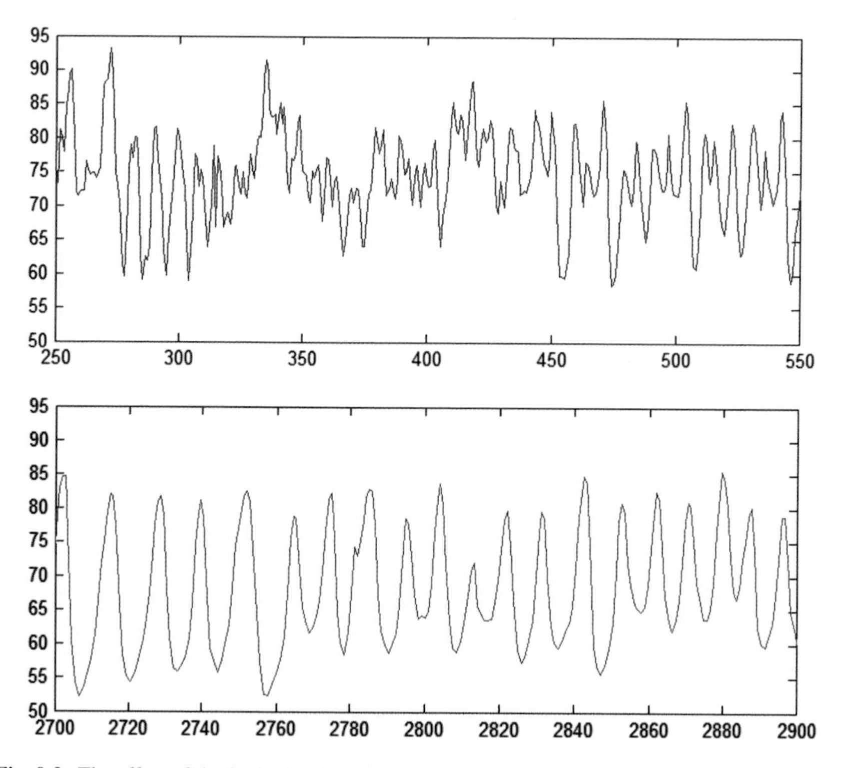

Fig. 9.3 The effect of rhythmic chanting: heart rate in beats/min as a function of time in seconds, initially (*upper plot*) and 45 min later (*lower plot*)

time resolution of measurements of interbeat intervals was 1 ms, as required for analysis of HRV. The subject was a 26-year-old healthy female with several years of experience of chanting. The upper plot represents interbeat variability at the beginning of chanting, while the lower plot represents the variability of the same subject after 45 min of chanting. A rhythmic, regular pattern of interbeat variability can be clearly seen on the lower plot. The pattern indicates slow, regular breathing, and strong coupling between breathing and heart rate. Precise quantification of the RSA would require precise measurement of lung volume that was not available at the time of experiment. Future work should include measurement of other physiological parameters, such as lung volume, blood pressure, and cardiac output.

9.4 Discussion and Conclusion

Increased coupling of bodily rhythms and conscious decoupling of the limbic system from the current content of consciousness may facilitate conscious exploration of the unconscious mind. We believe that the stabilization of basic physiological rhythms may serve as a foundation for altered states of consciousness that would facilitate insights and expansion of consciousness (Jovanov 1995, 2005; Rakovic et al. 1999).

The role of the limbic system is crucial for survival, since it allows activation of the high-priority body functions necessary during life-threatening events. During a journey into unconsciousness, the limbic system provides important indication about the emotional value of the experiences. However, during exploration the activation of the limbic system also influences mental content in our flow of thoughts. This results not only in a change in the stream of consciousness, but in the processing of conscious material. As we cannot consciously and directly control the functions of our limbic system, we cannot directly control our stream of consciousness. In a Buddhist philosophical context, this could account for the times when we are unable to stabilize the focus of our "wandering minds" on a particular issue. So, the critical question is that of how we can take advantage of the "message" of the emotional mind (limbic system) without significant modification of the stream of consciousness.

Undisturbed flow of consciousness might be attained by reflecting on the first of Patañjali's sutras (*yogah cittavrtti nirodhah*), on the cessation of movement in consciousness (Iyengar 1993). The sutra states that Yoga, or union of the individual and divine spirit, starts with the cessation of movements in consciousness.

We hypothesize that externally stabilized physiological rhythms can result in periods of uninterrupted conscious experience via the indirect stabilization of the limbic system. This stabilization can allow deep insights and a variety of integrating experiences to emerge. To paraphrase a frequently used metaphor, one can hardly see the bottom of the lake because of the constant chaotic ripples on the surface, whereas stable and regular waves allow one to see much deeper into the lake.

We also hypothesize that the entrainment techniques described here, including rhythmic drumming, slow breathing, and chanting, can stabilize and regulate basic body rhythms, which has therapeutic value in itself, and allow transcendent experiences to emerge. We hypothesize that these entrainment techniques provide extended control of the limbic system, offering one the chance to reduce "emotional noise" and settle the mind. We believe that externally stabilized physiological rhythms modulate control feedback between mental processes and the limbic system, producing periods of uninterrupted conscious experience and reflection, which allows deeper insights and facilitates spiritual growth. It would be interesting to simultaneously monitor subjects and drummer(s) to observe common patterns and synchronicity of changes. However, it is very important to understand the fundamental limitations of typical signal processing techniques used to analyze the changes of autonomic and central nervous system activity (Jovanov 2008). This is particularly important in the case of biofeedback techniques where processed physiological parameters are used in the entrainment process.

Future basic research in neurophysiology will greatly benefit from optical control of cells in animal experiments. This is a very promising approach used in molecular biology that will allow genetic modification of highly targeted cell assemblies and their control using optical stimulation at different wavelengths (Tononi and Koch 2008). This approach will allow researchers to "disconnect" at will very specific assemblies and test overall neural function in animal experiments.

Experimental research with human subjects is likely to include the development of experiments that will facilitate personalized selection of the monitored parameters and presentation modalities. Computerized monitoring and database acquisition of sessions correlated with personal experiences may facilitate mapping of uncharted territory of physiological correlates of states of consciousness; such a map would serve as a starting point for personal journeys to the realm of individual and collective consciousness.

Nunez and Shrinivasan conclude thus: "Maybe consciousness is a resonance phenomenon and only properly tuned brains can orchestrate beautiful music of sentience" (Nunez and Srinivasan 2006, p. 525). We believe that the current technology may facilitate proper "tuning" and self-exploration of the expanded realms of consciousness.

References

Banquet JP (1973) Spectral analysis of the EEG in meditation. Electroencephalogr Clin Neurophysiol 35:143–151

Basar E (1980) EEG brain dynamics. Elsevier, Amsterdam

Basar E (ed) (1988) Dynamics of sensory and cognitive processing by the brain. Springer, Berlin

Benson H (1984) Beyond the relaxation response. Berkley Books, New York

Cade GM, Coxhead F (1979) The awakened mind, biofeedback and the development of higher states of awareness. Delacorte, New York

Cahn BR, Delorme A, Polich J (2010) Occipital gamma activation during Vipassana meditation. Cogn Process 11:39–56
Chalmers D (2002) The puzzle of conscious experience in the hidden mind. Sci Am 12:90–100
Cook WH, Salinas J, McManus JG, Ryan K, Rickards CA, Holcomb JB, Convertino V (2006) Heart period variability in trauma patients may predict mortality and allow remote triage. Aviat Space Environ Med 2006:14–19
Cosic I, Cvetkovic D, Fang Q, Jovanov E, Lazoura H (2006) Human electrophysiological signal responses to ELF Schumann resonance and artificial electromagnetic fields. FME Trans 34:93–103
Crick FC (1994) The astonishing hypothesis. the scientific search for soul. Charles Scribner's Sons, New York
Csíkszentmihályi M (1990) Flow: the psychology of optimal experience. Harper and Row, New York
Drury N (1989) Elements of shamanism. Element Books, Shaftesbury
Egner T, Gruzelier JH (2003) Ecological validity of neurofeedback: modulation of slow wave EEG enhances musical performance. Cogn Neurosci Neuropsychol 14(9):1221–1224
Eliade M (1964) Shamanism: archaic techniques of ecstasy (trans: Trask WR). Princeton University Press, Princeton
Guyton AC, Hall JE (2006) Textbook of medical physiology, 11th edn. Sunders Elsevier, Philadelphia
Harner MJ (1990) The way of the Shaman: a guide to power and healing. Harper and Row, San Francisco
Heart rate variability: standards of measurement, physiological interpretation and clinical use (1996) Task Force of the European Society of Cardiology and the North American Society of Pacing and Electrophysiology. Circulation 93:1043–1065
Hirai T (1960) Electroencephalograpic study of Zen meditation. Psychiatr Neurol Jap 62:76–105
Iyengar BKS (1993) Light on the Yoga Sutras of Patanjali. Thorsons, San Francisco
Jovanov E (1995) On methodology of EEG analysis during altered states of consciousness. In: Rakovic D, Koruga D (eds) Consciousness: challenge of the 21st century science and technology. ECPD, Belgrade
Jovanov E (2005) On spectral analysis of heart rate variability during very slow yogic breathing. In: Proceedings of the 27th annual international conference of the IEEE engineering in medicine and biology society, Shanghai, China
Jovanov E (2008) Real-time monitoring of spontaneous resonance in heart rate variability. In: Proceedings of the 30th annual international conference of the IEEE engineering in medicine and biology society, Vancouver, Canada, pp 2789–2792
Jovanov E, Cox P, Saul P, Salinas J, Ryan K, Convertino VA (2006) A comparison of real-time performance of signal processing algorithms for minimum latency detection of hypovolemic states. In: Proceedings of the 28th annual international conference of the IEEE engineering in medicine and biology society, New York, Sept 2006, pp 1674–1677
Jung CG (1991) The archetypes and the collective unconscious. Routledge, London
Kreutz G, Ott U, Teichmann D, Osawa P, Vaitl D (2008) Using music to induce emotions: influences of musical preference and absorption. Psychol Music 36(1):101–126
Lane J et al (1998) Binaural auditory beats affect vigilance performance and mood. Physiol Behav 63(2):249–252
Lehman D, Grass P, Meier B (1995) Spontaneous conscious covert cognition states and brain electric spectral states in canonical correlations. Int J Psychophysiol 19:41–52
Lehmann D et al (2009) EEG microstates. Scholarpedia 4(3):7632
Lehrer P, Sasaki Y, Saito Y (1999) Zazen and cardiac variability. Psychosom Med 61(6): 812–821
Ludwig AG (1968) Altered states of consciousness. In: Prince R (ed) Trance and possession states. R.M. Bucke Memorial Society, McGill University, Montreal, pp 69–95

Maxfield MC (1990) Effects of rhythmic drumming on EEG and subjective experiences. Unpublished Doctoral Dissertation, Institute of Transpersonal Psychology, Menlo Park
Mormann F, Koch C (2007) Neural correlates of consciousness. Scholarpedia 2(12):1740
Neher A (1961) Auditory driving observed with scalp electrodes in normal subjects. EEG Clin Neurophysiol 13:449–451
Neher A (1962) A physiological explanation of unusual behavior in ceremonies involving drums. Hum Biol 34(2):151–160
Nunez PL, Srinivasan R (2006) Electrical fields of the brain. Oxford University Press, Oxford
Oster G (1973) Auditory beats in the brain. Sci Am 229:94–102
Ott U (2001) The EEG and the depth of meditation. J Meditat Meditat Res 1(1):55–68
Peng C-K, Mietus JE, Liu Y, Khalsa G, Douglas PS, Benson H, Goldberger AL (1999) Exaggerated heart rate oscillations during two meditation techniques. Int J Cardiol 70:101–107
Penrose R (1994) Shadows of the mind, a search for the missing science of consciousness. Oxford University Press, Oxford
Raichle ME (2010) The brain's dark energy. Sci Am 302:44–49
Rakovic D (1991) Neural networks, brainwaves, and ionic structures: acupuncture vs. altered states of consciousness. Int J Acup Electro Therap Res 16:89–99
Raković D (1995) Brainwaves, neural networks, and ionic structures: biophysical model for altered states of consciousness. In: Raković D, Koruga DJ (eds) Consciousness: scientific challenge of the 21st century. ECPD, Belgrade
Rakovic D, Tomasevic M, Jovanov E, Radivojevic V, Sukovic P, Martinovic Z, Car M, Radenovic D, Jovanovic-Ignjatic Z, Skaric L (1999) Electroencephalographic (EEG) correlates of some activities which may alter consciousness: the transcendental meditation technique, musicogenic states, microwave resonance relaxation, healer/healee interaction, and alertness/drowsiness. Informatica 23(3):399–412
Ray GC (1988) Higher stages of Rajayoga and its possible correlation with process of evolution. J Inst Eng (India) 68:37–42
Ray GC, Kaplan AY, Jovanov E (1999) Homeostatic change in the genesis of ECG during yogic breathing. J Inst Eng (India) 79(1):28–33
Saraswati SS (2004) A systematic course in the ancient Tantric techniques of Yoga and Kriya. Yoga Publication Trust, Munger
Schwartz MS, Andrasik F (eds) (2003) Biofeedback: a practitioner's guide. Guilford, New York
Song H-S, Lehrer PM (2003) The effects of specific respiratory rates on heart rate and heart rate variability. Appl Psychophysiol Biofeedback 28(1):13–23
Stevens L, Haga Z, Queen B, Brady B, Adams D, Gilbert J, Vaughan E, Leach C, Nockels P, McManus P (2003) Binaural beat induced theta EEG activity and hypnotic susceptibility: contradictory results and technical considerations. Am J Clin Hypn 45(4):295–309
Swami Muktibodhananda Saraswati, Swara Yoga (1984) The Tantric Science of Brain Breathing, Bihar School of Yoga, Munger, Bihar, India
Tang YY, Ma Y, Fan Y, Feng H, Wang J, Feng S, Lu Q, Hu B, Lin Y, Li J, Zhang Y, Wang Y, Zhou L, Fan M (2009) Central and autonomic nervous system interaction is altered by short-term meditation. Proc Natl Acad Sci USA 106(22):8865–8870
Tart CT (ed) (1972) Altered states of consciousness, 2nd edn. Anchor Books, New York
Tei S, Faber PL, Lehmann D, Tsujiuchi T, Kumano H, Pascual-Marqui RD, Gianotti LRR, Kieko K (2009) Meditators and non-meditators: EEG source imaging during resting. J Brain Topogr 22 (3):158–165
Thompson J (2006) Bio-Tuning®/sonic induction therapy manual – intensive course book. Center for Neuroacoustic Research, Encinitas. http://www.neuroacoustic.com
Tononi G, Koch C (2008) The neural correlates of consciousness – an update. Ann N Y Acad Sci 1124:239–261
Vaitl D, Birbaumer N, Gruzelier J, Jamieson G, Kotchoubey B, Kübler A, Lehmann D, Miltner WHR, Ott U, Pütz P, Sammer G, Strauch I, Strehl U, Wackermann J, Weiss T (2005) Psychobiology of altered states of consciousness. Psychol Bull 131(1):98–127

Vaschillo E, Lehrer P, Rishe N, Konstantinov M (2002) Heart rate variability biofeedback as a method for assessing baroreflex function: a preliminary study of resonance in the cardiovascular system. Appl Psychophysiol Biofeedback 27(1):1–27

Vaschillo E, Vaschillo B, Lehrer P (2006) Characteristics of resonance in heart rate variability stimulated by biofeedback. Appl Psychophysiol Biofeedback 31(2):129–142

Wallace RK, Benson H (1972) The physiology of meditation. Sci Am 226:84

Zhang JZ, Zhao J, He QN (1988) EEG findings during special psychical state (Qi Gong State) by means of compressed spectral array and topographic mapping. Comput Biol Med 18(6): 455–463

Zhang P, Tapp W, Reisman S, Natelson B (1997) Respiration response curve analysis of heart rate variability. IEEE Trans Biomed Eng 44(4):321–325

Chapter 10
States of Consciousness Beyond Waking, Dreaming and Sleeping: Perspectives from Research on Meditation Experiences

Frederick Travis

Abstract Three categories of meditation practices have been proposed: *focused attention* meditations, which involve voluntary and sustained attention on a chosen object; *open monitoring* meditations, which involve non-reactive monitoring of moment-to-moment content of experience; and *automatic self-transcending* meditations, which are designed to transcend their own activity. While *focused attention* and *open monitoring* meditations explore the nature of individual cognitive, affective, and perceptual processes and experiences, *automatic self-transcending* meditations explore the state when conscious processing and experiences are transcended, a state called pure consciousness. This paper reports unique phenomenological and physiological patterns during the state of pure consciousness, as experienced during Transcendental Meditation (TM) practice, a meditation in the *automatic self-transcending* category. These data support the description of pure consciousness as a fourth state of consciousness with unique phenomenological and physiological correlates. This paper also discusses the Junction Point Model that integrates meditation experiences with the three ordinary states of waking, sleeping, and dreaming. The Junction Point Model is supported by EEG data and provides a structure to integrate ordinary experience during waking, sleeping, and dreaming with meditation experiences and so can serve as a foundation for investigating the full range of human consciousness.

10.1 Introduction

Traditional meditation techniques are part of a subjective approach to gaining knowledge that parallels and complements the objective approach of gaining knowledge in the natural sciences. The objective approach in western science has

F. Travis (✉)
Center for Brain, Consciousness, and Cognition, Maharishi University of Management, Fairfield, IA 52557, USA
e-mail: ftravis@mum.edu

D. Cvetkovic and I. Cosic (eds.), *States of Consciousness*, The Frontiers Collection, DOI 10.1007/978-3-642-18047-7_10,

used instruments to objectively measure phenomena to identify the principles and laws that explain material and social interactions. Similarly, the subjective approaches in eastern traditions have used meditation techniques to directly experience and so understand the full range of human experience. Some meditation techniques are designed to explore the range of waking processes and experiences; others are designed to explore the nature of consciousness at the source of thought when mental processes and content are transcended. In this time when East meets West, scientists can objectively evaluate growth of subjectivity through meditation practice. Thus, meditation techniques can serve as scientific probes to fathom the range of human experience.

Lutz has divided meditation practices into two categories: *focused attention* meditations, which involve voluntary and sustained attention on a chosen object, and *open monitoring* meditations, which involve non-reactive monitoring of the moment-to-moment content of experience (Lutz et al. 2008). In *focused attention* meditations, attention is focused on a given object and regulative skills are developed to monitor the movement of attention – detecting distraction, disengaging attention from the source of distraction, and redirecting and refocusing on the object (Lutz et al. 2008). *Open monitoring* or mindfulness-based meditations refer to an alert and open mode of perceiving and monitoring mental content from moment to moment, including perception, sensation, cognition, and affect (Kabat-Zinn 2003). These meditation practices involve the non-reactive, dispassionate monitoring of the content of ongoing experience to become reflectively aware of the nature of emotional and cognitive processes.

Meditation techniques in these two categories explore the nature of waking processes and experiences. Waking experiences are characterized by subject/object duality. A subject, agent or experiencer observes and reflects on affective, cognitive, or sensory objects of perception that are separate from himself or herself – I am here observing the experience out there. Focusing attention on a specific object of experience or maintaining an orientation to monitor changing objects of experience uses and maintains the subject/object duality. One keeps the attention involved with the procedures of the technique.

A third meditation category has been proposed, *automatic self-transcending*, which includes meditation techniques designed to transcend their own activity (Travis and Shear 2010a). Meditation techniques in this category do not attempt to control the movement of attention or to monitor ongoing experience; rather they are designed to transcend their own activity – to allow a state of consciousness to emerge when mental activity and cognitive control has been transcended. Techniques in this third category necessary must be automatic, because any intention to control the attention would keep the mind active and not allow mental activity to settle to silence.

Meditation techniques in the *automatic self-transcending* category provide insight into the state of consciousness when thoughts have ceased, revealing a ground state of human consciousness. This paper explores the nature of this state, and then presents a model that integrates this state during meditation with those during waking, dreaming, and sleeping.

Note that these three categories are not necessarily mutually exclusive within a single session or over the course of a life-time of meditation practice. *Focused attention* and *open monitoring* are combined in Zen, Vipassana, and Tibetan Buddhism meditation traditions (Austin 2006; Gyatso and Jinpa 1995; Lutz et al. 2008). Also, with diligent practice over many years, *focused attention* meditations may lead to reduced cognitive control and could result in effortless concentration (Lutz et al. 2008; Wallace 1999).

A meditation technique within the *automatic self-transcending* category is the Transcendental Meditation® (TM®) technique. During TM practice, one appreciates a mantra at finer levels in which the mantra becomes secondary in experience and self-awareness becomes primary (Maharishi 1969; Travis and Pearson 2000). Ultimately, the mantra disappears and the subject-object relation that defines customary experiences is transcended. The subject, or the experiencer, finds him/herself awake to his/her own existence – called pure consciousness or a ground state of consciousness (Maharishi 1997). Pure consciousness is pure in the sense that it is free from the processes and contents of knowing. It is a state of consciousness in that self-awareness is maintained. Pure consciousness is a non-dual state of awareness – the self is both the subject and object of awareness. This would contrast with the end state of some Buddhist meditations that seek to lose the self in the object, such as during the practice of loving, kindness, and compassion (Lutz et al. 2008). While this is also a non-dual state, it is a state of object referral – the object alone is (see Travis and Shear 2010b).

The non-dual state of pure consciousness differs from the duality of conscious awareness or conscious experience. Conscious experience has a three-part structure – the experiencer, the object of experience, and the process of experience. These three components exist as separate even at the same time as they are unified in the conscious experience. In pure consciousness, the three-part structure of experiencer, object of experience, and process linking the two has been transcended. Now, the experiencer or subject is the both the subject and object of experience – it is described as a purely self-referral experience.

Table 10.1 presents a schematic of the qualitative shift of inner experience from sleeping to pure consciousness. This table presents a 2 $\times$ 2 grid with the presence or absence of affective, cognitive, or perceptual content as one axis, and presence or absence of sense of self as the other. As presented in this table, the waking state is characterized by the inner experience of a sense of self, the experiencer or doer and the experience of outer objects in the mind or in the environment. There is a clear separation between my inner reality and my outer experience.

Table 10.1 Phenomenological characteristics that differentiate waking, dreaming, sleeping and pure consciousness

		Sense of self is present	
		Yes	No
	Yes	Waking	Dreaming
Inner and/or outer perception is present	No	Pure consciousness	Sleeping

The sleep state is characterized by no sense of self and no awareness of any content. "Sleeping like a log" is a saying for having a good night's sleep. During deep sleep, there is no awareness of self or ongoing cognitive or perceptual experiences for large blocks of time.

Dreaming is arguably characterized by no sense of self and vivid dream images. This describes most dream experiences. Lucid dreaming, we argue, is meta-cognition within the dream state. A careful analyze of lucid dream content reveals that the dream ego and dream intellect make decisions that the waking ego and waking intellect would not make (see Travis 1994).

The fourth box in this 2 × 2 table is a sense of self without mental content. Before reading this paper, you along with the vast majority of today's scientists might say that state does not exist. The empty cell is simply an artifact of setting up a 2 × 2 grid. How can there be a sense of self without an object; without a sense of the body or the thinker or the thinking? William James, in his *Principles of Psychology*, observed:

> . . . it is difficult for me to detect in the activity any purely spiritual element at all. Whenever my introspection glance succeeds in turning round quickly enough to catch one of those manifestations of spontaneity in the act, all it can ever feel distinctly is some bodily process, for the most part taking place within the head (James 1950/1890, p. 300).

This conclusion is a valid conclusion if the experience of consciousness has been limited to waking experience, which includes sense-of-self (inner) and outer experiences. In waking consciousness, the self is never found without an object. However, the proposal put forth in this paper, is that meditation techniques uncover that state of pure consciousness and so make this seemingly anamolous state available for discussion and experimentation.

10.2 Phenomenological and Physiological Investigations of Pure Consciousness

Fifty-two college students who practiced the TM technique for a few months to over 8 years were asked to describe their deepest experiences during TM practice. They were asked to use their own words to describe their experiences, as though they were describing it to someone who did not meditate. A content analysis of these descriptions yielded three themes that were common to all reports – absence of time, absence of space, and absence of body sense (Travis and Pearson 2000). Time, space, and body sense are the framework that give meaning to waking experience. During deepest TM experiences, both the fundamental framework and the content of waking experience were reported to be absent. This suggests that the experience of pure consciousness may not be an "altered" state of waking. It is not described in terms of distorted content – strong emotions, strong visual, auditory or tactile sensations, or distorted sense of self. Rather, pure consciousness was described by the absence of the customary framework and characteristics that define waking experience. Phenomenologically, pure consciousness is distinct from experiences that characterize waking, dreaming, and sleeping.

Physiological, pure consciousness is also distinct from waking, dreaming, and sleeping. During pure consciousness, research reports higher EEG alpha coherence, and apneustic breathing – slow, extended inhalation from 10 to 20 s – with skin conductance orienting and a heart rate preparatory response at the onset of breath changes (Badawi et al. 1984; Travis and Pearson 2000; Travis and Wallace 1997). Apneustic breathing is not reported in normal populations (outside of meditation practices), and has never been reported in the literature with durations longer than 4–6 s (Plum and Posner 1980). The respiratory drive centers responsible for apneustic breathing (the parabrachialis medialis nuclei) are quiet during waking, dreaming, and sleeping, but become active during pure consciousness periods (Kesterson and Clinch 1989). Changes in the brainstem nuclei driving breathing, in autonomic functioning and brain state, with distinct phenomenological reports, supports the description of pure consciousness as a fourth major state of consciousness fundamentally different from waking, dreaming, or sleeping (Maharishi 1997).

10.3 Self-Referral Default Mode Network: Pure Consciousness Experiences Activate the Intrinsic Default State of the Brain

During TM practice, brain activity is reported to increase in the default mode network (DMN) (Travis et al. 2010). This network was first noted when comparing data from nine different neural imaging studies. Since neural imaging involves subtracting control from experimental images, higher activation in a control condition could lead to perceived "decreases" in the experimental condition. In these nine studies of unrelated and independent tasks, decreases in midline frontal and parietal cortices were consistently reported (Raichle et al. 2001); eyes-closed rest or simple fixation on a point were used as the control conditions. The researchers concluded that a default mode network exists that is an intrinsic, default property of the brain (Fox and Raichle 2007). Activation in this default mode is higher during low cognitive load periods, such as eyes-closed resting control periods, and is lower during goal directed behaviors requiring executive control (Gusnard et al. 2001; Raichle and Snyder 2007).

Further research into DMN activation reported higher activation during (1) self-referential mental activity (Gusnard et al. 2001; Kelley et al. 2002; Vogeley et al. 2001); (2) self-projection tasks; and (3) taking the viewpoint of others (Buckner and Carroll 2007). Activity in the default state is higher during eyes-closed experiences. When one closes the eyes, objects are reduced but sense-of-self remains. The person knows that they are sitting in space; they are there waiting for the next instruction. This is a predominately self-referral experience and DMN activity is reported to be high. When opens the eyes and attention streams through the senses and falls on an object – an object referral experience – DMN activity is reduced.

Relative to eyes-closed rest, DMN activation was higher during TM. This supports the description of pure consciousness during TM practice as being a fuller or higher sense of self-referral than just eyes-closed rest. Self can be written with a small and a capital "S". When "self" is written with a small "s" it denotes the self that thinks, feels, decides, and experiences – the self in a waking state; when "Self" is written with a big "S" it denotes that part of the individual that does not change and is the source of all streams of individual activity (Maharishi 1969). Thus, DMN activation during eyes-closed rest – small self-referral – would rise during the experience of pure consciousness – large self-referral.

10.4 Junction Point Model of Pure Consciousness, Waking, Sleeping, and Dreaming

A proposed Junction Point Model integrates meditation experience with waking, dreaming, and sleeping. This model helps to locate meditation experiences relative to those three states. It also provides a model for discussing higher states of consciousness. The Junction Point Model posits that waking, sleeping, and dreaming are not isolated states that interact, but are sequential expressions of an undifferentiated field – pure consciousness – that underlies them (Maharishi 1972; Travis 1994). This model starts with the observation that waking, sleeping, and dreaming are discrete states. This assumption is supported by unique brain stem activity (Siegel 1987), neurotransmitter balance (Hobson 1988), and EEG, EMG, and eye movement patterns (Niedermeyer 1997) during each state. The model suggests that one state must completely fade away before the next begins, and that between any two, a junction point can be located that will mark the end of one state and the beginning of the next. These junction points are windows into the field of consciousness posited to underlie waking, sleeping, and dreaming.

10.5 Research Testing the Existence of Pure Consciousness Between States of Consciousness

The prediction that pure consciousness can be located between states of consciousness is supported by two lines of research. First, similar EEG patterns have been reported during TM practice, which leads to pure consciousness between thoughts, as during the waking/sleeping transition. For instance, frontal alpha and slowing of peak EEG frequency by 1–2 Hz, reported during TM practice (Wallace 1970), were later independently reported (Santamaria and Chiappa 1987) during the waking/sleeping transition. This relation between EEG patterns during TM program and the waking/sleeping transition has also been experimentally investigated. EEG in 15 experienced TM subjects during TM practice was compared to EEG during the waking/sleeping transition in 15 non-meditating subjects matched for age, gender, and handedness. The raw EEG and the resulting power and coherence spectra were not significantly different between the TM sessions and the waking/sleeping transition (Travis 1990). However, the duration of these EEG patterns were different. During TM practice, they lasted for the entire 10-min TM session; during the waking/sleeping transition, they lasted for 3–5 min.

Other researchers have reported this similarity of EEG patterns during TM and during the waking/sleeping transition, and the fact that they last longer during TM. They concluded that TM practice balanced awareness between waking and sleeping (Fenwick et al. 1977; Stigsby et al. 1981; Wachsmuth and Dolce 1980), or that TM practice freezes the hypnagogic process (Pagano and Warrenberg 1983; Schuman 1980). The junction point model gives a more comprehensive interpretation of these findings. According to this model, EEG patterns would be similar during the waking/sleeping transition and during TM practice because both states involve a gradual minimizing of mental activity followed by pure consciousness periods between states in the first case, and between thoughts in the second. Also, this model would predict a longer duration of this pattern during TM practice because one continues to give an inward direction to awareness during TM, thereby cycling through pure consciousness many times in each session, in contrast to the natural transition between states of consciousness.

A second line of research directly compared EEG patterns during TM practice to those during the junction points between waking/sleeping, sleeping/dreaming, and dreaming/sleeping. In the subjects' power spectra, activity in each band, except 7–10 Hz, could be explained by known sleep mechanisms (Travis 1994). For instance, the rise and fall of 1–4 Hz power occurred during periods of slow wave sleep marked by high delta density and power; 13–16 Hz power was highest during Stage 2 sleep, reflecting sleep-spindle activity. In contrast, the rise and fall of 7–10 Hz activity (alpha1) occurred at the transitions between waking, sleeping, and dreaming in all subjects. Activity in this same band was seen in these subjects during their Transcendental Meditation program. In terms of the Junction Point Model, significant peaks in EEG power during the transitions between waking, sleeping, and dreaming and during TM practice suggests that a similar state might be available between states of consciousness and between thoughts.

10.6 Research Testing the Integration of Pure Consciousness with Waking, Sleeping and Dreaming

If pure consciousness underlies waking, dreaming and sleeping, can it be integrated with the three customary states of consciousness? If pure consciousness represents a fourth state of consciousness, then the integration of pure consciousness with waking, dreaming and sleeping will be a fifth state. This is the first stabilized state of enlightenment described in the Vedic tradition, called *turiyatit chetana* (Maharishi 1997). Since subjective experiences and states of consciousness have defining physiological characteristics, this proposed fifth state of consciousness should also have distinct physiological markers.

10.7 Research Testing the Integration of Pure Consciousness with Sleeping

Two research papers report EEG data that support the description of the experience of pure consciousness along with the body sleeping. Banquet and Sailhan (1974) recorded EEG during sleep in advanced TM subjects, and reported that alpha1 activity, seen during the TM practice, was superimposed over delta activity, seen during deep sleep. Although they used experienced TM subjects, they did not correlate this EEG pattern with self-reports of the integration of pure consciousness with sleep.

Mason tested this hypothesis more directly (Mason et al. 1997). She compared sleep EEG in 11 subjects reporting the integration of pure consciousness with sleep, to sleep EEG in 11 short-term TM subjects, who did not report this experience, and 11 non-meditating controls. Subjects reporting the integration of pure consciousness with sleep had simultaneous alpha1 and delta in their sleep records, which supports their subjective experience of self-awareness while the body rested deeply. Simultaneous alpha and delta during sleep, called alpha/delta sleep, has also been reported in clinical cases of subjects in pain (Moldofsky et al. 1983). However, these clinical subjects only reached Stages 2 and 3 during sleep. In contrast, the TM subjects did not complain of pain, discomfort, or problems during sleeping, and they had the same amount of Stage 4 sleep as normal subjects.

10.8 Research Testing the Integration of Pure Consciousness with Waking

A second line of research has investigated the integration of pure consciousness with waking tasks. EEG was recorded during simple and choice paired reaction time tasks in 17 long-term TM subjects, reporting the integration of pure consciousness with waking and sleeping, and compared to EEG patterns in 17 short-term TM subjects who did not report this experience, and 17 non-meditating controls. In individuals reporting the integration of pure consciousness with waking and sleeping, brain preparatory responses during the paired reaction time tasks were higher in simple but lower in choice trials, and alpha relative power and broadband frontal EEG coherence were higher during the challenging tasks (Travis et al. 2002). Increased alpha amplitude and coherence, characteristic of TM practice, appeared to become a stable EEG trait during challenging tasks in these subjects.

These individuals were also given a battery of personality and psychological tests including inner/outer orientation, moral reasoning, anxiety, and personality. Scores on these tests were factor analyzed. The first unrotated PCA component of the test scores yielded a "consciousness factor," analogous to the intelligence *g* factor. The individuals reporting the integration of pure consciousness with waking and sleeping had significantly higher consciousness factor scores – more inner directed, higher levels of moral reasoning, higher emotional stability, and lower anxiety. These same individuals had higher scores on the Brain Integration Scale (BIS) (Travis et al. 2004).

We can use a movie metaphor to give a sense of the growth of consciousness. Watching a movie, most individuals are "lost" in the movie. The movie is real. Emotions and thoughts are dictated by the ever-changing sequence of the film. This is a predominantly object-referral state that characterizes the waking state. The meditative experience of transcending – the repeated experience of pure, self-referral consciousness – alters this common movie-going experience. Subjectively, the individual begins to "wake up" to his/her own inner status. Although continuing to enjoy the movie, he/she gradually becomes aware that they exist independently of the movie. They experience a value of witnessing the activity around them. To these individuals, the ever-changing movie frames are a secondary part of experience because these frames are always changing. The most salient part of their every experience is pure self-awareness. What is "real" shifts with time from the movie to self-awareness, from the thoughts, feelings, and actions to the Self, from object-referral to self-referral awareness (Travis et al. 2004).

10.9 Other Research on the Brain Integration Scale

The Brain Integration Scale (BIS) was constructed from cross-sectional data of individuals reporting more frequent experiences of pure consciousness. A 3-month random assignment longitudinal study with college students supports the finding that TM practice leads to higher scores on this scale. After 3 months of TM practice, college students increased on brain integration scores and decreased in sympathetic reactivity (Travis et al. 2009). They also decreased in negative personality traits, such as total mood disturbance, anxiety, and depression, and increased in positive personality traits such as vigor, emotional intelligence, and behavioral and emotional coping (Nidich et al. 2009). Thus, the experience of pure consciousness during TM could to be a causal mechanism for increasing levels of brain integration over time.

In addition, BIS scores were explored in two groups of athletes: professional athletes who placed in the top ten in the Olympics, world games, or national games for 3 consecutive years, or control athletes who did not consistently place. The professional athletes who excelled had higher BIS scores, faster skin conductance habituation to loud tones, and higher moral reasoning and ego development than the controls (Harung et al. in press). The athletes were not practicing a meditation technique. Their level of brain integration reflects the sum of their lifestyle and life experiences to that point. However, this finding suggests greater success in life with those markers that could index higher consciousness.

10.10 Conclusion

Meditation techniques can serve as probes to investigate states of consciousness. Investigating the Transcendental Meditation technique, a technique designed to transcend its own activity, has led to phenomenological and physiological descriptions of a state called pure consciousness, a proposed fourth state of consciousness, and has generated a model, the Junction Point Model, which integrates waking, sleeping, and dreaming with meditation experiences. This model is supported by similar EEG patterns during the transitions between waking, sleeping, and dreaming and during TM practice. This model also suggests that the underlying field of pure consciousness can coexist with ordinary waking, sleeping and dreaming. This would be a fifth state of consciousness. Individuals reporting this experience were distinguished during slow wave sleep by the coexistence of alpha EEG, observed during TM, and delta EEG, observed during sleep, and during waking tasks by higher scores on the Brain Integration Scale and higher consciousness factor scores. The Junction Point Model could provide a structure to integrate ordinary experience with meditation experiences to help model and research the full range of human consciousness.

References

Austin JH (2006) Zen-brain reflections. MIT, Cambridge

Badawi K, Wallace RK, Orme-Johnson D, Rouzere AM (1984) Electrophysiologic characteristics of respiratory suspension periods occurring during the practice of the Transcendental Meditation program. Psychosom Med 46(3):267–276

Banquet JP, Sailhan M (1974) Quantified EEG spectral analysis of sleep and Transcendental Meditation. Electroencephalogr Clin Neurophysiol 42:445–453

Buckner RL, Carroll DC (2007) Self-projection and the brain. Trends Cogn Sci 11(2):49–57

Fenwick PBC, Donaldson S, Gillis L, Bushman J, Fenton GW, Perry I et al (1977) Metabolic and EEG changes during Transcendental Meditation: an explanation. Biol Psychol 51:101–118

Fox MD, Raichle ME (2007) Spontaneous fluctuations in brain activity observed with functional magnetic resonance imaging. Nat Rev Neurosci 8(9):700–711

Gusnard DA, Raichle ME, Raichle ME (2001) Searching for a baseline: functional imaging and the resting human brain. Nat Rev Neurosci 2(10):685–694

Gyatso T, Jinpa T (1995) The world of Tibetan Buddhism: an overview of its philosophy and practice. Wisdom Publications, Somerville

Harung HS, Travis F, Pensgaard AM, Boes R, Cook-Greuter S, Daley K, (2011) High Levels of Brain Integration in World-class Norwegian Athletes: Towards a Brain Measure of Performance Capacity in Sports. Scandinavian Journal of Exercise and Sport, 1:32–41

Hobson J (1988) The dreaming brain. Basic Books, New York

James W (1950/1890) The principles of psychology. Dover Books, New York

Kabat-Zinn J (2003) Mindfulness-based interventions in context: past, present, and future. Clin Psychol Sci Pract 10:144–156

Kelley WM, Macrae CN, Wyland CL, Caglar S, Inati S, Heatherton TF (2002) Finding the self? An event-related fMRI study. J Cogn Neurosci 14(5):785–794

Kesterson J, Clinch NF (1989) Metabolic rate, respiratory exchange ratio, and apneas during meditation. Am J Physiol 256(3 (Pt 2)):R632–R638
Lutz A, Slagter HA, Dunne JD, Davidson RJ (2008) Attention regulation and monitoring in meditation. Trends Cogn Sci 12(4):163–169
Maharishi Mahesh Yogi (1969) Maharishi Mahesh Yogi on the Bhagavad Gita. Penguin, New York
Maharishi Mahesh Yogi (1972) The science of creative intelligence. MIU, New York
Maharishi Mahesh Yogi (1997) Celebrating perfection in education, 2nd edn. Maharishi Vedic University Press, Noida
Mason LI, Alexander CN, Travis FT, Marsh G, Orme-Johnson DW, Gackenbach J et al (1997) Electrophysiological correlates of higher states of consciousness during sleep in long-term practitioners of the Transcendental Meditation program. Sleep 20(2):102–110
Moldofsky H, Lue FA, Smythe HA (1983) Alpha EEG sleep and morning symptoms in rheumatoid arthritis. J Rheumatol 10(3):373–379
Nidich SI, Rainforth MV, Haaga DA et al (2009) A randomized controlled trial on effects of the Transcendental Meditation program on blood pressure, psychological distress, and coping in young adults. Am J Hypertens 22:1326–1331
Niedermeyer E (1997) The normal EEG of the waking adult. In: Niedermeyer E, Lopes da Silva R (eds) Electroencephalography: basic principles, clinical applications and related fields. Urban Schwarzenberg, Baltimore, pp 301–308
Pagano RR, Warrenberg S (1983) Meditation: in search of a unique effect. In: Davidson JM, Schwartz GE, Shapiro D (eds) Consciousness and self-regulation: advances in research and theory. Plenum, New York
Plum F, Posner JB (1980) The diagnosis of stupor and coma. F.A. Davis, Philadelphia
Raichle ME, Snyder AZ (2007) A default mode of brain function: a brief history of an evolving idea. Neuroimage 37(4):1083–1090, discussion 1097–1089
Raichle ME, MacLeod AM, Snyder AZ, Powers WJ, Gusnard DA, Shulman GL (2001) A default mode of brain function. Proc Natl Acad Sci U S A 98(2):676–682
Santamaria J, Chiappa I (1987) The EEG of drowsiness. Demos Publishing, New York
Schuman M (1980) A psychophysiological model of meditation and altered states of consciousness: a critical review. In: Davidson JM, Davidson RC (eds) The psychobiology of consciousness. Plenum, New York
Siegel J (1987) Brain stem mechanisms generating REM sleep. In: Krye HH (ed) Principles and practice of sleep medicine. Raven, New York
Stigsby B, Rodenberg JC, Moth HB (1981) EEG findings during mantra meditation (TM): a controlled quantitative study of experienced meditators. Electroencephalogr Clin Neurophysiol 81:434–442
Travis F (1990) EEG patterns during TM practice and hypnagogic sleep. Soc Neurosci Abstr 15(1):244
Travis F (1994) The junction point model: a field model of waking, sleeping, and dreaming relating dream witnessing, the waking/sleeping transition, and Transcendental Meditation in terms of a common psychophysiologic state. Dreaming 4(2):91–104
Travis F, Pearson C (2000) Pure consciousness: distinct phenomenological and physiological correlates of "consciousness itself". Int J Neurosci 100:77–89
Travis F, Shear J (2010a) Focused attention, open monitoring and automatic self-transcending: categories to organize meditations from Vedic, Buddhist and Chinese traditions. Conscious Cogn 19(4):1110–1118
Travis F, Shear J (2010b) Reply to Josipovic: duality and non-duality in meditation research. Conscious Cogn 19(4):1122–1123
Travis F, Wallace RK (1997) Autonomic patterns during respiratory suspensions: possible markers of transcendental consciousness. Psychophysiology 34(1):39–46

Travis FT, Tecce J, Arenander A, Wallace RK (2002) Patterns of EEG coherence, power, and contingent negative variation characterize the integration of transcendental and waking states. Biol Psychol 61:293–319

Travis FT, Arenander A, DuBois D (2004) Psychological and physiological characteristics of a proposed object-referral/self-referral continuum of self-awareness. Conscious Cogn 13 (2):401–420

Travis F, Haaga DA, Hagelin J, Tanner M, Nidich S, Gaylord-King C et al (2009) Effects of Transcendental Meditation practice on brain functioning and stress reactivity in college students. Int J Psychophysiol 71(2):170–176

Travis F, Haaga D, Hagelin J, Arenander A, Tanner M, Schneider R (2010) Self-referential awareness: coherence, power, and eLORETA patterns during eyes-closed rest, Transcendental Meditation and TM-Sidhi practice. Cogn Process 11(1):21–30

Vogeley K, Bussfeld P, Newen A, Herrmann S, Happe F, Falkai P et al (2001) Mind reading: neural mechanisms of theory of mind and self-perspective. Neuroimage 14(1 (Pt 1)):170–181

Wachsmuth D, Dolce G (1980) Rechnerunterstützte Analyse des EEG wahrend Transzendentaler Meditation und Schlaf. ZEEG-EMG 11:183–188

Wallace RK (1970) Physiological effects of Transcendental Meditation. Science 167(926): 1751–1754

Wallace A (1999) The Buddhist tradition of Samatha: methods for refining and examining consciousness. J Conscious Stud 6:175–187

Chapter 11
Ethno Therapy, Music and Trance: An EEG Investigation into a Sound-Trance Induction

Jörg Fachner and Sabine Rittner

Abstract Music has been used since ancient times in healing rituals. It has been played for people to induce altered states of consciousness (ASC), which change the focus of attention, mood, and thoughts about the world and the self. Music and altered states are connected in various ways concerning context, personal set, socio-ecological setting, and cultural beliefs. Discussion is ongoing as to whether music itself induces the changes via a "trance mechanism" or whether the setting and rituals connected to music are responsible for the induction of ASC. The authors conducted an explorative attempt to represent interdependencies of set and setting, sound and trance through electrophysiological correlation in the topographic spontaneous EEG. They opted for a sound trance induction with the sound of a body monochord (a bed-like stringed instrument producing a droning sound) in the context of a group ritual. In comparison with uninfluenced rest, they found individual changes in spontaneous EEG. Trance reactions to sound were seemingly more determined by the person's susceptibility to hypnosis as measured by the Phenomenology of Consciousness Inventory than by sound alone.

11.1 Postmodern Intercultural Fascination

"World music is a hit now," said the cashier of the local music store, and music sales reveal that world music as a genre fills concert halls and makes profits. Witnessing debates on music and its therapeutic effects, many people with a classical western

---

J. Fachner (✉)
Finnish Centre of Excellence in Interdisciplinary Music Research, University of Jyväskylä, Jyväskylä, Finland
e-mail: jfachner@me.com

S. Rittner
Department of Medical Psychology, Centre of Psychosocial Medicine, University Clinic of Heidelberg, Heidelberg, Germany

D. Cvetkovic and I. Cosic (eds.), *States of Consciousness*, The Frontiers Collection,
DOI 10.1007/978-3-642-18047-7_11,

education appear to be disposed toward the traditions and music, healing rituals and lifestyles of foreign cultures. This reveals a fascinated and sometimes uncritical openness that dismisses their own existence and growth in the well-known paths of the west as limited and unsatisfactory, as superficial and short-sighted. Individuals who for whatever reasons are considered resistant to therapy can feel attracted by shamans. They project their hopes onto those who are perceived to be completely different, to have a different view of persons and whose culture is influenced by a different cosmology (understood as how they explain the origins of the world, mankind and the beyond). And do their wise shamans and medicine men not always play instruments in their healing rituals, at least a special drum or another traditional instrument? Are they not music therapists, too? Their music may, at first sight, not involve the technical sophistication of western music, but is rather a symbolic activity, representing long traces of knowledge and the energetic recharging of the instrument through generations and traditions of healers who use exactly this type of drum. And it is believed that its legacy and its potential will be passed on. Does an ethno music therapy require more than just the use of musical instruments and modes of play that are rather unknown in, or untypical of, our own culture? Does it celebrate rites that somewhere else are like our Sunday church-going tradition, which today is not attractive enough to our young people to lure them away from the computer screen?

11.1.1 Instruments in Traditional Rituals and Therapy

Instruments from an ethnic context can be used for music production in many ways, and instruments are just instruments, so we can use them in a traditional as well as in a modern way. But from anthropological research we know that the shaman has to find or even build his instrument, mostly a drum, and has to sanctify it in a ritual according to his cosmology and its sacred traditional knowledge (Eliade 1964; Rouget 1985). Therefore he has to load it with the energy, tradition and mythology which is needed for his shamanic journey. It is played constantly during the treatment process and the way it is played marks the stations on his shamanic journey. This stresses that ritual purposes and meaningful intentions are connected to it. It is not just specific properties of sound or certain tempos or rhythms that are inducing altered states, as discussed in the trance, rave and electronic music scene considerably more influence is attached to certain performances, actions and stations (Aldridge 2006a, b; Cousto 1995; Weir 1996).

Timmermann (2009) stresses that the improvisation and reception of modern western music therapy involve a variety of instruments from all over the world, independently of their traditional uses. Monochromatic sounds from instruments that are monotone, droning, rich in overtones, and pulsation instruments (e.g., rattles, shakers, drums) are used with notable frequency in ethnic healing rituals. Many of these instruments have specific functions within the rituals, which are rooted in the traditions, myths, cosmologies and musical practice of their sources.

There is an ongoing debate among trance researchers and clinicians about which is more significant: the intriguing sound qualities of the instruments and their effect on the body on the one hand, or their symbolic content and the context and procedures attached to them on the other (Dittrich and Scharfetter 1987; Eliade 1964; Rouget 1985; Ruud 2001).

The conviction that ethnic healing methods may induce changes in consciousness precisely because of their calling on a specific experience characterised by analogy, spirit worlds and an overarching irrationality, appears firmly established since the 1950s and has interested ethnologists from early colonial times.

11.1.2 Cross-Modalities and Altered States in Therapy

In their book on music and altered states, Fachner and Aldridge (2006) collected examples to demonstrate how music, musical instruments and modes of music-making are rooted in cultural traditions, how music in rituals focuses attention, and which role rhythmical body movements play in dancing and drumming with regard to altered states of consciousness The co-excitation of several senses in synaesthesia appears to be an important element in rituals. Intensified interactions of the sensory modes of smell, vision, touch, hearing and taste may produce a heightened interplay of perception and induce ritually evoked stereotype images and feelings (De Rios 2006).

Mastnak draws on the studies of Habib Hassan Touma as follows:

> The function of music within a healing ceremony may be analysed only if we include the interaction of all levels of communication – auditory, visual, tactile, olfactory, gustatory and ultimately psychical. Depending on the occasion, the focus is more on one component, either on the visual through gestures, mimics, dance of the magician, or on the auditory through sounds, drums, screaming or music. (Mastnak 2009, p. 292)

The deliberate processional change in the perception through music, meditation, hypnosis, psychoactive substances etc. is used for therapy purposes in many forms (Fachner 2006a, b). In many healing traditions it is an aim of the treatment to alter the habitual focus of perception. This is done to alter attention from being fixed on everyday life issues or on a narrowed scope of possible perspectives. Changing the focus of attention in a healing ritual may help to discover the personal meaning of a disease. It may open the client's eyes to personal potential and show how easily one can get stuck in habituated ways of perceiving the world. In 1902, William James experimented with nitrous oxide (laughing gas) to explore the limits of sensory perception and the socially and culturally determined selective processes of a consciously perceived world (James 1902). While under the influence of the substance, he interpreted the meaning or appearance of things and his body or person's action in a completely different way. In the late 1960s, the American psychologist Charles Tart (1969, 1975) introduced the psychological term "altered states of consciousness" for states that are perceived as different from normal. Much earlier, in 1902, William James wrote:

> Our normal waking consciousness, rational consciousness as we call it, is but one special type of consciousness, whilst all about it, parted from it by the filmiest of screens, there lies potential forms of consciousness entirely different. (James 1902, p. 228)

The term "altered states of consciousness" (ASC) implies that there is a consciousness that is unchanged, or "normal". Tart, as well as Dittrich (1996) or James (1902) discuss consciousness as a complex psycho-physiological system of states whereby our "normal" consciousness is only a specific construction in the sense of a specialized tool for everyday purposes (see Tart 1975, p. 3). From a neurophysiological perspective, Tassi and Muzet described the action of intended and therefore "evoked states of consciousness" gained by certain induction methods, as for instance sleep or sensory deprivation, drugs, meditation, trance dancing, etc. Further they discussed the range of "physiological states of consciousness", depending on spontaneously changing levels of vigilance, arousal and biological rhythm phenomena (Tassi and Muzet 2001, p. 185). Most of the latter would not be connoted as altered or extraordinary states of consciousness in the sense of Stanislaf Grof (1975), but would be recognized by each of us when we experience such moments spontaneously (Glicksohn 1993).

Altered states of consciousness may be perceived voluntarily by employing psychological and pharmacological triggers (Dittrich 1996). What is experienced in personal, meaningful ASCs can open up new perspectives and insights, and can raise questions about personal growth and the meaning of one's own life. It can provide an intensive experience of togetherness with others who share the same setting (e.g., on an outdoor 3-day rave party, a shamanic ritual or a rebirthing weekend). It can reveal hidden knowledge, which was not consciously present, but at the same time not unfamiliar. The content and insights of ASCs can change an individual's attitude to himself and his life. An artist may want to induce the same: to change the perspective, the feeling and the way of thinking about an object of art, or may want to show the pure sensual power of a plain colour. It is one of the aims of art as therapy to transcend the limits and obstacles erected by illness through an aesthetic variation of the perceptive context (Aldridge 2000). While improvising music with a music therapist, the scope may shift from producing an acceptably skilful performance to the experience of playing and the fun of self-expression. When the process and not the product of playing is at the focus, clients experience themselves as performers and might find analogies to their illness behaviour. Improvisation may open up a channel of communicating on a non-verbal level and offers the chance to communicate in an analogous mode to language. While doing this in a playful manner the clients may experience themselves and their communicative style in the process of interaction with others while creating music.

11.1.3 Experiential Systems and State-Dependent Cognition

A purposive therapy approach in the areas of music, consciousness, ritual and synaesthesia requires individual training according to established rules. In most ethnic traditions, such knowledge is passed on to disciples orally, in specific initiation rituals and in a close teacher-disciple relationship (Eliade 1964). The disciple embarks upon a spiritual shamanistic path of enlightenment where he achieves such knowledge in practical application and in close contact with his teacher (Castaneda 1998). Teaching and the terms explicated and used are combined with experience, the meaning of which is not immediately clear but becomes understood only in the course of training in a situational context. By frequent repetition it becomes perceived reality, as Aldridge (2006a, b) has shown for the example of traditional oriental music therapy.

Walter Freeman (2000) explored ways in which music and dance were related to the cultural evolution of human behaviour and forms of social bonding. Freeman discusses how complex knowledge like the mythology of a tribe is transmitted by using ASCs caused by chemical and behavioural forms of induction. The ASC serves to break through habits and beliefs about reality, but also to make one alert for new and more complex information. What are the preconditions for heightened perceptiveness or the modes for purposive new imprinting? To distinguish a normal waking state from an ASC, Glicksohn (1993) discusses personal modes of meaning during altered state cognition and stresses that ASCs are primary cognitive events. He denies a definition that reduces an ASC to vigilance changes. An ASC is:

> ... any mental state ... recognized ... as representing a sufficient deviation in subjective experience ... from certain general norms ... during alert, waking consciousness. (Ludwig in Glicksohn 1993, p. 2)

Ludwig (1966) described ASCs as changes in thinking and in time perception, loss of control, changes in emotionality, body scheme, perception and significance, feelings of the inexpressible, of renewal and rebirth, and hyper-suggestibility. Such purposive changes and the pertinent contents to be traded possibly resulted in the emergence of "initiated" groups and trust in the passing on of important findings. In times of primarily oral information transfer, memorization techniques were required that stimulated all senses for storing and processing that information. Musical abilities in particular seemed to be important for an effective transfer of knowledge.

11.1.4 Trance Paradigms

The question of how music induces ASCs, and states of trance and ecstasy in particular, is discussed on many levels. The following paradigms may assist in explaining the effects of music in therapeutic settings.

1. *Biomedical paradigm*: Music has an immediate and physically transmitted effect on consciousness. This effect is produced by certain sounds, instruments and modes of play that are intended to alter consciousness like drugs. ASCs are induced by additional factors (drugs, lack of sleep, dancing, fasting, pain stimuli etc.). In such states, music is perceived as altered in significance and sound.

This approach is mainly based on biomedical therapy concepts and attributes a music-induced healing action like that of a pharmaceutical agent.

2. *Communication paradigm*: Scenic presentation and contextualization of sound, joint play, symbolism and figuration within rituals produce changes in attention and cognitive function, and in consequence induce an ASC. Music as a relational development shapes intentional information from transmitter and receiver. Here the influence of music is reduced to contextual inter- and intrapersonal relations and symbolic meaning, but its physical information is interchangeable.
3. *Performance paradigm*: Rituals may integrate biomedical functions as well as relational structures into a temporal pattern of performative actions, ordered according to purpose. Music may mark and guide the ritual stages in their temporal sequence and may produce intensities. Here biological and relational aspects are interwoven.

Classical ethnomedical research approaches distinguish between disease, illness and sickness (Kleinman 1981). *Disease* is described as malfunction or maladaptation in the biological or psychological sense (e.g., an organic change in the patient). This is an explanation model based on complex mechanisms of generation. Disease comprises the opinion of an expert in health and illness and is mostly reduced to the perspective of biomedicine, that is, the scientifically based component. According to Kleinman (1978, p. 88) *illness*, in contrast, describes the patient's experience with deviations from a state he himself defines as healthy, and the significance he and his family and environment attach to these derivations, as well as personal explanations of the cause. *Sickness* is the general term covering both components of disease and illness. Sickness therefore stands for technical as well as personal, socially informed explanation models for being ill, as absence of health.

Ethnotherapy interventions use sonorous and pulsating instruments, but in a wider sense also trance induction and synaesthesia in rituals. Relating Kleinman's definition of disease as malfunction or maladaptation in the biological or psychological sense (e.g., an organic change in the patient) to ethno music therapy approaches, we recognize the influence of music and the physiological exertions caused by the ritual. This is where the debate on trance mechanism as discussed by Neher and Rouget plays a role, i.e., attempts are made to interpret the effects of music mainly from the perspective of biomedicine.

11.1.4.1 Trance Mechanics

Neher (1961, 1962) attempted to explain obsessional trance and the attributed epilepsy-like phenomena witnessed in ceremonial drumming and healing rituals by assuming that they arise causally from a certain sound and rhythm: a distinct frequency spectrum, and here especially a sound spectrum of drums dominated by bass frequencies, and moreover the repetition at a certain speed (beats per minute, bpm) of rhythmic patterns of drum beat sequences would cause obsession trance. Neher performed laboratory experiment to explore the complex phenomenon of entrainment (a coupling of inner rhythms through external timers), where a conformity of body movements, breath, heart beat and nerve activity is triggered and synchronized by rhythm. Neher calculated the number of drumbeats and their frequency and discerned an analogy with the EEG frequency pattern of trance states. Neher hoped to demonstrate auditory driving with the same EEG frequency range in his laboratory experiment, analogous to the epilepsy-inducing effect of photic driving (brain convulsions caused by rhythmic light emissions). However, he described drum beat frequencies performed in the range of theta waves (4–8 Hz; in the context of drums, strongly beaten, 4–8 beats per second, bps), whereas photic driving is in the range of alpha waves (8–13 Hz or flashes per second). It is difficult to achieve 8–13 bps on drums because it is so fast (Neher 1962, p. 153/154), but computer technology and new music hardware (sampler sequencer, sound modules, etc.) permit such modes of play. So Neher's ideas have been taken up again in rave culture to explain the trance states occurring in the context of techno music and rave parties through sound (bass frequencies), repetition (loops and sequences), and tempo (bpm) of rhythmic patterns (Hutson 2000; Weir 1996).

Rouget believed such attempts to explain a universal trance mechanism with reference to music alone to be incomplete, since the laboratory situation in Neher's experiment with trance cannot be compared to other settings. Ritual leaders and musicians do not enter trance states unintentionally. A person must be willing to fall into a trance and must know the pertinent cultural techniques of music, singing and trance. The person must have a specific aim and must be prepared for trance intellectually (Rouget 1985, pp. 315–326). In the case of obsessional trance, the obsessed individual must identify with the respective form of divine being pertinent to his culture and possibly attract the spirit through characteristic movements (ibid, pp. 35, 103, 105–108). Matussek (2001) writes that the contents of the cultural trance matrix and the physiological effects complement each other functionally, in order to produce a state of amnesia and a willingness to assimilate new information.

11.1.4.2 Trance and Healing Practice

Modern active music therapy practice, however, relies on communication between therapist, music and client. Even primarily receptive therapeutic approaches are directed to the individual and see the music therapy process as a joint, purposeful or open interpretation. Following a biomedical perspective of illness, the interpretation of the effects of music widens to become a joint interactive performance that in ethno music therapy also comprises the respective cultural traditions, their cosmologies and corresponding symbols and social experiences of participants. The ritual or ceremony with the performative act that is meant to cure a specific problem (sickness) is the treatment that integrates the senses to a wholeness experienced in the act (Aldridge 2004).

Individuals from western industrialised nations without personal experience of the cultures from which certain musical styles emerged appear fascinated by the exotic performance in such rituals. Considering that in addition to the triad of therapist, music and client, traditionally other transcendent and powerful entities form part of the interaction and exert their situational influence (Thomas 1927), what we find here is the fundamental search for transcendence, religiosity and spirituality described by Jung (1992, 1995). Such needs will certainly be met by therapeutic approaches that use music, trance and ecstasy in rituals and procedures evoking an experience of sensory wholeness.

The reception of the traditional and modern significance of instruments, tonal systems, rhythms and melodies – for example the Icaro songs in South American healing rituals (cf. De Rios 2006) or the polyrhythms of African healing ceremonies (cf. Maas and Strubelt 2006) – and their applicability in clinical practice is influenced by the potential for reflection and acceptance among clinicians. This depends on their personal social identity and also on the reception of their work by colleagues in the clinical and professional community. But ultimately it depends on clients and their concepts of healing. Patients want one thing, relief from suffering; how exactly this happens is not uppermost in the minds of patients who have suffered for a long time. But those with acute problems do not necessarily react with enthusiasm to shamans and strange rituals on their wards.

However, reviewing findings on music and altered states (Fachner 2006a, b) we can observe changes in attentive focus guided by music and its ritual context. The length of rituals, repetitions, monotonies, growing and fading volume and density of sound and rhythm produce altered levels of intensity, altered perception of time and space, and as a consequence, altered associations of musical parameters in the acoustic field of perception. Rhythm in particular in combination with dance, and also altered perception of time and space appear to be essential factors that trigger an ASC through music. In addition, music in combination with imagery techniques can make visual images more vivid through sensory dynamics, particularly where various sensory perceptions come together (synaesthesia).

Music offers a mental space where significant themes may be coded and decoded, depending on an individual's biography, socialisation and (health) belief system. Music focuses and directs attention and structures temporal occurrence and memories of internally and externally perceived events. It has the potential to support *therapeia* in the sense of the Greek root of "therapy": the accompaniment through a professional helper or even by itself as a guide during altered states of consciousness.

11.2 QEEG Study of a Trance Induction with Monochromatic and Pulsating Sounds

In view of the influence of music on the body, and of its communicated effects in a healing performance, the following experiment with an EEG accompanied trance induction procedure reviews and investigates some aspects discussed in Sect. 11.1.1

11.2.1 Sound-Trance Approaches in Music Therapy

Work with trance-inducing sounds has become an intrinsic part of receptive music therapy in practice. Therapeutically intended ASCs enable an intensification of perception and a weakening of the psychic barriers of everyday consciousness. They also facilitate access to association and imagination and other healing resources (Aldridge and Fachner 2006).

In the 1990s, music therapists started to experiment with the trance-inducing effect of monochromatic and pulsing sounds and formulated concepts on the basis of their experience. They mainly used the effects of monochromatic sounds of monochord, gong, didgeridoo, sound bowls, etc., and developed individual empirical concepts (Bossinger and Hess 1993; Oelmann 1993; Rittner 1997, 1998; Strobel 1988, 1999). These first approaches were revised, further developed and also reviewed critically (Hartogh 2001; Hess 1999; Hess et al. 2009; Rittner et al. 2009; Rittner and Jungaberle 2002). Although the brain plays a decisive role in the experience of trance states, there is an astonishing lack of music therapy research into psycho-physiological aspects of sound-induced trance. The function of sound in therapy is to induce, control and withdraw the ASC. In such interventions, music serves not only to induce altered states of alert consciousness but also to maintain, navigate and structure them. All this is necessary to make the healing potential of trance states available for the therapy process. Pulsating and monochromatic sounds have their effect through an "intensive rhythmic charge in the field of perception" and a "reduction of the field of perception and focusing"; the effects are either physiologically stimulating (ergotropic) or calming and contemplative (trophotropic) (Hess et al. 2009, p. 554).

Rouget, in his groundbreaking book *Music and Trance*, differentiates between trance and ecstasy. For him,

> trance is always associated with a greater or lesser degree of sensory over stimulation – noises, music smells, agitation – ecstasy, on the contrary, is most often tied to sensorial deprivation – silence, fasting, darkness. (Rouget 1985, p. 10)

He distinguished the characteristics of the two ASC types as shown in Table 11.1.

Rouget's definition serves as an example for the fact that there are many different and in part contradictory definitions of the terms "trance" (from Latin

Table 11.1 Differentiation of ecstasy and trance (Rouget 1985, p. 11)

Ecstasy	Trance
Immobility	Movement
Silence	Noise
Solitude	In company
No crisis	Crisis
Sensory deprivation	Sensory overstimulation
Recollection	Amnesia
Hallucination	No hallucination

transire for “passing through”) and “ecstasy” (from Latin *exstasis* for “to be out or stand out of one’s head”) in the literature (Fachner 2006a; Meszaros et al. 2002; Pekala and Kumar 2000; Rouget 1985; Winkelman 1986). Trance seems to have a more direct relationship to the body, its functions and vigilance states, while ecstasy seems to be more concerned with pure mental activity, like meditation, contemplation, etc. However, it should be kept in mind that the everyday connotation of both terms is very often used in the opposite sense or even to mean the same thing. In techno music, the genre “trance” stands for dance and excitation, and ecstasy goes along with it. As the term “trance” has been defined in various, sometimes contradictory ways in the literature, we use it here as a generic term for:

> various physical-mental alterations that may occur in persons at ASC independent of the cultural setting. Stimulus, techniques and ritual that induce and structure a trance depend on the socio-cultural context. (Rittner et al. 2009, p. 538)

So we use the term “trance” as a synonym for the experience of an altered state, whether it be an ergotropic (energy activation, arousal, vigilance, wakefulness, movement) or trophotropic (energy preservation, recreation, sleep induction) state of activity.

11.2.2 Trance Study

We present here an excerpt from an analysis of a trance induction in recumbent position on a resounding body monochord. In the pilot study, we analysed four different methods of sound trance in a multi-perspective approach. The research design comprised visual and quantitative evaluations of a spontaneous EEG as well as psychometric measurements using the questionnaire on extraordinary states of consciousness 5D-ABZ (Dittrich et al. 2002) and the Phenomenology of Consciousness Inventory (PCI) (Pekala 1991a, b). In addition, we did a qualitative, content-analytical evaluation of written reports by participants on their experience. Selected results have been published elsewhere (Fachner and Rittner 2003, 2007).

11.2.2.1 Body Monochord

The body monochord is based on a design from the early 1990s for a musical instrument for therapy purposes. It has the form of a double-walled wooden stretcher, with 26 strings of equal length and exactly the same tuning on the underside. A person lying on the instrument with eyes closed may perceive sound with the entire body, via skin, bones, vibration of body liquids, and auditory sense. Depending on the type of playing, various droning, monochromatic and pulsating “sound clouds” with overtones may be produced, based on a keynote.

11.2.2.2 Trance and EEG

No previous EEG analyses of trance experience on the body monochord have been found. Available studies on trance and pertinent alterations in the EEG did not focus on music and trance but reviewed individual differences in trance experiences between a variety of mainly verbal trance inductions (De Benedittis and Sironi 1985; Jaffe and Toon 1980; Meszaros et al. 2002; Oohashi et al. 2002; Park et al. 2002; Sabourin et al. 1990; Stevens et al. 2004). The research design of the more recent studies by Meszaros, Park and Oohashi come closer to the authentic situation, i.e., trance induction in situ.

The question of whether trance is a temporary state or a specific quality in persons addresses the state-trait discussion in psychology and has not been answered so far (see Meszaros et al. 2002, p. 500). Most studies assume differences in susceptibility to hypnosis, and subjects were psychometrically differentiated at the beginning. For example, Sabourin compared 12 persons with high and 12 with low susceptibility to hypnosis respectively. At rest as well as under hypnosis, persons with high susceptibility to hypnosis had higher theta amplitudes compared to persons with low susceptibility (Sabourin et al. 1990). For an overview and discussion of all these studies see Fachner (2006a).

Park et al. (2002) found changes in the EEG in the case of a Salpuri dancer in comparing rest, listening to music and memory (of a previous dance). Salpuri is a traditional dance performed by medicine men in Korea. In mentally recalling an altered state (ecstatic trance) of the dance, frontal and occipital low alpha (8–10 Hz) and theta frequencies increased, when comparing power values to rest. Theta increases were mostly obvious in the frontal midline (a set of three frontal electrodes in the middle of the forehead), an increase that is normally seen in relaxed concentration and heightened awareness and that can be elicited with anxiolytic medication. In the comparison of rest and listening to a piece of pop music, there was a highly significant increase in the frequency of high alpha frequencies (10–12.5 Hz) over the entire cortex. This may have indicated a difference between a primarily physical trance experience and the enjoyment felt in listening to music. The identified peak frequency in rest and memory of dance was 9.5 Hz, with an increase of amplitude and energy in the memory of dance. The peak frequency rose to 10 Hz in the process of listening to music, and high beta frequencies increased. Park supposes that the Salpuri dancer "reaches the altered state of ecstatic trance through suppression of frontal cortex functions and activation of subcortical functions" (Park et al. 2002, p. 961). This suggests that trance is characterised by the dominance of theta frequencies.

11.2.2.3 Aims of the Study

For the purpose of this study, we assume that a specific state of consciousness may be induced and that the characteristics emerging in the EEG may be differentiated.

> In musical psychotherapy with sound trance, music … is effective in two directions: 1. Physiologically stirring (ergotropic) towards *ecstasy* by intensified rhythm in the field of perception … or 2. Physically calming and internalizing (trophotropic) towards *enstasis* with reduced field of perception and focussing via monochromatic sounds. (Hess et al. 2009, p. 554)

An ergotropic state (*ecstasis,* or in Rouget's definition *trance*) in the sense of Fischer's mapping (Fischer 1971, 1976) is an alert, non-contemplative, wide-awake ASC. Fischer says:

> The mapping follows along two continua: the perceptive-hallucinatory continuum of increasing central-nervous (ergotropic) excitement, and the perceptive-meditative continuum of increasing (trophotropic) damping. (Fischer 1998)

> Along the two continua, the sensory/motoric ratio increases. This means: the further you go along one continuum, the less will it be possible to verify the sensory element through random motoricity. (ibid, p. 51)

Accordingly, a trophotropic state is characterized by a rather relaxed, contemplative, apparently sleepy state, by a rather inhibited movement profile, and reduced reaction and willingness to perform (*enstasis* in Eliade's or *ecstasis* in Rouget's definition). The EEG was expected to show the differences of a trophotropic state via EEG synchronisation, i.e., via deceleration of the main frequencies and increase of slower wave ranges: delta from 0.3 to 4 Hz, theta from 4 to 8 Hz and lower alpha frequencies of 8–10 Hz (cf. David et al. 1983; Schwendtner-Berlin et al. 1995).

Ergotropic states were expected to show an EEG desynchronisation and dominance of high-frequency waves (upper alpha waves at 10–12 Hz, beta-1 at 12–16 Hz, and beta-2 at 16–30 Hz).

Our research questions were:

1. Are there intra-individual alterations in the topographical spontaneous EEG compared to undisturbed rest?
2. Which inter-individual differences or common factors may be detected between the two test subjects?
3. Does our study reveal an increase in theta waves in persons highly susceptible to hypnosis?

11.2.2.4 Research Methods

The electroencephalogram (EEG), discovered by Berger in 1929, has lost nothing of its fascination (Berger 1991). Positron emissions tomography (PET), functional magnetic resonance tomography (fMRT) and other procedures produce highly revealing images of the living brain; but the spontaneous EEG with its temporal exactness is an ideal instrument to correlate and measure electrical processes in the brain, especially when listening to music. Topographical presentations of frequency ranges, coherence, power distributions and dynamics permit conclusions on functional interactions of brain regions and their levels of activity (Maurer 1989). Quantitative EEG (QEEG) involves computer-assisted imaging and statistical post-hoc analysis of the topographic EEG (Duffy 1986). It also contains FFT quantification of spectral components (frequency power and percentage), coherence and symmetry over selected epochs of ongoing spontaneous EEG from rest and activation. Also, it is possible to calculate statistical probability maps (SPM) of topographic differences between rest and activation with respect to certain frequency ranges (Coburn et al. 2006).

It is a well-known difficulty in physiological measurements that movements or body activities in the course of a therapy may interfere with the precision of measurement data. EEG brain mapping requires a limitation to receptive sound perception avoiding movement. Ideally, the test subject has to be recumbent or sitting still in order to avoid movement artefacts in measurement results. The mobile EEG BrainImager used here was designed for EEG on intensive wards where reliable data collection is essential.

Measurements were taken in a therapy group setting (N = 10). We attached specific importance to making measurements in the ritual setting of a group well known to the test subjects. Unlike an isolated laboratory situation, a group setting and familiarity with the experience ensure that supportive socio-physiological factors influence the sound-induced trance experience for all participants (naturalistic design). Most attempts to locate practical music therapy in a laboratory setting impair the authenticity of the situation. The documentation of significant moments in therapy on recording appliances in particular demands a sensitive approach. In the realization of such a qualitative electrophysiology study (see Fachner 2004) the measuring instruments must be integrated as close as possible into everyday practice in order to generate explorative data. Consequently, we collected our data in the immediate therapy situation and not under laboratory conditions. On the basis of such explorative data collection, later a test concept may be designed to review the tendencies revealed in exploration in a laboratory experiment under ideal technical conditions. For details and discussion, see Fachner (2001, 2004), Fachner and Rittner (2004) and Burgess and Gruzelier (1997).

In this study, topographical alterations of brain activities in two test subjects were measured in a ritualistic group setting at Heidelberg University Clinic.

We examined a total of four different, sound trance induction procedures, the trance-inducing effects of which on the participants were measured and compared in the following order:

1. Body monochord, designed by H.P. Klein (see Rittner 1997)
2. Singing of monochromatic vocal sounds (see Rittner 1998)
3. Peruvian whistling vessels, traditional, according to Statnekov (2003)
4. Ritual body posture with rattle induction (Goodman 1990)

Two turns were performed, with an interval in between. On each occasion a test subject was connected to the mobile brain imager in the group setting. The male test subject (T1) came first, the female (T2) second.

In this chapter, we report only the analysis of sound trance induction with the body monochord (the first of the above procedures). This instrument was played for 12 min while the test subjects were lying stretched out on it with their eyes closed.

The test subjects reported their experience in a written journal from which we quote. After the complete sequence, two psychometric questionnaires were filled out by the participants, the questionnaire on extraordinary states of consciousness 5D-ABZ (Dittrich et al. 2002) and the Phenomenology of Consciousness Inventory (PCI) (Pekala 1991a, b).

For this pilot study, we selected volunteers with trance experience who were acquainted with the induction methods. The intention was to ensure a high degree of familiarity with the ritual setting and as little irritation as possible with the research situation. We focussed mainly on the results for the test subjects T1 (male) and T2 (female) from the total of ten participants.

The EEG was recorded and analysed by using a mobile bedside-type quantitative EEG brain mapping device (NeuroScience BrainImager, Version 6.03, Florida, USA). This unit, designed for the neurological ward, records and samples 28 EEG traces at 12 bit. A dynamic range of 256 μV was applied. It calculates, stores and displays while recording average maps of 2.5 s EEG epochs. The data were fast Fourier transformed into the classic EEG frequency ranges (maps display delta at 0.3–4 Hz, theta at 4–8 Hz, alpha at 8–12 Hz, beta-1 at 12–16 Hz, beta-2 at 16–30 Hz, and spectral peak frequency in 2 Hz steps). A monopolar (referential) montage of 28 passive silver chloride electrodes were mounted with Grass paste in an ECI cap at 10/20 positions. Silver chloride earclips on left and right earlobes served as reference electrodes, which the unit computed as linked ears reference. Impedances were kept below 5 kΩ and no artefact channels for EOG or EMG were used.

A video mixer enabled an audio and video protocol of (1) the client's head and shoulders (filmed from above when lying on the monochord), (2) the sound produced with the monochord, and (3) the ongoing EEG traces (filmed from the EEG screen). This AV protocol was used for artefact control and sound examination.

The recording time of the closed-eyes rest and monochord EEG was 10 min for rest and 10 min for the monochord sound. Subjects were lying on the monochord in the rest and in the trance induction period. Eye movement artefacts were examined visually post hoc on the frontal EEG traces.

Statistical comparisons were performed with the mobile EEG unit's own statistical package (Version 7). Using a pre/post design, we compared artefact-free epochs and their means and standard deviations of undisturbed rest (baseline state) versus sound trance epochs (altered state induced by the body monochord). The averages of each condition (rest and monochord) were treated with a Student's T-test and thus significance maps of changes from baseline rest to altered state were produced.

11.2.2.5 Results

The PCI test that quantifies the occurrence of characteristic structures of ASC via 12 main dimensions and 14 subscales is a method to determine retrospectively an individual's degree of hypnotic susceptibility (Pekala 1991a, b). The hypnoidal score indicates how far the experience of a situation resembles the experience of highly suggestible persons during hypnotic induction. The predicted Harvard Group Scale (pHGS) revealed the male subject T1 to be moderately hypnotisable (with a score of 6.09) and the female subject T2 to be highly hypnotisable (7.78).

For test subject T1, compared to rest, the mean of the monochord EEG amplitude mapping showed a power decrease of the frontal delta waves and parietal theta and alpha and beta-2 waves. Topographically distributed standard deviations were most pronounced on alpha in the parietal and occipital cortex. However, in the trance condition the fast beta-2 frequencies (16–30 Hz) showed a marked increase in the frontal regions, most pronounced on the right frontal. Accordingly, the *T*-test comparison between rest and monochord revealed highly significant ($p < 0.001$) differences on the beta-2 frequency band over the whole frontal cortex. Highly significant changes also appeared in the frontal regions in the spectral map.

The NeuroScience BrainImager computes a fast Fourier transformed (FFT) spectral map based on the relative percentage of power in the EEG. The spectrum is displayed in 2 Hz bins and mapped according to their topographic occurrence (bins ranges 0.3–2, 2–4, 4–6 Hz, etc.). According to the BrainImager handbook, the relative amount of peak frequencies above 87% is topographically mapped in the frequency bins (NeuroScience 1992).

For the T1 spectrum, changes in frequency showed desynchronisation (increase of frequency) in right and left frontal regions compared to rest. While theta (6 Hz) and low alpha frequencies (8 Hz) dominated frontal regions at rest, the monochord epochs were characterised by desynchronisation, with a dominance of medium (10 Hz) and high (12 Hz) alpha frequencies. This change in frontal spectrum was highly significant.

For text subject T2, comparing rest and monochord EEG averages, a decrease of theta waves in temporal, midline and frontal regions occurred compared to rest, while alpha and beta (1 and 2) frequencies increased in parietal regions with a centre of gravity to left. In contrast to subject T1, a slowing of the EEG from central to frontal regions was observed in the spectral band during the monochord phase. In rest, high and middle alpha frequencies (10 and 12 Hz) dominated the frontal regions. In the monochord phase the frontal region was dominated from slow alpha (8 Hz) frequencies.

For T2, the *T*-test changes were highly significant ($p < 0.001$) all over the cortex on alpha on theta over temporal, occipital and parietal regions, but not in the left occipital. Moreover, highly significant changes occurred in occipital regions on both beta bands. In the spectrum, changes were high significant in the frontal midline (F3, Fz, F4) and on Cz and Pz.

11.2.2.6 Discussion

Both test subjects showed an increase in beta waves and a decrease in theta waves during the monochord phase. Diminutions in theta waves, specifically in temporal regions (where primary auditory centres are located), while subjects were listening to music, were also found in a further study using this topographical EEG method with the NeuroScience BrainImager. Fachner (2002) described this already in a comparison of rest and music. Measurements with a direct-current EEG also revealed and discussed amplitude decreases in temporal regions (Altenmüller and Beisteiner 1996; David et al. 1969). Our study appears to show that reaction to trance induction – in the case of the monochord in particular – is rather specific to the individual test subject.

Despite growing physical relaxation, the desynchronisation (rise of the EEG frequency as seen in the spectrum) and increase of high beta frequencies compared to rest suggests an active visual imagination in the case of the male test subject T1:

> A cloud rises in me and through the back of my head pulls me up to great heights . . . I am flying, surrounded by clouds . . . thousands of houses in green and lilac on a slope . . . a beautiful sight. . .

The simultaneous decrease of theta power suggests increasing physical–mental alertness and an intentionally controlled imagination. The male test subject T1 seems to influence his rather ecstatic experience deliberately. In contrast, the EEG of the female test subject T2 tends towards synchronization, induced by the alpha increases. For T2, the recumbent position on the monochord appears to promote a more relaxed and sensual body experience:

> . . . in me this turning movement . . . as if I hovered in this rotation . . . there were patches of haze or drifting veils and far away an unearthly music . . . a feeling of calmness, of being sheltered. . .

She seems to drift into an enstatic experience, a contemplation and deep physical relaxation. This was most obvious in the *T*-test in the highly significant change of the alpha frequency in the comparison of rest and trance. Accordingly, the trance phase EEG revealed an increase in alpha waves. Crawford (1994) underlined the correlation between theta activity, high suggestibility and a reaction of frontal and limbic regions. Function-related hippocampus and amygdale activities (in deeper layers of the central brain) suggest a complex pattern of facilitation and inhibition of neural interaction in the limbic system of persons highly suggestible to hypnosis (in Sabourin et al. 1990). Since our two subjects turned out to differ in their suggestibility to hypnosis as a test result, this seems to influence the experience regarding tendencies towards ecstatic or enstatic experience. For the female subject, the hypnoidal score of the pHGS indicated high hypnotisability (see Sect. 11.2.2.5); she reacts more in the sense of a trophotropic trance with increasing low-frequency waves, while the male subject reacts in the sense of an ergotropic trance with increasing high-frequency waves.

The personal written reports reflect the effects of the monochord on both test subjects in similar, subjective descriptions of hovering states, visions of cloud-like forms and changes in the body feeling. The male, T1, says: "My impression is that I drift through space lying on the monochord". And the female, T2, writes: "They were very slow, but even movements . . . as if I hovered in this rotating movement . . . a tremendous space in grey and white, through which veils were drifting".

The increase in high beta frequencies, while a subject listened to music reported in Park et al.'s (2002) trance study above is also known from other studies on music perception. Walker (1977) reported increased right-lateral activity while listening to classical music; Behne et al. (1988) reported occipital increases; Petsche (1993) found increases in posterior right-lateral coherence; and Bruggenwerth reported music-related, emotion-specific decreases or increases of posterior beta activity (Bruggenwerth et al. 1994). According to Petsche (1994), the beta frequency bands indicate differentiations of music-related cognitive activity.

The conspicuous reaction of the EEG in both trance inductions with an increase of high beta waves (16–30 Hz) seems to be a further indication of the influence of trance triggered by sounds. Beta frequencies indicate changes in emotional states. Isotani et al. (2001) explored hypnosis-induced states of relaxation and anxiety and discussed the conspicuous reactions with high beta frequencies and their EEG signature in emotional states.

Meszaros et al. (2002) interpreted the EEG for hemispheric differences and described a primarily right-hemispheric, parieto-temporal EEG reaction of the alpha and beta bands in persons highly susceptible to hypnosis; he concluded that in the "mainly emotion-focused hypnotherapies" (ibid, p. 511), as expected, right-hemispheric changes will be dominant. In his study, persons with high, medium and low susceptibility to hypnosis all experienced an ASC while listening to music in a relaxed position in an easy chair. Mescaros et al. categorized subjects' self-reported responses after an exposure to a computer generated drum-beat driven "shamanic journey" into several dimensions [relaxation, imagination/hallucination, alterations in attention, altered state experiences (describing feeling high, out-of-body, etc.) and depth]. These dimensions had to be rated again from the subjects on a scale from 0 to 7 points. Significant differences in this self-developed Altered State Index were only found for the areas imagination/hallucination (cf. ibid, p. 505/510).

Consequently, the beta changes we found for the monochord seem to illustrate the emotional reactions produced by the ritual and the sounds.

11.2.2.7 Study Conclusions

Returning to the research questions stated at the beginning of this study, we found inter- and intraindividual differences in comparing rest and trance induction. In this study with two test subjects, highly significant differences were found in the topographical EEG comparing rest and trance. The profile becomes irregular in the case of the monochord. The female subject recumbent on the monochord showed a more trophotropic trance, indicated by an increase in alpha waves, while the male subject showed a desynchronisation and an increase of beta waves, which suggests a more ergotropic trance. The data available do not explain whether these findings suggest gender-specific or mood-dependent differences, or differences in susceptibility to hypnosis. Moreover, the topographical EEG showed a distinct difference between rest and sound-induced activations. The temporal regions of both test subjects revealed a decrease on the theta band. In the monochord test we also found a tendency toward synchronisation and increased low frequencies in the easily hypnotisable subject.

11.3 Closing Remarks

Not everybody gets easily absorbed or entranced from music and changes his or her consciousness into an altered state when listening to droning and monotone sounds. It depends on personality and here foremost on his or her susceptibility to hypnosis. This has been found in hypnosis research before (Fachner 2006a) and seems to be true for ethnotherapeutic trance inductions as well. Therefore, before using an altered state induction for its possible benefits in a healing setting it is necessary to measure the subject's susceptibility. This might also help to reduce the recognition of trance and altered state induction methods as a dangerous weakening of the subject's will and self-control and enhance the compliance of clients treated with such approaches.

Which elements of an ethno therapy setting and which range of instruments may be used for the figurations of dance and masks of ethnic healing rituals in the clinic or in private practice? In most cases such applications can be realized only in contexts where there is a willingness to accept them. The discussion of shamanic roots in music therapy raises another question: How can we create a music therapy setting that does not alienate clients who are not familiar with such processes? Or, as Even Ruud remarks with regard to the believers in Indian music tradition and supporters of the healing biomedical effectiveness of certain ragas (which are certain tone scales played and performed, and believed in traditional Indian music to represent certain entities, or gods or spirits, that become active when the ragas are played in an intentional manner):

> I suppose that a raga without this perceived cosmology does not have much of an (therapeutic) effect upon any listener. (Ruud 2001)

However, Sumathy Sundar's (2007) research on the use of certain traditional Indian ragas as receptive music therapy in oncology may demonstrate that there is more to it than just being socialized and acculturated. In any case, the search for universal properties of sound for healing purposes will continue. Contextualization of music traditions and their cultural belief systems on what makes the music act therapeutically remains an exciting new field of experience and research for "reflective practitioners" (Aldridge 1996). We may conclude that the debate about the effects of music and its elements in healing rituals with an ethnic and cultural background resembles debates about music therapy theory. Is the music effective, is the relationship effective, or are both inseparable? The spontaneous answer is the latter, of course, but a closer look reveals that each approach has a different focus and underlines either one or the other element. The sociologist W.I. Thomas said:

> If people define situations as real, situational definitions are real in their consequences. (Thomas 1927)

This seems to suggest that a mutually shared system of beliefs and symbols ritualized in a healing setting can stimulate the activation of healing potentials.

Ethnologists confirm that traditions in other cultures may only be truly understood if we experience them in their original environment for longer periods. Maas and Strubelt (2006) underlined the significance of procedural details in ritual, and how important it is to live on site until such traditions have been absorbed. Well-meant imitations are often rather clumsy and even unwise. We will have to wait and see whether "world music" as part of the general global trend and aimed at greatest possible integration will appeal to a majority, and will sell, or whether it will only satisfy a certain section of the public with an open mind for this relatively new but at the same time very old style of music in therapy.

References

Aldridge D (1996) Music therapy and research in medicine: from out of the silence. Jessica Kingsley, London

Aldridge D (2000) Spirituality, healing, and medicine: return to the silence. Jessica Kingsley, London

Aldridge D (2004) Health, the individual and integrated medicine: revisiting an aesthetic of health care. Jessica Kingsley, London

Aldridge D (2006a) Music, consciousness and altered states. In: Aldridge D, Fachner J (eds) Music and altered states: consciousness, transcendence, therapy and addictions. Jessica Kingsley, London, pp 9–14

Aldridge D (2006) Performative health: a commentary on traditional oriental music therapy [Electronic Version]. Music Therapy Today 7(1):65–69. Retrieved 1 April 2006 from http://www.musictherapytoday.com

Aldridge D, Fachner J (eds) (2006) Music and altered states: consciousness, transcendence, therapy and addictions. Jessica Kingsley, London

Altenmüller E, Beisteiner R (1996) Musiker hören Musik: Großhirnaktivierungsmuster bei der Verarbeitung rhythmischer und melodischer Strukturen. In: DGMP (ed) Musikpsychologie: Jahrbuch der Deutschen Gesellschaft für Musikpsychologie. Florian Noetzel, Wilhelmshaven, pp 89–109

Behne KE, Lehmkuhl P, Hassebrauck M (1988) EEG-Korrelate des Musikerlebens, Teil II. In: Behne KE, Kleinen G, de la Motte-Haber H (eds) Musikpsychologie. Florian Nötzel, Wilhelmshaven, pp 95–106

Berger H (1991) Das Elektroenkephalogram des Menschen: Kommentierter Reprint des Erstdruckes aus dem Jahre 1938. PMI, Frankfurt am Main

Bossinger W, Hess P (1993) Musik und außergewöhnliche Bewusstseinszustände. Musiktherapeutische Umschau 14:239–254

Bruggenwerth G, Gutjahr L, Kulka T, Machleidt W (1994) Music induced emotional EEG reactions. EEG – EMG – Zeitschrift für Elektroenzephalographie. Elektromyographie und verwandte Gebiete 25:117–125

Burgess AP, Gruzelier J (1997) How reproducible is the topographical distribution of EEG amplitude? Int J Psychophysiol 26:113–119

Castaneda C (1998) The teachings of Don Juan: a Yaqui way of knowledge. University of California Press, Berkeley

Coburn KL, Lauterbach EC, Boutros NN, Black KJ, Arciniegas DB, Coffey CE (2006) The value of quantitative electroencephalography in clinical psychiatry: a report by the Committee on

Research of the American Neuropsychiatric Association. J Neuropsychiatry Clin Neurosci 18:460–500
Cousto H (1995) Vom Urkult zur Kultur: Drogen und Techno. Nachtschatten, Solothurn
Crawford HJ (1994) Brain dynamics and hypnosis: attentional and disattentional processes. Int J Clin Exp Hypn 42:204–232
David E, Finkenzeller P, Kallert S, Keidel WD (1969) Akustischen Reizen zugeordnete Gleichspannungsänderungen am intakten Schädel des Menschen. Pflügers Arch 309:362–367
David E, Berlin J, Klement W (1983) Physiologie des Musikerlebens und seine Beziehung zur trophotropen Umschaltung im Organismus. In: Spintge R, Droh R (eds) Musik in der Medizin: Neurophysiologische Grundlagen, Klinische Applikationen. Geisteswissenschaftliche Einordnung. Springer, Berlin, pp 33–48
De Benedittis G, Sironi VA (1985) Deep cerebral electrical activity during hypnotic state in man: neurophysiologic considerations on hypnosis. Riv Neurol 55:1–16
De Rios MD (2006) The role of music in healing with hallucinogens: tribal and western studies. In: Aldridge D, Fachner J (eds) Music and altered states: consciousness, transcendence, therapy and addictions. Jessica Kingsley, London, pp 97–101
Dittrich A (1996) Ätiologie-unabhängige Strukturen veränderter Wachbewusstseinszustände. Ergebnisse empirischer Untersuchungen über Halluzinogene I. und II. Ordnung, sensorische Deprivation, hypnagoge Zustände, hypnotische Verfahren sowie Reizüberflutung. Verlag für Wissenschaft und Bildung, Berlin
Dittrich A, Scharfetter C (1987) Ethnopsychotherapie: Psychotherapie mittels aussergewöhnlicher Bewusstseinszustände in westlichen und indigenen Kulturen. Enke, Stuttgart
Dittrich A, Lamparter D, Maurer M (2002) 5D-ABZ: Fragebogen zur Erfassung außergewöhnlicher Bewusstseinszustände. PSIN PLUS, Zürich
Duffy FH (1986) Topographic mapping of brain electric activity. Butterworths, Boston
Eliade M (1964) Shamanism: archaic techniques of ecstasy. Bollingen Foundation, New York
Fachner J (2001) Veränderte Musikwahrnehmung durch Tetra-Hydro-Cannabinol im Hirnstrombild. Witten/Herdecke University, Witten
Fachner J (2002) Topographic EEG changes accompanying cannabis-induced alteration of music perception: cannabis as a hearing aid? J Cannabis Ther 2:3–36
Fachner J (2004) Cannabis, brain physiology, changes in states of consciousness and music perception. In: Aldridge D (ed) Case study designs in music therapy. Jessica Kingsley, London, pp 211–233
Fachner J (2006a) Music and altered states of consciousness: an overview. In: Aldridge D, Fachner J (eds) Music and altered states: consciousness, transcendence, therapy and addictions. Jessica Kingsley, London, pp 15–37
Fachner J (2006b) Music and drug induced altered states. In: Aldridge D, Fachner J (eds) Music and altered states: consciousness, transcendence, therapy and addictions. Jessica Kingsley, London, pp 82–96
Fachner J, Rittner S (2003) Sound and trance in a ritualistic setting: two single cases with EEG Brainmapping. Brain Topogr 17:121
Fachner J, Rittner S (2004) Sound and trance in a ritualistic setting visualised with EEG brainmapping. MusicTherapyWorld.net, Accessed 18 Jan 2006 http://www.musictherapytoday.com
Fachner J, Rittner S (2007) EEG brainmapping of trance states induced by monochord and ritual body postures in a ritualistic setting. In: Frohne-Hagemann I (ed) Receptive music therapy: theory and practice. Reichert, Wiesbaden, pp 189–202
Fischer R (1971) A cartography of the ecstatic and meditative states. Science 174:897–904
Fischer R (1976) Transformations of consciousness. A cartography, II. The perception-meditation continuum. Confin Psychiatr 19:1–23
Fischer R (1998) Über die Vielfalt von Wissen und Sein im Bewusstsein. Eine Kartographie außergewöhnlicher Bewusstseinszustände. In: Verres R, Leuner HC, Dittrich A (eds) Welten des Bewußtseins. VWB – Verlag für Wissenschaft und Bildung, Berlin, pp 43–70

Freeman W (2000) A neurobiological role of music in social bonding. In: Wallin NL, Merker B, Brown S (eds) The origins of music. MIT, Cambridge, MA, pp 411–424
Glicksohn J (1993) Altered sensory environments, altered states of consciousness and altered-state cognition. J Mind Behav 14:1–12
Goodman FD (1990) Where the spirits ride the wind: trance journeys and other ecstatic experiences. Indiana University Press, Bloomington
Grof S (1975) Realms of the human unconscious: observations from LSD research. Viking, New York
Hartogh T (2001) Die Rezeption monotonaler Klänge. Eine empirische Untersuchung zur Klangwirkung des Monochords. Zeitschrift für Musik-, Tanz- und Kunsttherapie 12:111–119
Hess P (1999) Musiktherapie mit archaischen Klangkörpern. Musiktherapeutische Umschau 20:77–92
Hess P, Fachner J, Rittner S (2009) Verändertes Wachbewusstsein. In: Decker-Voigt HH, Knill P, Weymann E (eds) Lexikon Musiktherapie. Hogrefe, Göttingen, pp 550–557
Hutson SR (2000) The rave: spiritual healing in modern western subcultures. Anthropol Q 73:35–49
Isotani T, Tanaka H, Lehmann D et al (2001) Source localization of EEG activity during hypnotically induced anxiety and relaxation. Int J Psychophysiol 41:143–153
Jaffe JR, Toon JH (1980) EEG and polygraphic changes during hypnotic suggestibility. J Electrophysiological Technol 6:75–92
James W (1902) The varieties of religious experience. Modern Library, New York
Jung CG (1992) Psychology and religion. Yale University Press, New Haven
Jung CG (1995) Die transzendente Funktion. In: Gesammelte Werke: Die Dynamik des Unbewussten. Walter, Olten, Freiburg, pp 79–108
Kleinman A (1978) Concepts and a model for the comparison of medical systems as cultural systems. Soc Sci Med 12:85–93
Kleinman A (1981) Patients and healers in the context of culture: an exploration of the borderland between anthropology, medicine and psychiatry. University of California Press, Berkeley, Los Angeles
Ludwig AM (1966) Altered states of consciousness. Arch Gen Psychiatry 15:225–234
Maas U, Strubelt S (2006) Polyrhythms supporting a pharmacotherapy: music in the Iboga initiation ceremony in Gabon. In: Aldridge D, Fachner J (eds) Music and altered states: consciousness, transcendence, therapy and addictions. Jessica Kingsley, London, pp 101–124
Mastnak W (2009) Musikethnologie – Schamanismus – Musiktherapie. In: Decker-Voigt HH, Weymann E (eds) Lexikon Musiktherapie, 2nd edn. Hogrefe, Göttingen, pp 290–294
Matussek P (2001) Berauschende Geräusche. Akustische Trancetechniken im Medienwechsel. In: Hiepko A, Stopka K (eds) Rauschen Seine Phänomenologie zwischen Sinn und Störung. Königshausen & Neumann, Würzburg, pp 225–240
Maurer K (ed) (1989) Topographic mapping of EEG and evoked potentials. Springer, Berlin
Meszaros I, Szabo C, Csako RI (2002) Hypnotic susceptibility and alterations in subjective experiences. Acta Biol Hung 53:499–514
Neher A (1961) Auditory driving observed with scalp electrodes in normal subjects. Electroencephalogr Clin Neurophysiol 13:449–451
Neher A (1962) A physiological explanation of unusual behavior in ceremonies involving drums. Hum Biol 34:151–160
NeuroScience (1992) Neuroscience brain imager operators' manual. In. 7.0/7.32 ed. San Diego: Darox Corporation
Oelmann J (1993) Klang, Wahrnehmung, Wirkung. Zur therapeutischen Arbeit mit Gongs und Tam-Tams in rezeptiver Therapie. Musiktherapeutische Umschau 14:289–305
Oohashi T, Kawai N, Honda M et al (2002) Electroencephalographic measurement of possession trance in the field. Clin Neurophysiol 113:435–445
Park JR, Yagyu T, Saito N, Kinoshita T, Hirai T (2002) Dynamics of brain electric field during recall of Salpuri dance performance. Percept Mot Skills 95:955–962
Pekala RJ (1991a) Phenomenology of consciousness inventory (PCI) form I. MID-Atlantic Educational Institute, West Chester, PA

Pekala RJ (1991b) Quantifying consciousness: an empirical approach. Plenum, New York, NY
Pekala RJ, Kumar VK (2000) Operationalizing "trance" I: rationale and research using a psychophenomenological approach. Am J Clin Hypn 43:107–135
Petsche H (1993) Zerebrale Verarbeitung. In: Bruhn H, Oerter R, Rösing H (eds) Musipsychologie. Rowohlt, Reinbek bei Hamburg, pp 630–638
Petsche H (1994) The EEG while listening to music. EEG – EMG – Zeitschrift für Elektroenzephalographie. Elektromyographie und verwandte Gebiete 25:130–137
Rittner S (1997) Die Arbeit mit dem Ganzkörper-Monochord in der Musikpsychotherapie. In: Berger L (ed) Musik, Magie and Medizin. Junfermann, Paderborn, pp 110–117
Rittner S (1998) Singen und Trance: Die Stimme als Medium zur Induktion veränderter Bewußtseinszustände. In: Gundermann H (ed) Die Ausdruckswelt der Stimme. Hüthig, Heidelberg, pp 317–325
Rittner S, Jungaberle H (2002) Sounding systems: different approaches to the evaluation of music psychotherapy groups. In: Aldridge D, Fachner J (eds.) Music therapy in Europe. Proceedings of the 5th European music therapy congress in Castel Dell'Ovo, Naples, Italy, 20–24 Apr 2001. E-Book on Music Therapy Info CD-ROM ed. University Witten/Herdecke, Witten, pp 1514–1543
Rittner S, Fachner J, Hess P (2009) Trance. In: Decker-Voigt HH, Knill P, Weymann E (eds) Lexikon Musiktherapie. Hogrefe, Göttingen, pp 538–541
Rouget G (1985) Music and trance. A theory of the relations between music and possession. Chicago University Press, Chicago
Ruud E (2001). Music therapy: history and cultural contexts [Electronic Version]. Voices 1(3). Retrieved 13 Jan 2007 from https://normt.uib.no/index.php/voices/article/view/66/53
Sabourin ME, Cutcomb SD, Crawford HJ, Pribram K (1990) EEG correlates of hypnotic susceptibility and hypnotic trance: spectral analysis and coherence. Int J Psychophysiol 10:125–142
Schwendtner-Berlin H, Berlin J, David L et al (1995) Gesetzmäßigkeiten in der Psychologie veränderter Bewußtseinszustände. Curare 18:361–368
Statnekov DK (2003) Animated earth. North Atlantic Books, Berkeley, CA
Stevens L, Brady B, Goon A et al (2004) Electrophysiological alterations during hypnosis for ego-enhancement: a preliminary investigation. Am J Clin Hypn 46:323–344
Strobel W (1988) Klang – Trance – Heilung. Die archetypische Welt der Klänge in der Psychotherapie. Musiktherapeutische Umschau 9:119–139
Strobel W (1999) Reader Musiktherapie. Klanggeleitete Trance u.a. Beiträge. Reichert, Wiesbaden
Sundar S (2007). Traditional healing systems and modern music therapy in India [Electronic Version]. Music Therapy Today 8(3). Retrieved 9 Jan 2008 from http://www.musictherapytoday.com
Tart CT (1969) Altered states of consciousness; a book of readings. Wiley, New York
Tart CT (1975) States of consciousness. Dutton, New York
Tassi P, Muzet A (2001) Defining the states of consciousness. Neurosci Biobehav Rev 25:175–191
Thomas WI (1927) Situational analysis: the behavior pattern and the situation. Publ Am Sociol Soc 22:1–13
Timmermann T (2009) Ethnologische Aspekte in der Musiktherapie. In: Decker-Voigt HH, Weymann E (eds) Lexikon Musiktherapie, 2nd edn. Hogrefe, Göttingen, pp 123–126
Walker JL (1977) Subjective reactions to music and brainwave rhythms. Physiol Psychol 5:483–489
Weir D (1996) Trance: from magic to technology. Trans Media, Ann Arbor
Winkelman MJ (1986) Trance states: a theoretical model and cross-cultural analysis. Ethos 14:174–203

Chapter 12
States of Consciousness Redefined as Patterns of Phenomenal Properties: An Experimental Application

Adam J. Rock and Stanley Krippner

Abstract Although so-called states of consciousness have been the focus of considerable contemporary multi-disciplinary interest, this concept is neither well defined nor sufficiently understood. While definitions of "consciousness" usually distinguish it from its content, definitions of "states of consciousness" typically confuse consciousness and its contents by explicitly stating that a state of consciousness *is* the content (i.e., mental episodes) available to conscious awareness. In other words, the term "states of consciousness," along with the intimately related term "altered states of consciousness," rests on a conflation of consciousness and content whereby consciousness is erroneously categorized in terms of content rendered perceptible, presumably by consciousness "itself." This error, which we call the consciousness/content fallacy, may be avoided if one supplants "[altered] states of consciousness" with a new term, "[altered] pattern of phenomenal properties," an extrapolation of the term "phenomenal field."

12.1 Introduction

Chalmers suggested that "There is nothing we know more intimately than consciousness, but there is nothing harder to explain" (Chalmers 1995, p. 200). Indeed, psychologists and philosophers of mind remain engaged in an intricate debate regarding the concept of "consciousness" (e.g., Antony 2002; Rosenthal 2002; Silby 1998). Debates concerning consciousness are complicated by the fact that many meanings have been ascribed to the term, due to the erroneous conflation of very different concepts into a single one (Block 2002). For example, the key definitional elements of the term "consciousness" may be held to be conscious

A.J. Rock (✉)
Phoenix Institute of Victoria, Melbourne, VIC, Australia
e-mail: adam.rock@phoenixinstitute.com.au

S. Krippner
Saybrook Graduate School, San Francisco, CA 94111, USA

D. Cvetkovic and I. Cosic (eds.), *States of Consciousness*, The Frontiers Collection,
DOI 10.1007/978-3-642-18047-7_12, © Springer-Verlag Berlin Heidelberg 2011

awareness working in tandem with unconscious functioning (Krippner 1972) or simply conscious awareness, attention, and memory (Farthing 1992).

If one takes the term consciousness to refer to conscious awareness alone, then one may distinguish between (1) the *process* (i.e., conscious awareness) that renders objects perceptible, and (2) the *objects* (e.g., visual stimuli) that are being rendered perceptible. For example, imagine that you are attending an advanced screening of the latest blockbuster at your local cinema. When the film begins, it is, in part, the process of conscious awareness that allows you to perceive the faces of the actors and hear the dialogue. Clearly, your conscious awareness is not identical to the visual and auditory stimuli that are rendered perceptible. In other words, one may differentiate between consciousness (i.e., conscious awareness) and the content (i.e., objects rendered perceptible) of consciousness.

Based on a review of the literature, we will attempt to demonstrate that the key definitional elements of the term "consciousness" accurately reflect the aforementioned distinction between consciousness and the content of consciousness. At the same time, we will argue that the key definitional elements of the terms "states of consciousness" and "altered states of consciousness" indicate a theoretical confusion of consciousness and the content of consciousness – that is, objects (e.g., stimuli) are confused with the process that renders them perceptible. We will refer to this error as the consciousness/content fallacy.

The purpose of this chapter is, therefore, to elucidate and resolve the aforementioned fallacy, while demonstrating how our solution may be situated within an experimental methodology. We proceed by reviewing several definitions of consciousness, arguing that they all exemplify a commonly accepted distinction between consciousness and its content. Secondly, we examine the consciousness/content fallacy by analyzing the concept of [altered] states of consciousness, and propose a solution to the fallacy. Third, we summarize a recent experimental study, illustrating the utility of our proposal that so-called [altered] states of consciousness are more appropriately described as [altered] patterns of phenomenal properties. Finally, we explicate the implications of our proposal for current theory and research.

Before proceeding further, a qualifying statement is required. As previously stated, the present chapter is concerned with (1) the concept of consciousness as the process that renders objects (e.g., external events) perceptible, and (2) the fallacy that occurs when a shift from the term "consciousness" to "states of consciousness" is accompanied by a confusion of consciousness with its content. Consequently, for the purpose of this chapter, only the conscious awareness component of the concept of consciousness will be considered, although we understand that this awareness interacts with cultural and neurophysiological determinants that operate outside the scope of awareness.

12.2 Consciousness and Content

Forman (1996, p. 714) stated that the inherent difficulty associated with providing an adequate definition of consciousness is due, in part, to the multiplicity of meanings ascribed to the term. Block (2002) suggested that this multiplicity of meanings is due to the erroneous treatment of very different concepts as a single concept. For example, in an influential series of articles, Block (1995, 2002) distinguished between various notions of consciousness: phenomenal, access, self, and monitoring.

First, *phenomenal consciousness* (p-consciousness) refers to one's conscious awareness of one's phenomenology. P-consciousness is equated to experience; thus, p-consciousness properties constitute such experiential properties as "sensations, feelings and perceptions ... thoughts, wants and emotions." For Block, the "totality of the experiential properties of a state are 'what is it like' to 'have it'" (Block 2002, p. 206). For example, one may be p-conscious of the experiential properties of pain – that is, the "what is it like" to have pain.

Second, *access-consciousness* (a-consciousness) is a non-phenomenal notion of consciousness. An entity exemplifying a-consciousness is aware of information "poised for direct rational control of action" (Silby 1998, p. 3). For example, one may have learned various survival techniques useful for climbing Mount Everest. However, this information is not considered a-conscious until it is primed for the control of behaviors during a life-threatening event associated with ascending Mount Everest.

Third, *self-consciousness* (s-consciousness) is illustrated by "me-ishness." An s-conscious entity is aware of the concept of the self; furthermore, one's usage of this concept (explicitly or implicitly) in thinking about oneself also reveals s-consciousness.

Finally, consciousness may also be conceptualized as an *internal monitor* or observer of subjective experience – that is, monitoring consciousness (m-consciousness), as in "mindfulness meditation" and similar contemplative practices (Block 2002). An entity may be m-conscious of inner perceptions, internal scanning, and metacognitive thoughts, resulting in entering a particular cognitive state.

All the preceding concepts of consciousness either explicitly or implicitly suggest that "when people are conscious, they are always conscious of *something*. Consciousness always has an object" (Benjafield 1992, p. 58). For example, one may be p-conscious of phenomenal properties; a-conscious of information that may be invoked to control actions; s-conscious of one's self-concept; or m-conscious of, for instance, internal scanning. This contention is by no means atypical. Indeed, over a century ago Husserl (cited in Sartre 1958, p. ii) contended that "all consciousness ... is consciousness *of* something." Similarly, Sartre argued that consciousness always attends to a "transcendent object" (i.e., an entity other than consciousness) and is thereby precluded from being phenomenologically contentless (Sartre 1958, p. 629). This type of consciousness may be referred to as "positional self-consciousness," as in the following citation:

> All that there is of *intention* in my actual consciousness is directed toward the outside, toward the table; all my judgments or practical activities, all my present inclinations transcend themselves; they aim at the table and are absorbed in it. Not all consciousness is knowledge (there are states of affective consciousness, for example), but all knowing consciousness can be knowledge only of its object (Sartre 1958, p. iii).

A survey of the cognitive psychology literature further supports the above contentions. In brief, cognitive psychologists (e.g., Matlin 1998; Nairne 1997;

Solso 2001) tend to define consciousness as the awareness of internal and external events (e.g., mental phenomena and stimuli in the environment, respectively). In contrast, others (e.g., Westen 1999, p. G-4) limit the definitional boundary of consciousness to "the subjective awareness of mental events." These assertions may indeed constitute the core of consciousness concepts among cognitive psychology today. Commenting on the definition of consciousness as awareness of something (i.e., internal and/or external events), Natsoulas wrote: "It is difficult to emphasize sufficiently the fundamental importance of consciousness in the present sense. It is arguably our most basic concept of consciousness, for it is implicated in all the other senses" (Natsoulas 1978, p. 910).

The salient point exemplified by the preceding descriptions of consciousness is the distinction between consciousness and the content of consciousness. For example, Block's p-consciousness is not composed of experiential properties (contents of p-consciousness) such as sensations (e.g., sight) or perceptions (e.g., recognizing what one sees) (Block 2002). Rather p-consciousness refers to one *being p-conscious* of experiential properties such as sensations and perceptions (for example, being aware of seeing an object and perceiving the shape of that object, discriminating it from similar shapes).

However, we acknowledge that not all experiences are characterized by a distinction between consciousness and its content. Indeed, one notable exception is an unmediated form of mystical experience referred to as the "pure consciousness event" (PCE) (e.g., Almond 1982; Bucknell 1989a; Franklin 1990; Kessler and Prigge 1982; Matt 1990; Perovich 1990; Prigge and Kessler 1990; Rothberg 1990; Woodhouse 1990). Forman (1990a, p. 8) defines the PCE as "a wakeful though contentless (nonintentional) consciousness." A substantial body of evidence has been produced to support this claim. For example, Chapple (1990, p. 70) reported that descriptions of *kaivalyam* in the Samkhya system and *samadhi* in the *Yoga Sutras* are suggestive of the "attainment of a purified consciousness that is beyond characterization." Griffiths (1990, p. 78) surveyed the Indian Buddhist tradition and found evidence for a condition referred to as the attainment of cessation (*nirodhasamapatti*), which is defined as "the non-occurrence of mind and mental concomitants." Bucknell (1989b, p. 19) has suggested that the "third non-material *jhana*" encountered in Buddhist meditation is analogous to the introvertive mystical experience "in which both the thought-stream and sensory input have ceased, leaving zero mental content." Forman (1990b, p. 112) examined the mystical theology of the Christian mystic Meister Eckhart and concluded that Eckhart considered one's encounter with the Godhead to be "phenomenologically contentless."

12.3 Confusing Consciousness and Content

As previously stated, consciousness is often defined as awareness of internal and external events (e.g., Solso 2001) or merely awareness of something (e.g., Natsoulas 1978). In contrast, a so-called "state" of consciousness (SoC) tends to be defined as "[the set] of mental episodes of which one can readily become directly aware" (Natsoulas 1978, p. 912). That is to say, definitions of the term "consciousness" emphasize that consciousness is a *process* that renders perceptible objects such as visual stimuli corresponding to a painting that one is viewing in a gallery. In contrast, definitions of the term "states of consciousness" emphasize that states of consciousness are *sets of objects* that might themselves include objects such as visual stimuli corresponding to a painting that one is viewing in a gallery. While, as previously argued, definitions of consciousness typically distinguish consciousness from its content, the preceding definition of SoCs represents a theoretical confusion of consciousness and its contents by explicitly stating that a SoC is itself the content (i.e., mental episodes) available to conscious awareness. That is, when the qualifier "state of" is affixed to consciousness, "it" [consciousness] is held to be content. Consequently, the term *states of consciousness* rests on a conflation of consciousness and content whereby consciousness is erroneously categorized in terms of content rendered perceptible, presumably, by "itself". We refer to this error as the consciousness/content fallacy.

In this context, it is noteworthy that Tart (1975) contended that terms such as *states of consciousness* "have come to be used too loosely, to mean whatever is on one's mind at the moment" (p. 5). Consequently, Tart coined the term *discrete states of consciousness* (d-SoC) in an attempt to address this issue. Tart defined a discrete state of consciousness (d-SoC) as a "unique configuration or system of psychological structures or subsystems ... that maintains its integrity or identity as a recognizable system in spite of variations in input from the environment and in spite of various (small) changes in the subsystems" (p. 62). According to Tart, psychological structures include, for example, sensory qualities and body image (i.e., content of consciousness). It is evident that Tart is suggesting that a d-SoC is *not* a process (i.e., conscious awareness) that renders a system of psychological structures (content, e.g., body image) recognizable – but rather the actual system of psychological structures (content) that is rendered recognizable. Thus, when Tart affixes the qualifiers "discrete state of" and "state of" to the concept of consciousness he confuses consciousness with "its" content (i.e., a recognizable system of psychological structures).

Implicit in the consciousness/content fallacy is the erroneous notion that, during a SoC, consciousness may observe "its" own qualities. That is to say, a privileged observer (i.e., one who has first-person access to a particular experience) would only be conscious of the fact that he or she was experiencing a particular SoC (i.e., that consciousness exemplified state-like properties), if consciousness could observe its own properties. However, one cannot directly experience the process of being consciously aware, CA_1, which functions to render an object perceptible because this would require the postulation of a second conscious awareness process, CA_2, necessary to render CA_1 a perceptible object, thus, committing one to a vicious regress.

Indeed, various scholars (e.g., Feinberg 2001; Kant 1781) have argued that consciousness cannot directly experience "itself" as a perceptible object, for then it would cease to be the subject (i.e., a process of subjectivity that renders objects perceptible). Wilber (1993) stated that the circumstance is analogous to a sword that cannot cut itself, an eye that cannot see itself, a tongue that cannot taste itself, or a finger that cannot touch its own tip. This argument has been reiterated in Baladeva's commentary to the *Vedanta-sutras of Badarayana*, in which he wrote, "If the Self could perceive His own properties, He could also perceive Himself; which is absurd, since one and the same thing cannot be both the agent and the object of an action" (Vasu 1979, p. 331). Similarly, in a Hindu philosophy text, the *Brihadaranyaka-Upanishad*, it is stated, "You cannot see the seer of sight, you cannot hear the hearer of sound, you cannot think the thinker of the thought, you cannot know the knower of the known" (Swami and Yeats 1970, p. 138).

It is perhaps noteworthy that decades ago, Krippner and Meacham (1968) suggested that, "it may make more sense to speak of the 'objects' of consciousness than to speak of the 'states' of consciousness" (p. 150). It is salient, however, that this recommendation was not arrived at via recognition of the consciousness/content fallacy, but rather the methodological difficulties associated with "searching" for a particular state of consciousness.

12.4 The Consciousness/Content Fallacy with Reference to Altered States of Consciousness

During the early stages of consciousness studies, various definitions of the term "altered states of consciousness" (ASCs) were developed (Krippner 1972; Ludwig 1966; Tart 1969). These decades-old definitions have remained standard among ASC scholars.

Ludwig (1966) asserted that ASCs may be defined as

> any mental state(s), induced by various physiological, psychological, or pharmacological manoeuvres or agents, which can be recognized subjectively by the individual himself (or by an objective observer of the individual) as representing a sufficient deviation in subjective experience or psychological functioning from certain general norms for that individual during alert, waking consciousness (p. 225).

Ludwig also identified a variety of general characteristics purportedly exhibited by ASCs: alterations in discursive thought, subjective time, body image, emotional expression, perception, and significance or meaning; loss of volitional control; a sense of ineffability; rejuvenation; and hypersuggestibility. However, one shortcoming of Ludwig's definition is that it does not operationalize a "sufficient deviation in subjective experience" (Ludwig 1966, p. 225). Additionally, the "general norms" purportedly associated with ordinary waking consciousness are not explained. That is, the "general norms" for one person or culture might differ considerably from the "general norms" for another person or culture.

In contrast, Krippner's (1972) ASC definition ostensibly resolves the problems associated with operationalizing the qualifier "sufficient." Krippner defined an ASC as "a mental state which can be subjectively recognized by an individual (or by an objective observer of the individual) as representing a *difference* in psychological functioning from the individual's 'normal' alert state" (p. 1). However, Krippner neglected to specify whether it is the pattern and/or the intensity of the psychological functioning that is required to be different relative to the individual's "normal alert state" for an ASC to be inferred. A related shortcoming is that Krippner neglected to operationalize "mental state" and "normal alert state." Additionally, Krippner's definition seems to suggest that any "difference in psychological functioning" compared to a "normal alert state" is sufficient for one to infer an ASC (p. 1). However, it seems reasonable to assume that one's normal alert state may vary from moment to moment and from one context to another. Thus, if one were to accept Krippner's definition, one might suggest that the state(s) experienced each moment subsequent to an initial moment constitutes ASCs relative to the "normal alert state" experienced in that initial moment.

Similarly, Tart (1969) defined an ASC for a particular person as a "*qualitative* shift in his pattern of mental functioning, that is, he feels not just a quantitative shift (more or less alert, more or less visual imagery, sharper or duller, etc.), but also that some quality or qualities of his mental processes are *different*" (p. 1). Qualities may include, for example, visual imagery, subjective time distortion (i.e., dilation or contraction), and changes in body image. The utility of Tart's definition is undermined by the fact that he does not specify how pronounced the qualitative shift must be, or precisely how many qualities of one's mental processes must be different, before an ASC may be inferred. Furthermore, when Tart referred to a qualitative and quantitative shift, he neglected to outline what might constitute an appropriate comparison group.

It is noteworthy that the preceding definitions postulate that it is the shifts, deviations, or differences in subjective experience (Ludwig 1966), psychological functioning (Krippner 1972), or mental functioning (Tart 1969) that constitute an ASC. If one accepts the definition of an altered state of *consciousness* as shifts, deviations, or differences in subjective experience, psychological functioning, or mental functioning, then it would seem to follow that ordinary consciousness is the baseline subjective experience, psychological functioning, or mental functioning. Furthermore, it is arguable that, if an ASC did constitute shifts, deviations, or differences in subjective experience, psychological functioning, or mental functioning, then a privileged observer would not be conscious of such shifts on the ground that to be conscious of such shifts would necessitate that consciousness could observe changes in its own properties – that is, alterations held to constitute an ASC. The above-listed citations nonetheless emphasize that an ASC may be *subjectively recognized* by a privileged observer (i.e., an individual who has first-person access to the experience). Consequently, if these three authors are using "ASC" as a subsidiary part of the notion of consciousness as one being conscious of something (e.g., an internal or external event), then they have confused consciousness and the content of consciousness on the ground that consciousness is implicitly held to be both (1) the cognizer of shifts in subjective experience, and (2) the shifts in subjective experience themselves. If "ASC" is *not* being used as a subsidiary part of the aforementioned notion of consciousness, then the definition of consciousness that has been used to extrapolate a definition for ASC needs to be explicitly stated.

If one defines consciousness as being consciously aware of something, then it would seem to follow that during an ASC it is the phenomenal properties that consciousness may be aware of that are altered (e.g., mental imagery, body image, time sense), rather than the *state* of consciousness. It may be, however, that phenomenal properties do not encapsulate the variety of mental phenomena that may be objectified by consciousness. For example, as previously discussed, Block (2002) formulated the notion of a-consciousness, whereby an entity is held to be conscious of non-phenomenal mental objects, or information primed for the rational control of one's actions (Silby 1998). Similarly, O'Brien and Opie (1997) suggested that "phenomenal experience" does not refer to objects associated with self-consciousness and access-consciousness (e.g., self-concept and information that may be invoked to control actions, respectively), but rather the "what is it like" of experience (p. 269). However, for the purposes of this chapter, Reber and Reber's definition of phenomenal field as "absolutely anything that is in the total momentary experiencing of a person, including the experience of the self" (Reber and Reber 2001, p. 532), will be adopted and applied to "phenomenal properties." Consequently, we will define "phenomenal properties" as the qualities of "absolutely anything that is in the total momentary experiencing of a person, including the experience of the self," which means that the phrase "altered pattern of phenomenal properties" encapsulates what has been referred to by Block (1995) and others (e.g., Lormand 1996) as phenomenal and non-phenomenal objects of conscious awareness – that is, the content that a privileged observer may be aware of during what earlier researchers (Krippner 1972; Ludwig 1966; Tart 1969) referred to as an ASC. One may then recommend that the term "altered state of consciousness" be supplanted by the new term, "altered pattern of phenomenal properties." It would seem that, by reconceptualizing the notion of an ASC in this manner, the confusion of consciousness with the content of consciousness is avoided. In the next section we will explain how one may use a retrospective phenomenological assessment instrument, the Phenomenology of Consciousness Inventory, to investigate patterns of phenomenal properties experimentally.

12.5 Experimentally Investigating Patterns of Phenomenal Properties

A pattern of phenomenal properties is held to be altered relative to a baseline pattern of phenomenal properties – that is, what is generally referred to as normal waking consciousness or an ordinary waking state. One may use a retrospective phenomenological assessment instrument referred to as the Phenomenology of Consciousness Inventory (PCI) (Pekala 1991) to investigate patterns of phenomenal properties. In fact, the PCI is held to quantify "both the major contents of consciousness, and the processes or means by which these contents are 'illuminated,' cognized, perceived, and so forth by consciousness" (Pekala 1991, p. 82). The former is denoted by our use of the term "phenomenal properties."

The PCI is a 53-item questionnaire consisting of 12 major dimensions (altered state, rationality, positive affect, arousal, self-awareness, memory, inward absorbed attention, negative affect, altered experience, volitional control, vivid imagery, and internal dialogue) and 14 minor dimensions (joy, sexual excitement, love, anger, sadness, fear, body image, time sense, perception, meaning, visual imagery amount, vividness, direction of attention, absorption). Minor dimensions are constituents of major dimensions. For example, the major dimension of negative affect consists of the minor dimensions of fear, anger, and sadness. Major and minor dimensions are scored on a 7-point Likert scale where 0 signifies "no or little" increased intensity ratings and 6 indicates "much or complete" (Pekala and Wenger 1983; Pekala et al. 1985).

At this point a qualifying statement is required. The PCI's "altered state" major dimension consists of questionnaire items such as "I felt in an extremely different and unusual state of consciousness" (Pekala 1991, p. 355). However, when a respondent answers the aforementioned item, clearly they are not reflecting on the properties of their SoC because, as argued earlier (Feinberg 2001; Kant 1781; Swami and Yeats 1970; Vasu 1979; Wilber 1993), consciousness cannot observe "its" own properties. Consequently, it is arguable that when respondents address the item they are, in fact, reflecting on the extent to which their *phenomenology* is unusual or altered.

The PCI possesses respectable psychometric properties. For example, it has good internal consistency as evidenced by coefficient alphas ranging between 0.7 and 0.9 on the various dimensions (Pekala et al. 1986). Furthermore, it is reported that participants who were administered different stimulus conditions recorded significantly different PCI scores. This finding suggests that the PCI can successfully distinguish between qualitatively different SoCs and, thus, supports the scale's criterion validity. Previous research has used the PCI to quantitatively map and diagram patterns of phenomenal properties associated with, for instance, deep abdominal breathing (e.g., Pekala et al. 1989), listening to drumming (e.g., Maurer et al. 1997), progressive relaxation exercises (e.g., Pekala et al. 1989), hypnotic induction (e.g., Pekala and Kumar 1984, 1986), and sitting quietly with one's eyes closed (e.g., Pekala and Kumar 1989).

PCI data may be used to construct graphs referred to as "psygrams" that map or diagram the patterns of relationships between pairs of PCI major dimensions (i.e., phenomenal properties) associated with a particular stimulus condition (Pekala 1991). A psygram pictorially represents two types of information: (1) mean phenomenological intensity values for each PCI major dimension, and (2) strength of "coupling" or association among the various PCI major dimensions (Pekala and Kumar 1986). One creates a psygram by first producing a correlation matrix consisting of the 12 PCI major dimensions. The non-significant correlations ($p > 0.05$) are removed and the significant r obtained values are converted to r^2 values, that is, coefficients of determination. Subsequently, the r^2 values are converted to percentages. Each line linking a pair of major dimensions constitutes 5% of the r^2 or variance in common.

By way of illustration, one may consider an experimental study conducted by Rock et al. (2008) that investigated differential phenomenological effects of a control or baseline condition (i.e., sitting quietly with one's eyes open) and two composite activity conditions (i.e., mental imagery cultivation while listening to monotonous drumming or being exposed to an altered sensory environment). More specifically, this study consisted of a between-subjects design with three conditions: (1) a control condition consisting of sitting quietly with eyes open for 15 min (S-group); (2) listening to guided imagery instructions followed by 15 min of white noise, while viewing diffuse red light (G-group); and (3) listening to guided imagery instructions followed by 15 min of listening to monotonous drumming at 8 beats/s (D-group). One aim of the study was to use Pekala's PCI to investigate – with the aid of psygrams – the pattern of relationships between pairs of phenomenal properties associated with the S-group, the G-group, and the D-group, respectively.

Figure 12.1 depicts a psygram of the pattern of relationships between pairs of phenomenal properties associated with the S-group. This group reported *little* negative affect; *mild* alterations in altered state of awareness, positive affect, arousal, and altered experience; an *average* amount of change in vivid imagery and inward absorbed attention; and *moderate* alterations in rationality, self-awareness, memory, volitional control, and internal dialogue. With regard to pattern analysis, altered state is strongly coupled with rationality, positive affect, self-awareness, memory, altered experience, and volitional control. Altered experience is strongly coupled with rationality, positive affect, memory, negative affect, and volitional control; rationality is strongly coupled with memory, inward absorbed attention, volitional control, and internal dialogue; inward absorbed attention is strongly coupled with memory and vivid imagery; and volitional control is strongly coupled with self-awareness and memory.

Figure 12.2 represents the pattern of relationships between pairs of phenomenal properties reported by the G-group. In comparison to the baseline condition (Fig. 12.1), there is an *increase* in altered state, altered experience, and vivid imagery; a decrease in rationality, arousal, volitional control, and internal dialogue; and *no change* to positive affect, self-awareness, memory, inward absorbed attention, or negative affect. With regard to pattern relationships, the configurations evident in Fig. 12.2 are noticeably different from those depicted in Fig. 12.1. The strength of the relationships between altered experience and altered state and positive affect, altered state and self-awareness and positive affect, and self-awareness and volitional control has *increased*; and the strength of the relationships between altered state and rationality and volitional control has *decreased*. Additionally, positive affect and vivid imagery; altered experience and self-awareness and inward absorbed attention; negative affect and internal dialogue and arousal; vivid imagery and rationality and memory; and self-awareness and rationality are all now strongly coupled. Furthermore, it is evident that many of the pattern relationships present in the S-group are no longer present in the G-group.

Figure 12.3 depicts the pattern of relationships between pairs of phenomenal properties reported by the D-group. With the exception of arousal (which has *increased*), intensity ratings across the 12 major PCI dimensions are *equivalent* to

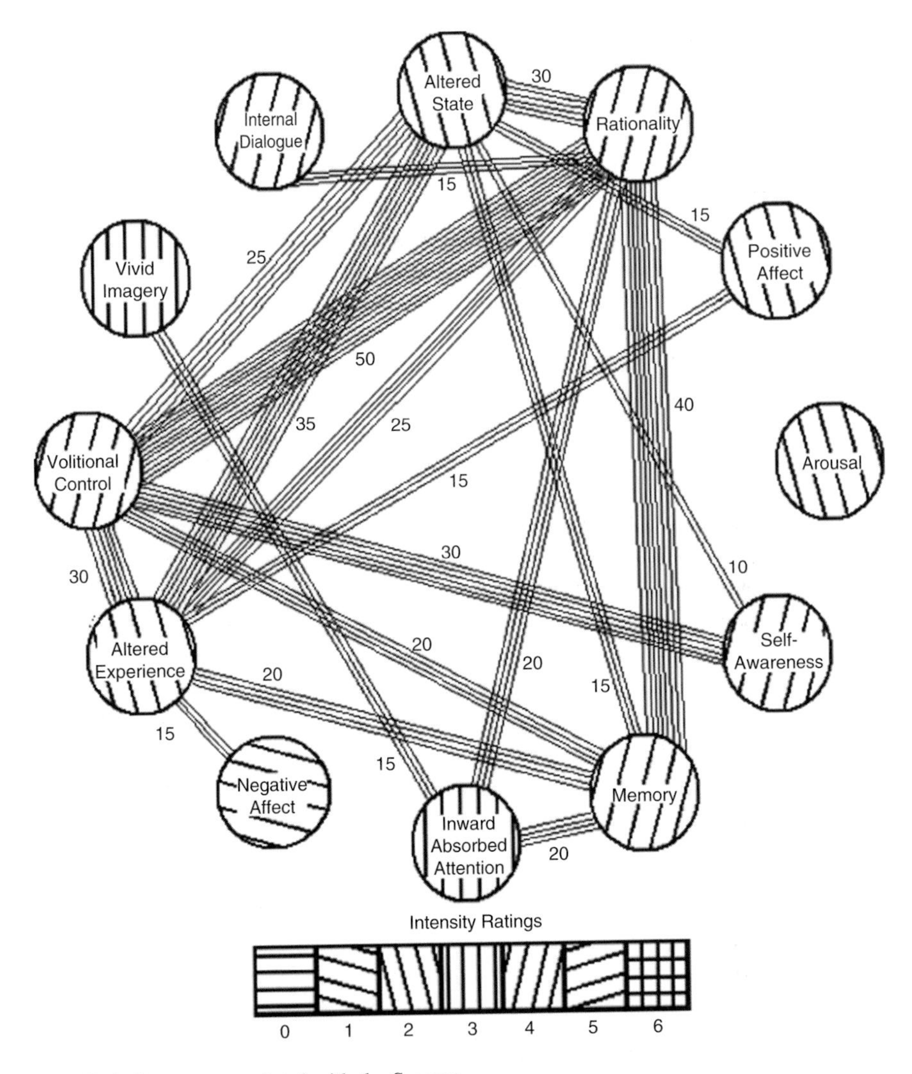

Fig. 12.1 Psygram associated with the S-group

the G-group. Despite this finding, several changes in pattern relationships are evident. The strength of the relationships between altered state and altered experience has *increased*; and the strength of the relationships between altered state and positive affect and self-awareness, and positive affect and vivid imagery has *decreased*. Altered state is now strongly coupled with vivid imagery, inward absorbed attention, and memory. Additionally, altered experience is now strongly coupled with vivid imagery and volitional control, and self-awareness is now strongly coupled with negative affect and inward absorbed attention.

Thus, inspection of the psygrams for the S-group, G-group, and D-group reveals many differences between conditions with regard to the patterns of relationships

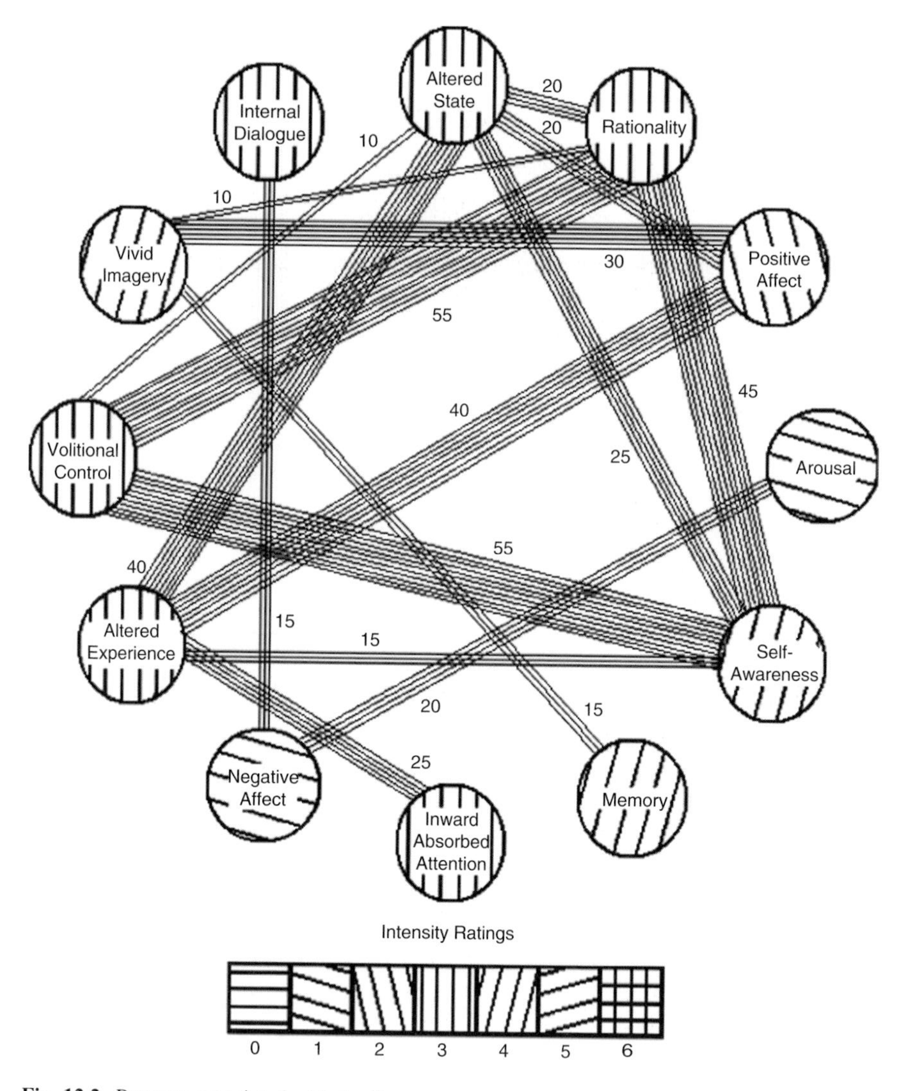

Fig. 12.2 Psygram associated with the G-group

between pairs of phenomenal properties. Various statistical methods such as the Jenrich test and box test (Pekala 1991) may be used to determine if there is a *statistically significant* difference between the "pattern structures" (i.e., patterns of relationships between pairs of PCI major dimensions or phenomenal properties) of psygrams. Rock et al. (2008) reported no statistically significant difference between psygrams. This result suggests that, compared to the S-group, the G-group and D-group were not associated with a "major reorganization in pattern structure that is hypothesized by Tart (1975) to be associated with" what has typically been referred to as "an altered state of consciousness" (Woodside et al. 1997, p. 84). Thus, this

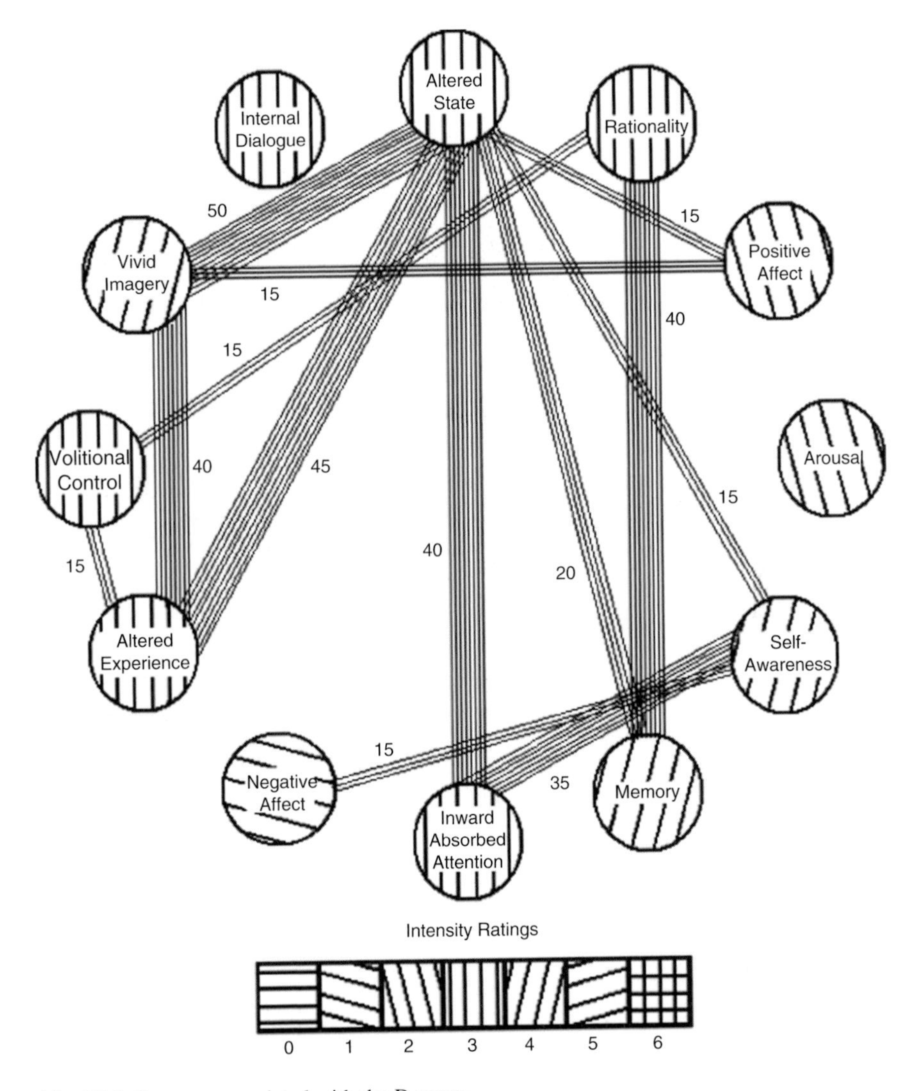

Fig. 12.3 Psygram associated with the D-group

methodology has the advantage of allowing one to "operationally define, map and diagram" [altered] patterns of phenomenal properties (Pekala 1985).

12.6 Concluding Remarks

In this chapter we reviewed several definitions of the term "consciousness," arguing that they all share the implicit distinction between consciousness and the content of consciousness. We further suggested that definitions of the terms "states of consciousness" and "altered states of consciousness" erroneously conflate consciousness and content by explicitly defining SoCs as the content (i.e., mental episodes) available to conscious awareness. That is, when the qualifier "state of" is affixed to consciousness, "it" [consciousness] is held to be content. We referred to this error as the consciousness/content fallacy, contending that this fallacy may be avoided by reconceptualizing an altered state of consciousness as an altered pattern of phenomenal properties. Finally, we described a recent experimental study illustrating the utility of our proposal that so-called [altered] states of consciousness are more appropriately described as patterns of phenomenal properties.

The consciousness/content fallacy has numerous theoretical implications. Theories of SoCs and ASCs, for example, would be enhanced by supplanting the term "[altered] states of consciousness" with "[altered] patterns of phenomenal properties." Theories containing the consciousness/content fallacy would need to be revised to avoid fallacious contentions such as that consciousness is simultaneously (1) the "cognizer" of shifts in, for instance, "subjective experience," and (2) the shifts in "subjective experience" themselves. If a particular SoC or ASC theory did not incorporate the term "[altered] states of consciousness" as a subsidiary of the concept of consciousness as conscious awareness of something, then this would need to be explicitly stated. Fundamentally, SoC and ASC theories would need to be reformulated such that the phenomenon being explained is patterns of phenomenal properties rather than states of consciousness.

In addition, the consciousness/content fallacy has implications for quantitative and qualitative research. A researcher who is aware of this fallacy and wishes to develop a survey instrument to quantitatively measure, for example, meditation experiences, would construct items pertaining to alterations in phenomenal properties rather than alterations in consciousness. For instance, items such as "I experienced an extremely unusual state of consciousness" would be omitted in favor of items addressing a range of phenomenal properties (e.g., "My subjective time sense seemed to slow down"; "My visual imagery became extremely vivid"; "I felt great joy"). Similarly, consider a research situation in which, for example, a qualitative study of meditative experiences is conducted using semi-structured interviews for the purpose of obtaining non-numerical data that may be organized into comprehensive constituent themes. A researcher who is mindful of the consciousness/content fallacy would not pose open-ended questions about "meditative states of consciousness" or "alterations in consciousness." Instead, open-ended questions pertaining to phenomenal properties would be asked, as in, "Can you tell me about the most important aspects of your most profound meditation experience?" The authors hope that the present chapter's elucidation and proposed resolution of the consciousness/content fallacy will encourage consciousness theoreticians and researchers from diverse backgrounds to address its implications.

Acknowledgements Preparation of this chapter was supported by the Chair for the Study of Consciousness, Saybrook University, and by the Director, Phoenix Institute of Victoria.

References

Almond P (1982) Mystical experience and religious doctrine. Mouton Publishers, New York

Antony MV (2002) Concepts of consciousness, kinds of consciousness, meanings of "consciousness". Philos Stud 109:1–16

Benjafield JG (1992) Cognition. Prentice-Hall International, Englewood Cliffs

Block N (1995) On a confusion about a function of consciousness. Behav Brain Sci 18: 227–247

Block N (2002) Concepts of consciousness. In: Chalmers D (ed) Philosophy of mind: classical and contemporary readings. Oxford University Press, New York, pp 206–218

Bucknell R (1989a) Buddhist *Jhana* as mystical experience. In: Zollschan GK, Schumaker JF, Walsh GF (eds) Exploring the paranormal: perspectives on belief and experience. Prism Press, Bridgeport, pp 131–149

Bucknell R (1989b) Buddhist meditation and mystical experience. Paper presented to the 14th annual conference of the Australian association for the study of religions, Perth

Chalmers D (1995) Facing up to the problem of consciousness. J Conscious Stud 2:200–219

Chapple C (1990) The Unseen seer and the field: consciousness in Samkhya and Yoga. In: Forman RKC (ed) The problem of pure consciousness. Oxford University Press, New York, pp 53–70

Farthing GW (1992) The psychology of consciousness. Prentice-Hall, Englewood Cliffs

Feinberg TE (2001) Altered egos: how the brain creates the self. Oxford University Press, New York

Forman RKC (1990a) Introduction: mysticism, constructivism, and forgetting. In: Forman RKC (ed) The problem of pure consciousness. Oxford University Press, New York, pp 3–49

Forman RKC (1990b) Eckhart, *Gezucken*, and the ground of the soul. In: Forman RKC (ed) The problem of pure consciousness. Oxford University Press, New York, pp 98–120

Forman RKC (1996) What does mysticism have to teach us about consciousness? Revised version of a paper delivered to "Toward a science of consciousness 1996," Tucson II, April 1996

Franklin RL (1990) Experience and interpretation in mysticism. In: Forman RKC (ed) The problem of pure consciousness. Oxford University Press, New York, pp 288–304

Griffiths PJ (1990) Pure consciousness and Indian Buddhism. In: Forman RKC (ed) The problem of pure consciousness. Oxford University Press, New York, pp 71–97

Kant I (1781/1933) The critique of pure reason, 2nd edn. Macmillan Press, London (Original work published 1781)

Kessler G, Prigge N (1982) Is mystical experience everywhere the same? Sophia 21:39–55

Krippner S (1972) Altered states of consciousness. In: White J (ed) The highest state of consciousness. Doubleday, Garden City, pp 1–5

Krippner S, Meacham W (1968) Consciousness and the creative process. Gifted Child Quart 12:141–157

Lormand E (1996) Nonphenomenal consciousness. Nous 30:242–261

Ludwig AM (1966) Altered states of consciousness. Arch Gen Psychiatry 15:225–234

Matlin MW (1998) Cognition, 4th edn. Harcourt Brace College Publishers, Fort Worth

Matt DC (1990) *Ayin*: the concept of nothingness in Jewish mysticism. In: Forman RKC (ed) The problem of pure consciousness. Oxford University Press, Oxford, pp 121–159

Maurer RL, Kumar VK, Woodside L, Pekala RJ (1997) Phenomenological experience in response to monotonous drumming and hypnotizability. Am J Clin Hypn 40:130–145

Nairne JS (1997) Psychology: the adaptive mind. Brooks/Cole, Pacific Grove

Natsoulas T (1978) Consciousness. Am Psychol 33:906–914

O'Brien G, Opie J (1997) Cognitive science and phenomenal consciousness: a dilemma, and how to avoid it. Philos Psychol 10:269–286

Pekala RJ (1985) A psychophenomenological approach to mapping and diagramming states of consciousness. J Relig Psych Res 8:199–214

Pekala RJ (1991) Quantifying consciousness: an empirical approach. Plenum Press, New York
Pekala RJ, Kumar VK (1984) Predicting hypnotic susceptibility by a self-report phenomenological state instrument. Am J Clin Hypn 27:114–121
Pekala RJ, Kumar VK (1986) The differential organization of the structures of consciousness during hypnosis and a baseline condition. J Mind Behav 7:515–540
Pekala RJ, Kumar VK (1989) Phenomenological patterns of consciousness during hypnosis: relevance to cognition and individual differences. Aust J Clin Exp Hypn 17:1–20
Pekala RJ, Wenger CF (1983) Retrospective phenomenological assessment: mapping consciousness in reference to specific stimulus conditions. J Mind Behav 4:247–274
Pekala RJ, Wenger CF, Levine RL (1985) Individual differences in phenomenological experience: states of consciousness as a function of absorption. J Pers Soc Psychol 48:125–132
Pekala RJ, Steinberg J, Kumar CK (1986) Measurement of phenomenological experience: phenomenology of consciousness inventory. Percept Mot Skills 63:983–989
Pekala RJ, Forbes E, Contrisciani PA (1989) Assessing the phenomenological effects of several stress management strategies. Imagination Cogn Pers 88:265–281
Perovich AN (1990) Does the philosophy of mysticism rest on a mistake? In: Forman RC (ed) The problem of pure consciousness. Oxford University Press, New York, pp 237–253
Prigge N, Kessler GE (1990) Is mystical experience everywhere the same? In: Forman RKC (ed) The problem of pure consciousness. Oxford University Press, New York, pp 269–287
Reber AS, Reber E (2001) The Penguin dictionary of psychology, 3rd edn. Penguin, London
Rock AJ, Abbott G, Childargushi H, Kiehne M (2008) The effect of shamanic-like stimulus conditions and the cognitive-perceptual factor of schizotypy on phenomenology. N Am J Psychol 10:79–98
Rosenthal DM (2002) How many kinds of consciousness? Conscious Cogn 11:653–665
Rothberg D (1990) Contemporary epistemology and the study of mysticism. In: Forman RKC (ed) The problem of pure consciousness. Oxford University Press, New York, pp 163–210
Sartre JP (1958) Being and nothingness: an essay on phenomenological ontology. Philosophical Library, New York
Silby B (1998) On a distinction between access and phenomenal consciousness. http://www.def-logic.com/articles/silby011.html. Accessed 2 Feb 2005
Solso RL (2001) Cognitive psychology, 5th edn. Allyn and Bacon, Boston
Swami SP, Yeats WB (1970) The ten principal Upanishads. Faber and Faber, London
Tart CT (1969) Introduction. In: Tart CT (ed) Altered states of consciousness. Wiley, New York, pp 1–7
Tart CT (1975) States of consciousness. Dutton, New York
Vasu RBSC (1979) The Vedanta Sutras of Badarayana with the commentary of Baladeva, 2nd edn. Oriental Books Reprint Corporation, New Delhi
Westen D (1999) Psychology: mind, brain, and culture, 2nd edn. Wiley, New York
Wilber K (1993) The spectrum of consciousness, 2nd edn. Quest, Wheaton
Woodhouse MB (1990) On the possibility of pure consciousness. In: Forman RKC (ed) The problem of pure consciousness. Oxford University Press, New York, pp 254–268
Woodside LN, Kumar VK, Pekala RJ (1997) Monotonous percussion drumming and trance postures: a controlled evaluation of phenomenological effects. Anthropol Conscious 8:69–87

Index

D. Cvetkovic and I. Cosic (eds.), *States of Consciousness*, The Frontiers Collection, DOI 10.1007/978-3-642-18047-7, © Springer-Verlag Berlin Heidelberg 2011